L.W. SUMNER, JOHN G. SLATER, and FRED WILSON are members of the Department of Philosophy at the University of Toronto.

The essays in this volume are connected with the main areas of Thomas A. Goudge's research: Peirce studies and the philosophy of science, especially the philosophy of biology. There are two sections, each opened with an essay on Goudge's contribution to the field.

The first section deals with C.S. Peirce. Essays examine major features of his thought: his proof of pragmatism, his logic of relations, his semiotics, his theory of truth, his philosophy of induction, and his Scotist metaphysics of individuals. Other essays in this group relate pragmatism and pragmaticism to the work of others, in particular Collingwood, Descartes, and James and Santayana. Each involves substantial dialogue with Peirce, and a philosophical evaluation of his work and the work of others.

The second is concerned with the philosophy of science. Some of the essays study the nature of scientific explanation: whether there is a logic of discovery, and whether life and teleology could be fitted into the mechanistic view of the universe that arose in the early modern period of philosophy. Various problems specifically concerning the synthetic theory of evolution are also discussed: for example, is it a theory somehow intrinsically different in kind from theories in physics? does it yield special forms of explanation? how is it related to modern genetic theory? Other essays deal with problems arising out of an evolutionary view of life and man: for example, emergence, the place of consciousness in nature, man's knowledge of necessary truths.

This is an important collection for those working in Peirce studies or the philosophy of biology, and a substantial contribution to the history of Canadian philosophy.

Thomas A. Goudge
Photograph by Kenneth Quinn

Edited by
L.W. SUMNER
JOHN G. SLATER
FRED WILSON

Pragmatism and Purpose: Essays Presented to Thomas A. Goudge

UNIVERSITY OF TORONTO PRESS
Toronto Buffalo London

© University of Toronto Press 1981
Toronto Buffalo London
Printed in Canada
Reprinted in 2018
ISBN 0-8020-5481-1
ISBN 978-1-4875-8247-0 (paper)

Canadian Cataloguing in Publication Data
Main entry under title:

Pragmatism and purpose

Contents: Pt. 1. Savan, D. The unity of Peirce's thought. – Burbidge, J.W. Peirce on historical explanation. – Fisch, M.H. The 'proof' of pragmatism. (etc.). Pt. 2. Wilson, F. Goudge's contribution to philosophy of science. – Hull, D.L. Historical narratives and integrating explanations. – McRae, R. Life, vis inertiae, and the mechanical philosophy. (etc.).
ISBN 0-8020-5481-1
1. Goudge, Thomas A., 1910– Addresses, essays, lectures. 2. Peirce, Charles S., 1839–1914 – Addresses, essays, lectures. 3. Pragmatism – Addresses, essays, lectures. 4. Evolution – Addresses, essays, lectures. 5. Biology – Philosophy – Addresses, essays, lectures. I. Goudge, Thomas A., 1910– II. Sumner, L.W., 1941– III. Slater, John G. IV. Wilson, Fred, 1937–
B832.P72   144'.3   C80-094471-2

Contents

PREFACE vii
ACKNOWLEDGMENTS xi
ABBREVIATIONS xiii

PART ONE: THE THOUGHT OF C.S. PEIRCE

The Unity of Peirce's Thought 3
DAVID SAVAN

Peirce on Historical Explanation 15
J.W. BURBIDGE

The 'Proof' of Pragmatism 28
MAX. H. FISCH

Peirce's Remarkable Theorem 41
HANS G. HERZBERGER

Scientific Inference: Peirce and the Humean Tradition 59
EDWARD H. MADDEN

The Labeling Problem 75
DON D. ROBERTS

Peirce and Descartes: Doubt and the Logic of Discovery 88
PETER A. SCHOULS

James, Santayana, Tarski, and Pragmatism 105
T.L.S. SPRIGGE

Peirce on Truth 121
H.S. THAYER

Peirce's Conception of an Individual 133
MANLEY THOMPSON

PART TWO: THE ASCENT OF LIFE

Goudge's Contribution to Philosophy of Science 151
FRED WILSON

Historical Narratives and Integrating Explanations 172
DAVID L. HULL

Life, *Vis Inertiae*, and the Mechanical Philosophy 189
ROBERT McRAE

The Ascent of Consciousness 199
ROLAND PUCCETTI

The Interaction of Evolutionary and Genetic Theory 207
ALEXANDER ROSENBERG

Philosophical Aspects of the Darwinian Revolution 220
MICHAEL RUSE

Evolution and the Necessities of Thought 236
BARRY STROUD

The Autonomy of a Logic of Discovery 248
PAUL R. THAGARD

Emergence Revisited 261
R.E. TULLY

Is Biology a Different Type of Science? 278
MARY B. WILLIAMS

NOTES 293
LIST OF WORKS CITED 317
THE PUBLISHED WORKS OF THOMAS A. GOUDGE 329
NOTES ON CONTRIBUTORS 337
INDEXES 341

Preface

When Thomas A. Goudge retired from the University of Toronto in 1976 the event marked a decisive step in a remarkable career. For nearly forty years he was one of the leading members of Canada's largest and most important philosophy department, serving as teacher to dozens of philosophers who are now established figures. During many of those years he also bore administrative duties, including a term as chairman (1963-9) during a difficult time of expansion and change. His influence, moreover, was not confined within the boundaries of his own university, for he was an active member of many organizations of professional philosophers. But it is safe to say that the source of his greatest impact on the world at large has been his writing. Of the many fruits of his scholarly labours two merit special attention. *The Thought of C.S. Peirce* (1950) led to Goudge's election in 1955 to the Royal Society of Canada; his second book, *The Ascent of Life* (1961), won the Governor-General's Award for academic non-fiction. Both books were recognized from the first as outstanding contributions to their fields: the philosophy of Peirce and the philosophy of biology. They established their author as one of the foremost figures in the history of Canadian philosophy. It is in order to celebrate the career of this outstanding teacher, administrator, and scholar that this volume of essays has been assembled.

Thomas Anderson Goudge was born in Halifax in 1910. He received his BA from Dalhousie University in 1931 and his MA from the same institution the following year. His association with the University of Toronto began in 1932 when he registered as a PHD student. He was awarded his doctorate in 1937 for his thesis, 'The Theory of Knowledge in Charles S. Peirce.' While writing his thesis he spent two years, 1935-7, as a special student in Harvard University where he had access to Peirce's unpublished papers. In 1934 he taught a term at Waterloo Lutheran University (now Wilfrid Laurier University), substituting for a regular member who was ill. The next year he was appointed a Fellow in Philosophy in Queen's University, returning there after his Harvard studies as a lecturer in 1937–8. So it was with considerable teaching experience that, in the summer of 1938, he joined the University of Toronto Philosophy Department as a lecturer and its sixth member.

His appointment to the department was clearly intended by Professor G.S. Brett, then its head, to bring in a promising young scholar ready and willing to teach recent work in British and American philosophy of a non-idealistic kind. A story told by Professor A.H. Johnson of the University of Western Ontario, who was also awarded a PHD from Toronto in 1937, indicates the direction of Brett's thinking. Bertrand Russell came to Toronto in the early 1930s on a lecture tour. At a meeting with Russell, Brett asked him if he thought there should be a course in symbolic logic offered for graduate students. Russell answered that there should be if there was a man on staff competent to teach it. One cannot help speculating that Russell's reply may have been instrumental in securing Goudge his appointment at Toronto.

Goudge introduced, with Brett's blessing, a graduate course in pragmatism and logical positivism in 1939–40 and one in symbolic logic in 1940–1, both firsts in the history of the department. These two courses are representative of the majority of the courses, both graduate and undergraduate, he offered during his years at Toronto, although he did considerable teaching of the history of philosophy, notably for several years a graduate seminar on the philosophy of Kant. He also pressed both before and during his chairmanship for a redress of the balance between systematic and historical courses – a reform long delayed at Toronto where philosophy was (or so it seemed to many) equated with the history of philosophy. Because of his position in this department and the place of this department in Canada, Goudge was influential in changing the direction which philosophy in Canada has taken since the war. An almost exclusive interest in the history of philosophy has given way to a primary concern with contemporary problems, while preserving sufficient attention to their history to avoid repeating past errors.

In addition to his teaching Goudge undertook more than his share of administrative duties. He was graduate secretary for a number of years, and for most of the time during Fulton Anderson's long headship (1946–63) Goudge was second only to Anderson in seniority and so was frequently called upon to act for Anderson when the latter was away or ill. When Anderson reached the mandatory retirement age, Goudge was the obvious choice to succeed him. During his chairmanship (the title changed but not the duties or responsibilities), and in keeping with his temperament and beliefs, he steered the department from autocracy to democracy. When he retired from office in 1969, the department was probably the most democratically governed of any in the Faculty of Arts and Science.

The most important duty Goudge had to discharge during his long service at Toronto was finding candidates for the large number of new appointments he was required to make during his tenure as chairman. The task was made more difficult by the fact that every other university system in the English-speaking world was also expanding at the same time. To find candidates of high calibre he had to turn to the large graduate departments in the United States and Great Britain, for there were simply too few doctoral candidates being trained in Canada to offer an adequate selection. It was in those days a standard practice for Canadians, after earning baccalaureate degrees in Canada, to go on to graduate work in the United States or Great Britain. After completion of their studies, many applied for teaching positions in the University of Toronto. Goudge repatriated a large number of them, following the policy that when academic qualifications were equal, the Canadian citizen was to be favoured for appointment.

It seems certain that Goudge holds, in the world of academic philosophy, the record for having made the largest number of teaching appointments. During the six years of his chairmanship he made thirty-seven tenure-stream appointments – a number larger than the size of any other North American department. After the usual resignations and denials of tenure, there remain twenty-five professors of philosophy at Toronto who owe their initial appointments to Goudge. Because most of his appointees were young, he has effectively determined the character of the department for the rest of this century. And since this department is the largest in the country, his ultimate influence on the development of philosophy in Canada cannot fail to be enormous.

Goudge was also active in promoting the profession in Canada and abroad. He joined the American Philosophical Association and the Association for Symbolic Logic very early in his career. He was a founding member and later president (1958 and 1959) of the Charles S. Peirce Society. From 1965 to 1969 he served as a member of the Council for Philosophical Studies, a very active arm of the Carnegie Corporation in New York. One of the founding members of the Canadian Philosophical Association, he served on its board for a number of years before being elected president for 1964–5. He aided the new association's journal, *Dialogue*, not only by sending it several articles and contributing frequent reviews, but also by securing its financial support from the University of Toronto.

It was during these busy years of teaching and administering that he continued to write and to publish. As befits a festschrift, the focus in this volume is on Goudge's scholarly work. The overall plan of the volume was determined by his two dominating interests in the philosophy of Peirce and the philosophy of biology. The editors commissioned a lead essay in each field to analyse and evaluate some of the leading principles of Goudge's own work. The remaining papers are original contributions to one or the other of the areas which have been illuminated by Goudge's scholarly attention. The volume also contains a complete bibliography of his published writings.

We have tried to satisfy a number of desires in selecting contributors. Established workers in these fields are, of course, represented, but we also invited younger philosophers whose work is just getting under way on the ground that a fitting way of marking one outstanding career is to provide a forum for a new generation of scholars. We also decided to invite some philosophers whose reputations have been established in other fields to turn their minds toward problems in an area relatively new to them. Specialization tends to produce work both deep and narrow, and the conceptions and frameworks of outsiders can be as refreshing as they are novel. Since Goudge spent forty years teaching philosophy, it seemed appropriate to invite contributions from some of his former students and colleagues, and, since all of that teaching was in Canada, we strove as well to present a large sample of the philosophical work now being done in Canada. Half of the contributions report the work of philosophers currently based in Canada. For the high quality of philosophical scholarship current in this country Goudge may justly claim an important part of the credit.

Acknowledgments

This book has been published with the help of grants from the Canadian Federation for the Humanities, using funds provided by the Social Sciences and Humanities Research Council of Canada, and the Publications Fund of University of Toronto Press.

Throughout the preparation of the volume we enjoyed the enthusiastic support of our colleagues in the University of Toronto Philosophy Department, and especially of our former chairman, David Gauthier. We are indebted to Barbara Ritomsky and Willa Sobel for textual editing, to Catherine Sherlock for bibliographic work, and to David Winiewicz for compiling the index. We wish especially to thank Myrna Friend for her professional guidance in assembling the bibliography. Linda Rothman, Barbara Anno, and Joyce Wright organized and provided indispensable clerical assistance.

Our largest debt is to our contributors, who were usually faithful in meeting their deadlines and invariably understanding when we failed to meet ours.

Finally, we would like to record a special thanks to Tom and Helen Goudge – to Tom for having been such a patient and co-operative subject, and to Helen for having conspired with us in those enterprises which it would have been indiscreet to reveal to Tom.

Abbreviations

Notes which merely provide citations have, wherever possible, been incorporated into the main text of the essays. The form of such citations has been standardized and is keyed to the List of Works Cited. Special abbreviations have been employed for the following frequently recurring items.

AL   Thomas A. Goudge *The Ascent of Life: A Philosophical Study of the Theory of Evolution* Toronto: University of Toronto Press and London: George Allen & Unwin Ltd. 1961

CP   Charles Hartshorne and Paul Weiss, eds *Collected Papers of Charles Sanders Peirce* volumes I–VI. Cambridge: Harvard University Press 1931–5
Arthur W. Burks, ed *Collected Papers of Charles Sanders Peirce* volumes VII and VIII. Cambridge: Harvard University Press 1958

PP   The Peirce Papers in the Houghton Library, Harvard University. The standard catalogue is Richard S. Robin *Annotated Catalogue of the Papers of Charles S. Peirce* University of Massachusetts Press 1967

TP   Thomas A. Goudge *The Thought of C.S. Peirce* Toronto: University of Toronto Press 1950

PART ONE: THE THOUGHT OF C.S. PEIRCE

DAVID SAVAN

The Unity of Peirce's Thought

It is now more than twenty-five years since the publication of Thomas A. Goudge's *The Thought of C.S. Peirce*. The principal thesis of the book, that there is an irreconcilable conflict between naturalist and transcendentalist tendencies in Peirce's thought, is important and provocative. Morton White, for example, has accepted Goudge's thesis and takes it to be a major clue to the understanding of the history of philosophy in the United States, back to its beginnings in Jonathan Edwards (White 1972). Nevertheless, Goudge's two-Peirce thesis has not yet received careful critical examination. It is such an examination that I will attempt in this paper.

The publication of the first six volumes of Peirce's *Collected Papers* began to make evident his immense range, diversity, originality, and profundity. One of the first responses among philosophers was to search for the unity in this variety. Paul Weiss and Charles Hartshorne found the unifying centre in Peirce's pervasive categories, Firstness, Secondness, and Thirdness. Justus Buchler argued that Peirce advanced a logical empiricism more comprehensive but looser in texture and argument than the logical positivism that flourished between the two world wars. James Feibleman tried to demonstrate that Peirce was a thoroughly systematic philosopher who had developed a realist metaphysics.

It is this assumption of unity that Goudge questioned. Much as Peirce may have tried to develop a systematically unified philosophy, Goudge argued, he did not succeed. He failed because he worked with two conflicting and irreconcilable sets of premisses, one naturalist and the other transcendentalist. Goudge proposes the conflict of naturalism and transcendentalism as an hypothesis, to be tested through an examination of Peirce's work as it appears in the six volumes of the *Collected Papers*. It is Goudge's adjunct hypothesis that Peirce's naturalism is dominant, and is clearest in his theory of inquiry, his work in mathematics and semiotics, in deductive, inductive, and abductive logic, and in his theory of scientific or critical metaphysics. On the other hand, his transcendentalism is most evident in his cosmology, ethics, and theology. This division results in a flowing and orderly analysis, sympathetic and fair, but at the same time critical and penetrating.

After twenty-five years and the appearance of many further excellent studies of Peirce, Goudge's book remains one of the most useful and most frequently cited. This is in itself evidence in support of the central hypothesis on which the book is based. Furthermore, Murray Murphey (1961) and, more recently, Gerd Wartenberg (1971), drawing on the unpublished manuscripts, have amply confirmed the strong transcendentalist strain in Peirce's thought.[1] Prima facie there is a tension in Peirce's thought.

What remains unexamined, however, is Goudge's contention that Peirce did not succeed in reconciling the conflict between his naturalism and his transcendentalism.[2] It is this omission which I wish to remedy.

Goudge sets out the conflict in eight pairs of contrasting propositions which delineate what Goudge means by *naturalism* and *transcendentalism*, as follows. I will label the naturalist propositions N and the transcendentalist propositions T.

1N: 'Scientific method ... is the only means of obtaining knowledge.'
1T: 'The appeal to feeling, sentiment, or instinct is a more important source of knowledge than the appeal to reason or science.'
2N: 'Theory and practice are interdependent and inseparable components of human life.'
2T: 'Theory and practice have no intrinsic connection with one another.'
3N: 'Conceptual precision achieved by logical analysis is a primary desideratum in philosophy.'
3T: '... No precise definition (of the most important philosophical ideas) is possible.' And
3'T: 'Metaphysical construction is a more significant activity than logical analysis.'
4N: 'There is no good reason for accepting a Platonic Realism with regard to universals.'
4T: 'There are convincing reasons for espousing a Platonic realism of universals.'
5N: A 'philosophical account of [the] world requires the use of non-subjective or non-personal categories.'
5T: We 'may properly interpret [the universe] in terms of highly personal or anthropomorphic categories.'
6N: 'All *a priori* modes of thought about matters of fact are illegitimate.'
6T: '*A priori* modes of reasoning are both necessary and legitimate in dealing with such matters as the origin and destiny of man and the universe.'

5 The Unity of Peirce's Thought

7*N*: 'The attempt to construct a "system" of philosophy in which the nature of the world is "explained" should be regarded with suspicion.'
7*T*: 'The aim of philosophical thought is to produce a comprehensive system of knowledge which will be ultimate and final.'
8*N*: 'The conclusions of philosophy must be broadly compatible with the ideas of common sense.'
8*T*: The conclusions of philosophy 'must inevitably differ from the superficial opinions of common sense.'

The examination of these eight pairs (3*T* and 3'*T* may be regarded as a conjunction) of conflicting theses will be facilitated if they are grouped as follows. (A) Science and Scientific Method (1 and 2); (B) Philosophy and Philosophical Method (3, 5, 6, 7, 8); and (C) Realism (4).

A. SCIENCE AND SCIENTIFIC METHOD

It should be noted at the outset that the oppositions presented by 1 and 2 fall short of outright contradiction. Scientific method is a means by which knowledge may be obtained. The sources from which that method draws its hypotheses may well include sentiment and instinct, as well as empirical observation. Again, theory and practice may be interdependent and inseparable, like two close friends, without being intrinsically connected. The question posed, however, is a fundamental one for Peirce's philosophy. Does he hold a consistent and integrated view of the relation of instinct, feeling, and practical action to scientific method?

Goudge speaks of 1 and 2 as premisses. This is misleading. Peirce does not present his analysis of scientific method dogmatically. Rather, he offers an extended argument. It is central to Peirce's conception of science that it is not an alien intrusion into the web of human activity. He attacks Descartes for supposing that science and philosophy can make a sharp break with common belief by means of the purgative of radical systematic doubt.

Among man's basic instinctive needs three are most important – feeding, breeding, and security. Around these needs cluster our primary feelings and emotions. We have a vague dread of doubt, indecision. and uncertainty. Instinctively, we trust those habits of action, in seeking our objectives, which we inherit from our ancestors through the long centuries of biological and social evolution.[3] We cling to pleasing and encouraging beliefs tenaciously, especially if they are familiar. When our unaided efforts are not sufficient to secure our beliefs and habits against surprise and disruption, we rely upon social institutions like the state and the church to hold them fast. If even these powerful external aids fail us, we change our beliefs and habits slowly and minimally. In matters which are of vital and instinctive importance to the survival of the individual and the species, belief is most intense and doubt most disturbing. Belief in such matters is the expression of habits of action and is inseparable from such habits. In moral, religious, and social life, doubt delays, inhibits, and destroys the direct expression of our habits and beliefs.

Scientific method, according to Peirce, arises out of this context of natural instinct, feeling, belief, and action. Our conception of the beliefs and concepts of science is the same as our conception of the practical interpretants of moral, religious, and political beliefs and concepts in habits of action together with attendant sensory perception. Since scientific beliefs are rarely of *vital* importance for the individual, they do not carry the same passionate force that permeates our moral, religious, and social beliefs. Some might, therefore, prefer to speak of scientific opinions and questions, leaving the language of belief and doubt to those other graver concerns (CP 5.394). Still, the aim of science is to secure its theories and observation reports ultimately against doubt. In this respect, it is like those more personal, instinctive, and vital beliefs. Like them too, scientific belief cannot be separated from habits of action. In this sense, it too is practical. There is one more important link joining our primary and personal beliefs with scientific theory. In his search for explanatory hypotheses, the scientist must of course consider observation reports and other hypotheses – those that have been accepted as well, perhaps, as some which have been rejected. But in addition, to hit upon the fruitful and successful hypothesis the scientist must turn to his own fertile imagination, which must be sensitive to the possible ways in which masses, forces, organisms, and societies might act. This sensitivity, or as it is sometimes called, *il lume naturale*, must rest, according to Peirce, upon that primary instinctive adaptation to nature which our need for food, reproduction, and security has formed. 'Unless man [has] a natural bent in accordance with nature's, he has no chance of understanding nature at all' (CP 6.477).

However, all other methods of fixing belief suffer from a fundamental defect from which the scientific method is free. Unscientific methods fix beliefs and change them through caprice and accident. Ruling authorities, or personal taste and emotion are the primary determinants of pre-scientific belief and change of belief. There is no promise of security, no hope that random change of belief will terminate in a set of stable, publicly accepted, true beliefs.

Only the method of science rests on the faith that there is a real world which exists independently of man's wishes, dreams, and imperatives. Hence it is only the scientific method which can make a clear distinction between true and false belief. Peirce is not saying that there is a purely factual and a purely logical component in scientific beliefs. He is saying rather that scientific

7 The Unity of Peirce's Thought

beliefs must be justified by a public and universal standard which is independent of private feeling and of personal practical imperatives. Peirce opposes theory not only to practice, in the sense of moral and social objectives of vital personal importance. He opposes theory also to feeling, taste, and emotion. As usual, Peirce is making a triadic and not a dyadic distinction.

Peirce's conception of the relation of theory and science to feeling and instinct is rather complex, and it has not received the attention given to other aspects of his thought (but see Ayim 1974). As I have pointed out above, Peirce maintains that theoretical inquiry grows out of our basic needs and feelings connected with feeding, breeding, and security. We are strongly inclined to believe what is pleasing and are moved by a vague dread of doubt. Our success in hitting upon successful scientific theories depends in part at least upon our natural sympathy and familiarity with our physical and social surroundings. Further, Peirce contends, scientific inquiry rests upon certain sentiments. It is a matter of faith or trust that there is a real world which is independent of what any man or finite group of men may think it to be. It is a matter of hope that this independent reality can be known eventually through the long painstaking process of formulating, clarifying, and sifting our theoretical beliefs about it. Beyond the two sentiments of faith and hope, the sicentist must also be moved by the love of truth, that is to say, by a willing sacrifice of personal short-term achievement for the long-run approximation to an ideally ultimate and stable truth, agreed upon by the scientific community. It is Peirce's view that sentiment is necessary because there can be no assurance, not even an assessment of probability, concerning our success in arriving at any one truth upon which the scientific community will permanently agree. Peirce can, in his own way, echo Gorgias: there may not be any reality; and if there is, we may never know it; and if we know it, we may not know that we know it.

But theory may serve the mistress of feeling as it may the master of practice – and with equally disastrous results. Thought, Peirce writes, 'may serve to amuse us ... and among *dilettanti* it is not rare to find those who have so perverted thought to the purpose of pleasure that it seems to vex them to think that the questions upon which they delight to exercise it may ever get finally settled; and a positive discovery which takes a favourite subject out of the arena of literary debate is met with ill-concealed dislike' (CP 5.396). In such a case, thought is 'similar to the development of taste; but taste, unfortunately, is always more or less a matter of fashion' (CP 5.383). Beliefs which thus serve the mistress of taste, fashion, and pleasure aim at 'pleasing and encouraging visions, independently of their truth; and thus, upon unpractical subjects, natural selection might occasion a fallacious tendency of thought' (CP 5.366).

Science and theory can flourish only when practice is the servant and interpretant of thought. Theoretical thinking requires the subordination of short-term personal and social goals to a succession which is potentially infinite. Particular cases are significant primarily through what they contribute to our ultimate understanding of a law or theory. When practice is master, however, the situation is reversed. Long-term investigations are subordinated to the need for results now, or in the immediate future. Each practical case is unique and unrepeatable. If my child in the next room cries out in terror, I know what to do – but not through critical scientific research. The point is not merely that time is too short. The point is that what counts is success in saving *this* child *now*. Theoretical correctness in assessing probabilities is worthless in practice if I do not succeed here and now. Since the case is for practical purposes unique, fallacious reasoning which succeeds in saving my child is vastly more valuable than sound reasoning which fails.

It is sometimes argued against Peirce that both feeling and practice, especially the pressure of long-lasting social institutions, ensure a greater security of belief than does science, with its frequent changes of theory, The argument is misdirected. Peirce's point is that the success of practice rests upon accident, and is consequently inherently unstable. Science, on the other hand, rests on a method which systematically discovers its own errors, errors both of method and of result, one by one, and corrects them. If there is any true answer to our questions, it is science alone which offers any hope that it will ultimately be discovered.

The virtues required of a scientist – honesty, willing admission and correction of his errors, communicating and sharing his results – follow from the subordination of practice to science. On the other hand, when practice is master dishonesty results. The conclusions of inquiry are predetermined on practical grounds. Paper doubt is followed by arguments designed to reestablish those beliefs which we cling to on practical grounds. This is one reason Peirce pours scorn upon theologians and seminarians in philosophy. In his own work for the US Geodesic survey he had some personal experience of the effects of subordinating scientific investigation to political considerations.

I submit, then, that there is a tension between science, practice, and feeling or instinct, but that Peirce succeeded in achieving a unified reconciliation of the three. It is not too much to say that Peirce's troubled life was exemplary in its subordination of practice and feeling to scientific and philosophical inquiry.

B. PHILOSOPHY AND PHILOSOPHICAL METHOD

The theses numbered 3, 5, 6, 7, and 8 concern philosophical method, systems, and conclusions, and it will be simplest to treat them together. As in the case

9 The Unity of Peirce's Thought

of the first two pairs of propositions, 1 and 2, it should be noted that 3, 6, 7, and 8 fall short of outright contradiction. Only 5, as Goudge formulates it, is a flat contradiction.

It has already been pointed out that it is Peirce's position that the ideal limit of theoretical inquiry, a limit which may never be reached in fact, differs fundamentally from theory at any prior stage. It is this distinction, applied specifically to philosophy, which resolves the conflict expressed in 3 and 7. As to 6 and 8, any apparent contradictions dissolve once certain ambiguities in their formulations are clarified. In 6 Goudge presumably means by '*a priori* modes of reasoning' inference whose validity can be established without appeal to matters of fact. The question raised by 6 is, then, whether independently of empirically established premises Peirce claims to establish categorical assertions of fact concerning the origin and destiny of man and the universe. I shall argue that Peirce makes no such claim. In 8 the ambiguity lies in the words 'broadly' and 'superficial opinions.' The conclusions of the theory of colour vision, for example, may differ from the *superficial opinions* of common sense while remaining *broadly* compatible with them.

There is no question that Peirce regarded careful logical analysis as a primary philosophical desideratum. For example, the analytical power which he claimed for his system of existential graphs seemed to him a major point in its favour. However, whatever is meant by 'significant' in 3'*T*, Peirce saw the systematic construction of theory, whether in metaphysics or science, as part and parcel of logical analysis. Is the construction of comprehensive scientific theories, like the theory of evolution, or Newtonian mechanics, more significant than the analysis and clarification of the major concepts which figure in such theories? Certainly, Peirce considers the development of theory in science and metaphysics to be of major significance. But the abductive formulation, deductive elaboration, and inductive testing of these theories is inseparable from the analysis and clarification of their concepts.

No theory and no concept, when first formulated, is perfectly precise and exact. The *objects* of our beliefs and theories are initially *general*. In Peirce's published papers of 1867–8 and 1877–8, notably *Upon Logical Comprehension and Extension* and *The Fixation of Belief*, he begins his account of how scientific and philosophical inquiry renders such objects more specific and precise. The *meaning* and sense of our theories and concepts is initially *vague*. In the same two series of papers he begins his account of how to make our ideas and concepts *clear*, less vague. The cognitive meaning or *information* of a symbol is 'the sum of synthetical propositions in which the symbol is subject or predicate' (CP 2.418). 'We can have no conception of wine except what may enter into a belief, either (1) That this, that, or the other, is wine; or, (2) That wine possesses certain properties' (CP 5.401).

10 David Savan

On the other hand, Goudge agrees that it is Peirce's naturalist fallibilism which leads him to conclude that 'there are three things to which we can never hope to attain by reasoning, namely, absolute certainty, absolute exactitude, absolute universality' (CP 1.135). 3*T* is as much part of Peirce's naturalism as is 3*N*. Only in an ideally complete and comprehensive body of confirmed theory could the concepts of science and philosophy be perfectly determinate. Peirce does not claim, *pace* Quine (1960, 20), that there *is* a unique, determinate, and ultimate truth.[4] Peirce insists throughout that we can have no reason for such a supposition. And he gradually came to the conclusion, several times repeated, that our hope must accommodate itself to an inevitable residue of indeterminacy, vagueness, and unanswered questions (CP 4.79, 8.43).

Peirce considers philosophy a science. If its methods have rarely in the past been scientific that is because it has too often been subordinated to feeling and practical purposes. Through his own work and that of others, Peirce believes that in the future philosophy in all its divisions will be based on careful observation and hypothetico-deductive-inductive methods. The pragmatic maxim is one important part of the movement toward a scientific philosophy. Philosophy must begin, then, with a broad but critical examination of the vague and general ideas of common sense (cf 8*N*). It must regard with suspicion any claim to an ultimate system explaining the nature of all things (cf 7*N*). Its ties with common sense may well lead, initially, to anthropomorphic accounts of its categories (cf 5*T*). Ontological arguments which conclude to the existence of matters of fact from premises none of which affirms any matter of fact are illegitimate (6*N*).

Nevertheless, as philosophy develops and tests its hypotheses it will inevitably move away from the superficial opinions of common sense (8*T*). Its categories will be analysed and characterized in terms that are less personal, less anthropomorphic (5*N*). Its aim, perhaps never to be fully attained, is to become more comprehensive, more precise, and clearer (7*T*) (Thompson 1953 164ff).

Peirce writes, in 1868, that the fact acknowledged by common sense 'with which we must commence our inquiry must be one whose existence is indubitable, and whose laws are best known ... This is no other than the process of valid inference' (CP 5.267). He makes it clear that he means abductive and inductive as well as deductive inference. Early in his logical studies, in the mid-1860s, Peirce arrived at the conclusion that the key to the understanding of concepts, propositions, and inference is the triadic sign relation. The three terms or relatives of the sign relation, Ground, Object, and Interpretant, are the three categories, Firstness, Secondness, and Thirdness. The conception of the semeiotic triad is the central unity in Peirce's philosophy, from beginning to end. The critical study of Peirce's semeiotic idealism remains to be written (see Savan 1952 and Savan 1977).[5]

For our purpose in this paper it must suffice to point out that Peirce's basic philosophical hypothesis is that everything is a virtual sign, and that anything actually thought about is an actual sign. Peirce's vast output, both in the natural sciences and in philosophy, is an immense inductive exploration of the possibility of interpreting logic, man, and nature through the semeiotic hypothesis and its categories.

The categories are non-subjective, non-personal. So far 5N represents Peirce's position correctly. But a mind, a person, a man, is a sign. It is this conception of man as a developing hierarchy of interrelated interpretants that Peirce presents in his earliest papers. At the end of his life he writes to Victoria Lady Welby that he still despairs of explaining it adequately and convincingly to his readers. Our sensations, feelings, emotions, actions, and thoughts are interpretants whose objects are the external independent events and processes of nature.[6] The contemplation of nature – what Peirce calls 'musement' – evokes an emotional interpretant. This emotional interpretant is hypothetical or abductive in its inferential form, and it interprets nature as a series of signs whose final interpretant is a quasi-mind. It is this that Peirce is trying to convey when he writes that the heart, that is, the emotions and sentiments, is also a perceptive organ through which we see God (CP 6.493). The dynamic or active interpretant of this awareness is that way of life which loves and pursues truth (CP 6.490). It is this hypothesis of nature as a virtual sign which becomes an actual sign only when it is interpreted that leads Peirce to his anthropomorphic account of the categories (5 T).

I turn finally to 6T. Did Peirce hold that a priori reasoning, alone and without reference to empirically known matters of fact, was necessary and legitimate in dealing with questions concerning the origins and destiny of man and the universe? He nowhere proposes or defends such a position. What Goudge appears to have chiefly in mind are (1) Peirce's seemingly Kantian position that ontology and metaphysics draw upon logic, and (2) Peirce's evolutionary cosmology.

(1) Peirce's conception of the dependence of ontology upon logic is not transcendentalist but semeiotic. He is concerned with the logical analysis of such words as 'being,' 'real,' 'exists,' 'actual,' and 'possible.' 'The conception of being is ... a conception about a sign' (CP 5.294). The word 'reality' particularly concerns the logic of scientific inquiry, since scientific inquiry aims to arrive ultimately at theories which are true, and true not of a fictitious world but of the real world (CP 5.405ff). Peirce would not jib, I think, at saying that to *exist* is to be the value of a variable. It is always important to

remember that when Peirce speaks of the dependence of ontology and metaphysics upon logical analysis he intends such analysis to include abductive and inductive as well as deductive inference.

(2) Peirce recognizes that his evolutionary cosmology is a supposition, an hypothesis, an abduction (CP 1.408ff, 6.189ff). Like any scientific cosmology it must eventually be made sufficiently precise and clear that it can be checked empirically. Although Peirce uses the Hegelian term, *objective logic*, he repeatedly criticizes Hegel for following a purely a priori method, following his intellectual inclinations without taking account of brute existent singularities (CP 5.79, 6.218). Peirce is quite clear that a priori reasoning alone cannot give us any information concerning particular laws or about the general structure of the universe. 'The relative probability of this or that arrangement of Nature is something which we should have a right to talk about if universes were as plentiful as blackberries ... But, even in that case, a higher universe would contain us, in regard to whose arrangements the conception of probability could have no applicability' (CP 2.684).

The matters of fact involved on Peirce's cosmological hypothesis are (a) that 'at any assignable date in the past, however early, there was already some tendency toward uniformity' (CP 1.409), (b) that at any assignable date in the future there will be some aberrancy from law (CP 1.409), and (c) that the direction of change is from vague singularities in which detailed microscopic information approximates to zero toward an entropic state in which microscopic information approximates to its maximum. Because the universe is infinite, novel singularities are constantly introduced, so that the final state of dynamic equilibrium – what Peirce calls 'dead matter' (CP 6.585) – required by the principle of entropy is never reached.

Such a vague and general theory, Peirce realized, must be made more precise and clear if empirical considerations are to count for or against it. He did in fact make some efforts to do just that. And he considered his existential graphs to be an admittedly inadequate model for such a cosmological hypothesis. That the theory is not such a transcendental one as Goudge claims may be seen by comparison with the much more detailed and specific cosmology recently proposed by Layzer (1975).[7] I believe that Peirce would have welcomed Layzer's work as one of several possible clarifications of his own much vaguer theory.

What of the figurative and mythological language in which Peirce writes of the evolution of the cosmos? Does that not demonstrate his deep-seated transcendentalism? The answer is that it does. None can deny the strong transcendentalist strain in Peirce's thought and leanings. Just as his mathematician father, and various physicists, described God as a divine mathematician, Charles Peirce thinks of God figuratively as a divine semeioticist, a divine scriber of existential graphs. I hope to have shown, however, that Peirce succeeded in reconciling his transcendentalist and naturalist tendencies through his semeiotic philosophy of science.

C. REALISM

There remains the contradiction between 4N and 4T. Was Peirce a Platonic realist with regard to universals? The question has continued to exercise students of Peirce's philosophy, notably Boler (1963), Fisch (1967), and Roberts (1970). I will confine myself here, however, to the question of the contradiction between Peirce's naturalist rejection and transcendentalist acceptance of universals.

If Platonic realism is taken as asserting that universals are single or individual entities of a special kind, then Peirce is clearly opposed. Such a theory treats what is general as a singular, and in Peirce's eyes it is a disguised nominalism. The realism which he defends and tries to define is a realism of general laws. Such a realism Peirce prefers to call *Scotist* or *Scholastic* rather than Platonic. When the distinction between a Platonic realism of singular universals and a Scotist realism of general laws is kept in mind, the apparent contradiction between 4N and 4T vanishes.

What is the difference between a random uniformity such as the chance truth that all the people at a public lecture have surnames beginning with 'S,' and a law or general truth – for example, that if a stone is released from the hand it will fall? According to Peirce, the nominalist denies that there is any general difference between these two cases. Uniformities in nature are all random regularities of single unconnected events. If causality is a non-random regular sequence, as Peirce maintains, then such a chance uniformity is not causal. If a world of pure chance were possible, agreement between two or more observers, or between two or more interpretants, would also be a matter of chance. There could be no reliable way in which any procedure could be counted on to bring about a publicly agreed belief. There could be no scientific – or other – methods for bringing about an eventual agreement on what is true. Given Peirce's semeiotic theory of reality, this is equivalent to saying that in a world of pure chance nothing is real.

Peirce's Scotist realism is thus integral to his semeiotic philosophy. Unless some general laws, as distinct from chance regularities, were operative in nature, there could be no hope of communicable agreement in belief and so no hope of truth. In order that information be reliably communicated, there must be a theoretically possible way of deciding concerning two or more interpretants whether or not they are true of the same object. The object must have some minimal efficiency in bringing about this result (CP 5.431). It is because Peirce maintains that every real natural object is a virtual sign that he attacks Platonic realism and defends Scotist realism.

I conclude that Goudge is indeed right – there is an important tension in Peirce's thought between naturalism and transcendentalism. I have tried to indicate, however, that there is no such fundamental *contradiction* as Goudge imputes to Peirce. Peirce's semeiotic idealism is the basic unity, never adequately worked out, never clearly elucidated, which mediates and resolves the tension.

J.W. BURBIDGE

Peirce on Historical Explanation*

Professor Thomas A. Goudge, in initiating a discussion of Peirce's naturalism, refers to his analysis of human inquiry.

Few men were as well equipped as Peirce to undertake such a study. His first-hand acquaintance with the procedures employed in mathematics and logic, his training in chemistry, his practical experience in astronomy, experimental psychology, geodesy, and optics, provided him with a wealth of material for examination. His great generalizing power enabled him to use this material for the formulation of principles exhibited in all types of inquiry. And his naturalistic orientation led him to insist that since inquiry does not take place in a void, it must be exhibited as a mode of human activity (TP 11).

What Goudge does not mention in the list of sciences, however, is Peirce's interest in ancient history and in the procedures of historical inquiry. Yet in the Lowell Lectures of 1903 Peirce referred to five occasions when he 'had an opportunity of testing my Abductions about historical facts, by the fulfillment of my predictions in subsequent archeological or other discoveries; and on each one of those five occasions my conclusions, which in every case ran counter to that of the highest authorities, turned out to be correct' (CP 7.182 n7). And at the National Academy of Sciences in November 1901 he read the abstract of a long paper on 'The Logic of Drawing History from Ancient Documents.'[1] Its argument develops themes already contained in a series of studies of Hume's essay on miracles written at the request of Samuel P. Langley.[2] Since Peirce not only claimed success in historical investigations but also studied the methods of historical inquiry, therefore, there is reason to

---

*I first tackled this topic in a graduate course taught by Professor Goudge in the fall of 1968.

include history in the list of sciences 'which provided him with a wealth of material for examination' (TP 11). Indeed many of his conclusions bear directly on recent debates concerning the nature of historical explanation.

Some passages, for example, resemble the thesis put forward by Carl Hempel (1949). With reference to a historical question concerning Pythagoras Peirce writes: 'An explanation is a syllogism of which the major premiss, or rule, is a known law or rule of nature, or other general truth; the minor premiss, or case, is the hypothesis or retroductive conclusion, and the conclusion, or result, is the observed (or otherwise established) fact' (CP 1.89).[3] This parallels Hempel's claim that explanation involves subsumption under a law:

The scientific explanation of the event in question [E] consists of
(1) a set of statements asserting the occurrence of certain events $C_1, \ldots C_n$ at certain times and places,
(2) a set of universal hypotheses, such that
    (a) the statements of both groups are reasonably well confirmed by empirical evidence,
    (b) from the two groups of statements the sentence asserting the occurrence of event E can be logically deduced (Hempel 1949, 460).[4]

Both maintain that explanation requires a logical inference from a general truth (or universal hypothesis) and a specific case (or determining conditions) to the event in question.

At the same time, however, other passages echo themes in Collingwood (1946). 'It is a primary hypothesis underlying all abduction,' Peirce writes, 'that the human mind is akin to the truth in the sense that in a finite number of guesses it will light upon the correct hypothesis' (CP 7.220; cf 6.530). Earlier he had noted: 'If we subject the hypothesis, that the human mind has such a power in some degree, to inductive tests, we find that there are two classes of subjects in regard to which such an instinctive scent for the truth seems to be proved.' One of these is 'in regard to the ways in which human beings and some quadrupeds think and feel' – the science of psychics (CP 6.531). The suggestion that the mind can relate instinctively to the way other humans think and feel is analogous to, if not identical with, Collingwood's claim that 'history is nothing but the re-enactment of past thought in the historian's mind' (Collingwood 1946, 228).

On the basis of these superficial comparisons, then, we appear to be faced with a dilemma. For Peirce, is the method of historical inquiry to subsume an event under a general law, or is it to reproduce in thought the events which are being investigated? Does Peirce's philosophy of history support the thesis of

17 Peirce on Historical Explanation

the covering law theorist, or does it accept the claim that history has its own unique methods of explanation? Can he successfully overcome the apparent discrepancy which results from combining Hempel and Collingwood? The following reflections will attempt to resolve the dilemma.

I

Peirce does not adopt the straightforward positivist view of general law outlined by Hempel. In his various drafts of the article on Hume and in the long paper on history Peirce explicitly examines such a conception and finds it wanting when applied to historical events.

Hempel defines a general law or universal hypothesis as that which asserts 'a regularity of the following type: In every case where an event of a specified kind C occurs at a certain place and time, an event of a specified kind E will occur at a place and time which is related in a specified manner to the place and time of the occurrence of the first event' (Hempel 1949, 459). Peirce, while avoiding the suggestion of causal connection which Hempel introduced with the symbols 'C' and 'E,' renders explicit

> two common characters of all truths called laws of nature. The first of these characters is that every such law is a generalization from a collection of results of observations, *gathered* upon the principle that the observing was done so well to conform to outward conditions; but not *selected* with any regard to what the results themselves were found to be ...
>
> The second character is that a law of nature is neither a mere chance coincidence among the observations on which it has been based, nor is it a subjective generalization, but is of such a nature that from it can be drawn an endless series of prophecies, or predictions, respecting other observations not among those on which the law was based (Wiener 1958, 289–90).[5]

The first character requires that each result observed be independent of all others. Such independence is implicit in Hempel's formulation of the law, for the symbols 'C' and 'E' isolate classes of events (Hempel 1949, 460) which simply include a collection of discrete individuals. In addition, while he recognizes that 'the complete description of an individual event ... would require a statement of all the properties exhibited by the spatial region or the individual object involved, for the period of time occupied by the event in question' (Hempel 1949, 460–1), Hempel does not discuss how they are to be combined. The independence of the data, however, serves as a sufficient condition for formulating laws mathematically as probabilities. Therefore the most accurate explanation of an event with many properties would involve computing its probability simply by multiplying the quantities cited in the various laws.[6]

Whether or not these assumptions are appropriate to the methods of natural science, Peirce points out that they are singularly inappropriate to questions of history. When, for example, a historian has in front of him a document reporting an unfamiliar state of affairs he cannot simply compute the probability of a witness reporting an event falsely against the probability of the event's actual occurrence. For the required independence is not present. Whether or not a witness tells the truth about an event depends on what the character of that event is and on the subjective expectations of the witness vis-à-vis that character. 'The notion that a man tells a falsehood once in so often, *independently* of what the subject matter is,' notes Peirce, 'is a theory too far from the truth to be of the slightest service' (Wiener 1958, 313; cf CP 7.176).

Hempel's failure to recognize this point leads to further complications. The lack of independence makes it more difficult to isolate any property such that it can be formulated precisely in the specific terms required by Hempel's definition of universal hypotheses. For independence is a necessary condition for quantification without distortion. Laws now are more imprecise simply because the terms used cannot be subsumed into a mathematical formula. Hempel, indeed, is forced to recognize that such vagueness is present in historical explanation. Even though he wants to argue that, ideally, all laws can be stated in terms of mathematical probability, he admits that 'it would often be very difficult to formulate the underlying assumptions explicitly with sufficient precision and at the same time in such a way that they are in agreement with all the relevant empirical evidence available' (Hempel 1949, 464).[7]

Hempel adds, however, another reason for the lack of precision. 'The universal hypotheses in question frequently relate to individual or social psychology, which somehow is supposed to be familiar to everybody through his everyday experience; thus, they are tacitly taken for granted' (Hempel 1949, 464). Peirce focuses on this practice and relates it to the lack of precision. Because the general laws cannot be formulated mathematically and therefore tested by means of random samples, those who appeal to the positivist theory substitute familiar suppositions for empirical laws. Instead of leading to a more rigorous explanatory procedure, however, this leaves the researcher free to rely on subjective generalizations – beliefs and expectations which reflect the limitations of his own intellectual experience. Such a method Peirce caricatures with reference to the 'higher critic' for whom 'probability'

19 Peirce on Historical Explanation

means 'the degree to which a hypothesis in regard to what happened in ancient Greece recommends itself to a professor in a German university town ... A probability, in that sense, is nothing but the degree to which a hypothesis accords with one's preconceived notions; and its value depends entirely upon how those notions have been formed, and upon how much objectivity they can lay a solid claim to' (CP 7.177; cf Wiener 1958,313). Unable to perform a rigorous calculation of mathematical probabilities, the historian substitutes the method of balancing likelihoods,[8] which can in its extreme form ignore documents and testimonies simply because they are unlikely. Ironically, the result is a surrender to 'the method of empathetic understanding' which it is Hempel's purpose to reject (Hempel 1949, 467).

What opens the door in practice to this reliance on unverifiable assumptions is the failure to recognize the lack of independence among the various events and phenomena to be explained. Far from indicating a procedure which would make history more rigorous, therefore, Hempel's model fails to take account of the specific logical features involved. The practising historian cannot convert his 'explanation sketches' to 'explanation' simply because the required independence cannot be imposed on the data. Paradoxically, then, if the historian takes Hempel seriously he is prevented from explaining events in a way which retains adequate testing procedures.

In history documents are related to events not simply by means of external, observed correlations, but through the conscious reaction of witnesses to the events, and of participants in events to other events. These reactions are not simply a matter of cause and effect, but of reason and consequence. This means that an adequate reconstruction and explanation of the historical event requires some statement of rational connections and of dependent relations. It was precisely to establish this point that Collingwood argued his case for history as the re-enactment of past experience (cf Collingwood 1946, 282ff).

II

Even though Peirce agrees with Collingwood that the data to be explained by the historian are not independent events and therefore resemble mental functions, he suggests that thought alone is not sufficient for a real explanation.

'The processes of nature,' writes Collingwood, 'can ... be properly described as sequences of mere events, but those of history cannot. They are not processes of mere events but processes of actions, which have an inner side, consisting of processes of thought; and what the historian is looking for is these processes of thought' (Collingwood 1946, 215). Thought is able to re-enact the actions to be explained because it is able to discover the continuity which binds them together into a coherently related whole. 'The peculiarity of thought,' adds Collingwood, 'is that, in addition to occurring here and now in this context, it can sustain itself through a change of context and revive in a different one' (Collingwood 1946, 297).[9] Thought, then, is not an immediate presence but a mediating relation. Therefore, turning his back on the scissors and paste method which takes each document as an independent given, the contemporary historian endeavours to mediate between events by re-enacting the thought implicit in them. It is this structure of internal relations which is ignored by the covering law theory of Hempel.

For Peirce as well, the purpose of explanation is to discover the continuity which underlies events. Discreteness and independence present themselves most forcefully to the researcher in the moment of surprise. Whenever we expect a regularity or relation and do not find it, or whenever we expect a complete lack of relation and discover instead an evident pattern, we are stimulated to develop an explanation.[10] The moment of surprise reflects this conflict between a perceived or expected continuity and an event which appears to be independent of that continuity. 'An isolated fact,' says Peirce, 'is precisely what a demand for an explanation proper never refers to; it always applies to some fact connected with other facts which seem to render it improbable' (CP 7.200; cf 8.270). The historian who seeks to explain events, then, must establish a continuity between events by supplying 'a proposition which, if it had been known to be true before the phenomenon presented itself, would have rendered that phenomenon predictable, if not with certainty, at least as something very likely to occur' (CP 7.192).

As with Collingwood, this explanatory endeavour is a process of rational thought that presupposes a fundamental continuity between the activities of our mind and the structures to be explained. Peirce writes: 'We are ... bound to hope that, although the possible explanations of our facts may be strictly innumerable, yet our mind will be able, in some finite number of guesses, to guess the sole true explanation of them' (CP 7.219). This expectation has received some measure of empirical confirmation. 'If we subject the hypothesis, that the human mind has such a power in some degree, to inductive tests, we find that there are two classes of subjects in regard to which such an instinctive scent for the truth seems to be proved ... In fact, the two great branches of human science, physics and psychics, are but developments of that guessing-instinct under the corrective action of induction' (CP 6.531).[11]

The fact that Peirce includes physics with psychics suggests, however, a basic disagreement with Collingwood. For Collingwood history is an intellectual discipline absolutely distinct from the natural sciences.[12] For Peirce the two cannot be so contrasted. For natural science has the same structure. 'Every scientific explanation of a natural phenomenon is a hypothesis that

21 Peirce on Historical Explanation

there is something in nature to which the human reason is analogous; and that it really is so all the successes of science in its applications to human convenience are witnesses' (CP 1.316).

To this disagreement is related another: that of the terms used. Peirce, for example, talks about the mind 'guessing' possible explanations, whereas Collingwood speaks of the 're-enactment' of what actually occurred.[13] This opens a double question: How, according to Peirce, are possible explanations which are the result of guesses transformed into *real* ones? (Cf CP 8.270) On the other hand, what procedure does Collingwood propose to prevent guesses or 'likelihoods' being substituted for adequate re-enactment?

We turn to the latter question first. For Collingwood, the testing of historical explanations takes place within the mind: 'The historian not only re-enacts past thought, he re-enacts it in the context of his own knowledge and therefore, in re-enacting it, criticizes it, forms his own judgement of its value, corrects whatever errors he can discern in it. This criticism of the thought whose history he traces is not something secondary to tracing the history of it. It is an indispensable condition of the historical knowledge itself' (Collingwood 1946, 215). Although the historian relies on evidence presented, this datum simply becomes the occasion for questioning. The crucial development of an explanation takes place after the evidence has been noted when thought rigorously investigates the meaning of what is given and puts it under scrutinizing cross-examination. Critical reflection on the evidence previously given is the procedure advocated by Collingwood.

However, the illustration Collingwood provides for the historical method – the detective scenario 'Who killed John Doe' – contains an operation the significance of which has been overlooked. Consider the final paragraph of the Inspector's monologue:

'Well, well: there's a pretty story for you; but how can we tell whether it's true or not? There are two questions we've got to ask. First: can we find the ashes of those gloves? And the metal buttons, if they are like most of his gloves? If we can, the story is true. And if we can find a lot of writing-paper ash as well, the blackmail bit is true, too. Second: where is that jacket? Because if we can find the tiniest speck of John Doe's paint on it, there's our case' (Collingwood 1946, 273).

This passage belies Collingwood's claim that 'every step in the argument depends on asking a question' (Collingwood 1946, 273). For a different operation was introduced: The ash bin was sifted to find metal buttons, and the gift of the jacket to a deserving parishioner was discovered.

Collingwood would claim that this simply adds more evidence to be explained.[14] But there is a difference between the evidence from which the investigation starts and the evidence which is discovered in the attempt to verify a hypothesis. For Peirce, the distinction is central: 'Augustus De Morgan very simply demonstrated that, taking any selection of observations whatever, propositions without number can always be found which shall be strictly true of all those observations ... and yet no one of them likely to be true of any other observations which the same principle of selection might add to the collection' (Wiener 1958, 290). When thought refers simply to evidence given in the past, then, it cannot assume that the explanation proposed will extend to other data not yet observed. The Inspector had to find a generalized explanation which would not only cover the facts known, but would also anticipate new facts to be discovered.

It is this procedure which transforms a *possible* explanation into a *real* one. A possible explanation, hypothetically proposed, must certainly cover the facts already perceived. But it only becomes a real explanation when it successfully anticipates other facts. This reference back to the evidence of immediate experience in the future is the critical test against which the proposed constructions of thought, whether 'guessed' or 'imagined' are measured. Verification by successful prediction is a procedure common to both physics and psychics.[15]

Because Collingwood does not explicitly include this reference to successful prediction in his analysis of the historical method, he leaves open, once again, the possibility of preconceived notions overwhelming the critical question of truth. Subjective contingency will determine what questions are asked, and for how long, in the process of critical reflection. However, the purpose of explanation is not to provide thought with a comfortable generalization; it seeks to reconstruct the objective state of affairs.

From Peirce's point of view, then, Collingwood's theory fails because it does not introduce any criterion for assessment which is in some sense independent of the process of thought itself. On this question he agrees with Hempel, who states that the method of empathetic understanding 'may sometimes prove heuristically helpful; but its use does not guarantee the soundness of the historical explanation to which it leads. The latter rather depends upon the factual correctness of the empirical generalizations which the method of understanding may have suggested' (Hempel 1949, 467).

III

In a footnote Peirce defines what he means by 'reasonable':

By reasonableness, I mean, in the first place, such unity as reason apprehends – say generality ... However, Generality, as commonly understood, is not the whole of my

23 Peirce on Historical Explanation

'reasonableness.' It includes *Continuity*, of which, indeed, Generality is but a cruder form. Nor is this all. We refuse to call a design reasonable unless it be feasible ... These, then, are the naked abstract characters that must be recognized in the 'reasonableness' of a law of nature.[16]

To generality, then, he has added two further features: continuity and feasibility. The first clarifies more precisely the nature of the understanding process, the second the relation to reality.

This passage bears directly on the fundamental question in the debate between Hempel and Collingwood: the nature of universals. While Hempel assumes that each universal is denotative, defining a set of independent entities having the same property, Collingwood holds that the universals relevant to history are the mediating continua articulated by thought. The independence of entities in Hempel's theory grounds the reliability of objective empirical verification of the general law. The structure of continuity stressed by Collingwood requires the critical questioning of thought to establish a coherent framework within which all the evidence can be encompassed. In his definition of reasonableness, Peirce places a foot in each camp. He agrees with Collingwood in equating universality and continuity, and with Hempel in relating universality and feasibility. To cope successfully with both continuity and feasibility, however, Peirce's theory must do two things. On the one hand it must establish the structure of generality as continuity; on the other it must permit, and require, a procedure for testing explanatory proposals empirically. We will consider each of these questions in turn.

(1) Earlier in the paper we acknowledged a parallel between Hempel's claim that in explanation an event's occurrence should be deduced from a general law and a set of specific conditions and the position of Peirce. There is, however, a critical difference.

As we have seen, Peirce defines explanation as 'a syllogism of which the major premiss, or rule, is a known law or rule of nature, or other general truth; the minor premiss, or case, is the hypothesis or retroductive conclusion, and the conclusion, or result, is the observed (or otherwise established) fact' (CP 1.89). The known law relates the middle term to the predicate of the conclusion, while the hypothesis proposes that the subject be characterized by the middle term. Since Peirce allows that the major premiss 'is not actually thought, though it is in the mind habitually' (CP 8.65), it is not the purpose of explanation to focus directly on the 'general law.' The critical question, then, is the relation between the conclusion 'S is P' and the minor premiss, 'S is M.' Since 'S is P' is given, the task is to discover an 'M' which we already believe on other grounds to be related to 'P.'

Another of Peirce's definitions throws more light on this structure. 'Explanation ... is the replacement of a complex predicate, or one which seems improbable or extraordinary, by a simple predicate from which the complex predicate follows on known principles' (CP 6.612). The middle term, '*M*,' then, is proposed as a hypothesis because it unifies rationally a complex set of predicates which have been discovered to be true of the subject.

Peirce himself provides an illustration:

I see a woman of forty. Her countenance is so sinister as scarcely to be matched among a thousand, almost to the border of insanity, yet with a grimace of amiability that few even of her sex are sufficiently trained to command: – along with it, those two ugly lines, right and left of the compressed lips, chronicling years of severe discipline. An expression of servility and hypocrisy there is, too abject for a domestic; while a certain low, yet not quite vulgar, kind of education that is evinced, together with a taste in dress neither gross nor meretricious, but still by no means elevated, bespeak companionship with something superior, beyond any mere contact as of a maid with her mistress. The whole combination, although not striking at first glance, is seen upon close inspection to be a very unusual one. Here our theory declares an explanation is called for; and I should not be long in guessing that the woman was an ex-nun. (CP 7.196)

In this example the proposed explanation, 'ex-nun,' integrates the complex set of predicates 'about forty,' 'sinister countenance,' 'subject to severe discipline,' etc. From it we can infer with some measure of probability not simply to the likely presence of each of the alternative predicates but, more significantly, to their being combined in just this fashion. There is something about being a nun in the nineteenth century which shapes and forms the person such that these specific predicates become appropriate. Because they are deducible from the proposed mediating predicate 'ex-nun,' the various descriptions can be shown not to be independent of each other.[17]

This act of guessing a hypothesis, called 'abduction,' is central to the historical enterprise. The historian has presented to him a number of testimonies and monuments which relate to a particular event or set of events. To explain them he must propose a general term from which the actual combination of testimonies and monuments can be inferred on the basis of known principles.[18] On other occasions, when a particular observed predicate does not conform to expectations based on known facts, the researcher, by guessing, proposes a middle term which would relate the unexpected predicate to the rest of his evidence, and thus reintroduce continuity.

Hempel, because he took universals to be simply names for sets of independent events, concentrates on the major premiss or general law in his theory of

25 Peirce on Historical Explanation

explanation. Peirce, however, shows that, while the major premiss is essential to a successful explanation, it need not be explicit. Indeed, there are an indefinite number of such principles present in the mind of the investigator upon which he may call. The critical task is to determine the right middle term, for it alone will integrate the set of observed characters within a coherent continuum. This is why Peirce does not follow Hempel in dismissing as incomplete historical explanations which 'fail to include an explicit statement of the general regularities they presuppose' (Hempel 1949, 464). Since the primary task of explanation is to provide the minor premiss and middle term, the major premiss is not central.

(2) Abduction only proposes possible explanations. It does not establish them as real. If the connection between the middle term and the predicates were a causal correlation of independent events, then an experimental procedure could, by establishing the relative frequency of such a correlation, test the reliability of the proposal. In the model discussed, however, the middle term integrates the various predicates. In contrast to correlation, integration establishes an intrinsic relation of mutual dependence and thereby destroys both independence and linear causal sequence. How, then, can the hypothesis be verified?

Once again, Peirce answers the question with an example:

Suppose we wish to test the hypothesis that a man is a Catholic priest, that is, has all the characters that are common to Catholic priests and peculiar to them. Now characters are not units, nor do they consist of units, nor can they be counted, in such a sense that one count is right and every other is wrong. Characters have to be estimated according to their significance. The consequence is that there will be a certain element of guesswork in such an induction; so that I call it an *abductory induction*. (CP 6.526)

The characters in question are the predicates to be applied to the man as subject. The term 'Catholic priest' is the proposed middle term which is to be tested. Therefore this example fits the model Peirce has proposed for explanation. Instructive, then, is his description of the method of testing. One deduces some other characters that belong to Catholic priests besides those already noticed in this individual; one then searches for, and finds, evidence of such characters. This procedure, Peirce notes, 'is an induction, because it is a test of the hypothesis by means of a prediction, which has been verified' (CP 6.526). But because the sample is not determined randomly, nor is any specific test numerically equivalent to any other in value, this inductive procedure does not follow the canons of probability calculus. Indeed, because 'characters have to be estimated according to their significance' an element of reflective evaluation is introduced. The testing procedure is not, then, strictly objective. At each stage there remains an element of guessing which relies on the instinctive capacity of thought to which Peirce has applied the adjective 'abductory.'

Later in the same year Peirce elaborated on this genus of induction. He contrasts it with the sampling of independent units to establish probabilities, and with the use of persistent reference to experience over a period of time in such a way that errors will be corrected, as in the testing of natural uniformities.[19] The third type 'tests a hypothesis by sampling the possible predictions that may be based upon it. Predictions are not units; for they may be more or less detailed. One can say roughly that one is more significant than another; but no approach to actual weighing of their significance can, in most cases, be made' (CP 7.216). As a result no particular test will be either strongly positive or decisively negative. Nor indeed can it be quantified without distortion. However, evidence from the testing may confirm, disconfirm, or suggest modifications to the original hypothesis.

The procedure for testing whether a proposed middle term will in fact explain a situation or not, then, involves deducing from that term other possible predicates which could be ascribed to the subject, sampling that collection of possibilities, and investigating to see whether such predicates can in fact be discovered. Any results are evaluated reflectively in light of other principles and contexts. It was this procedure that the Inspector followed when, in Collingwood's illustration, he tested his hypothesis that the rector killed John Doe (Collingwood 1946, 273).[20] He inferred that the rector would have burned his gloves and disposed of his jacket. These possible implications then became the basis for further investigation to validate the hypothesis. Since the confirming evidence was strong, the two tests were sufficient.

In historical research, unlike judicial investigation, it is not necessary to reach a definitive conclusion within a specified period of time (see Collingwood 1946, 268). Indeed, Peirce says, historical inquiry 'never will be completely closed,' for 'scientific interest lies in finding what we roughly call generality or rationality or law to be true, independently of whether you and I and any generations of men think it to be so or not' (CP 7.185, 186; cf 5.589). In this, science is sharply distinguished from practice, for 'practice requires something to go upon' (CP 5.589). It cannot wait until complete certainty has been established. It believes certain things and as a result decides and acts. Peirce allows that the more established a scientific hypothesis is, the more reliable it is for decision: 'In other words there is now reason to believe in the theory, for belief is the willingness to risk a great deal upon a proposition. But this belief is no concern of science, which has nothing at stake on any temporal

27 Peirce on Historical Explanation

venture but is in pursuit of eternal verities (not semblances to truth) and looks upon this pursuit, not as the work of one man's life, but as that of generation after generation, indefinitely' (CP 5.589). This long-term perspective for research in history (as well as in any science) anticipates that any wrong results which stem from the faulty weighing of evidence will in time be corrected through experience. The contingency introduced in the evaluation of confirming evidence, although unavoidable, is thus not vicious. And history can take its place as a discipline whose results can be empirically validated.

Peirce thus demonstrates that the universals required for historical explanation combine the concept of continuity with an element of feasibility or predictability. While his model may need elaboration and refinement in light of more recent discussions, there is sufficient evidence from the preceding dialogue between Peirce, Hempel, and Collingwood to confirm Professor Goudge's conclusion: 'Peirce's subtle, original, and wide-ranging intellect was struggling to break loose from the patterns of thought typical of his own century, and to point the way to the leading conceptions of the next.' As with his other work, Peirce's writing on the methods of historical inquiry provides 'an enormously rich and stimulating *point de départ* for contemporary thought' (TP 345).

MAX H. FISCH

The 'Proof' of Pragmatism*

There are very few topics of Peircean scholarship that are not dealt with in *The Thought of C.S. Peirce* or in other publications by Thomas A. Goudge. One of those few is the elusive 'proof' of pragmatism.

When Herbert Spencer died in December 1903, Peirce was giving a series of lectures at the Lowell Institute in Boston. He began his next lecture with a tribute to Spencer in the course of which he said: 'When philosophy becomes an adult science, as it will before the twentieth century is half over, the first question to be asked in weighing the importance of any philosopher will be what important truth did he *prove*, in the sense in which truths of philosophy can be proved' (PP 470, 146).

Less than a year later, in the article 'What Pragmatism Is,' he called his own form of it 'pragmaticism' and promised a second article on some of its applications and a third containing 'a proof that the doctrine is true.' That proof, it seemed to him, was 'the one contribution of value' that he had to make to philosophy. 'For it would essentially involve the establishment of the truth of synechism.'[1]

Still another year later, sketching his proof in a letter to the Italian pragmatist Mario Calderoni, he wrote that the 'one contribution to philosophy' he had already made was his 'New List of Categories,' thirty-eight years earlier; and he went on to show how that contribution functioned in the contribution he had still to make, the proof of pragmatism (CP 8.213).

The 'New List' had been confined to what Peirce later called universal categories, as distinguished from such particular categories as those of Hegel's Encyclopedia. In the end Peirce concluded that no final list of particular

---

*Presented to the Society for the Advancement of American Philosophy at Chicago, 24 April 1975; to the Department of Philosophy at Pennsylvania State University, 29 April 1975; and to the Washington Philosophy Club, 11 October 1975.

29 The 'Proof' of Pragmatism

categories is possible, but he believed that he had proved that there are three universal categories and not more, and that he had accurately identified them. It was chiefly on that proof that the proof of pragmatism rested (CP 5.43; see Fisch 1975, 172, 178–88).

Our first question is, in what sense *can* truths of philosophy be proved? Not in the sense in which the theorems and problems of mathematics can be proved. For that kind of proof Peirce reserved the term *demonstration*. He produced several such demonstrations. But the question of the truth of pragmatism is a question of what *is* true, not of what *would be* true under an arbitrary hypothesis, such as those of pure mathematics. The proof of matters of fact consists in bringing them to the test of indubitable experience. 'When I say *indubitable*, I mean of course indubitable today for me. Nothing can be imagined more absolutely satisfactory than that, being indubitable to me, it is equally so to you, for your doubting it would cause me to do so.'[2] 'Now, proof does not consist in giving superfluous and superpossible certainty to that which nobody ever did or ever will doubt, but in removing doubts which do, or at least might at some time, arise' (CP 3.432).

Our second question is, in what does the pragmatism that Peirce is proving consist? What are the matters of fact that it asserts? For a first approximation, we go back to his first published account in the 'Illustrations of the Logic of Science' in the *Popular Science Monthly* in 1877 and 1878, and particularly to the focal paragraph in 'How To Make Our Ideas Clear': 'It appears, then, that the rule for attaining the third grade of clearness of apprehension is as follows: Consider what effects, which might conceivably have practical bearings, we conceive the object of our conception to have. Then, our conception of these effects is the whole of our conception of the object' (CP 5.402).

Construing this paragraph in the light of all six of the published 'Illustrations,' we gather that pragmatism consists in affirming (1) that there is a third grade of clearness of ideas beyond the two discerned by Descartes and Leibniz; (2) that this third grade of clearness is characteristic of the scientific way of fixing beliefs, as distinguished from such other ways as tenacity, authority, and apriority; (3) that the scientific way of fixing beliefs, including the application of this rule for attaining the third grade of clearness, is applicable to logic and metaphysics; (4) that the above formulation of the rule is accurate and adequate; and (5) that once we have applied the rule to a given conception and thus attained the third grade of clearness of apprehension of it, there will remain nothing in it to be brought into the clear by applying any different or additional rule.

Assuming that pragmatism, at a first approximation, consists in these five affirmations, our third question is, did the "Illustrations' contain a proof of it?

Yes – of a sort. At least the greater part of one. It includes everything in the first two papers up to the focal paragraph. It includes the remainder of the second paper, in which Peirce brings us to the third grade of clarity concerning the rule itself by applying it to four important conceptions: hard, weight, force, and reality. It includes the third paper in which he goes on to apply the rule to a fifth and more difficult conception, that of probability. It is continued in the three following papers, in which he shows, or begins to show, that pragmatism yields an adequate account of the logic of science. But the six 'Illustrations' were to have been followed by others; the whole series was then to be brought out in book form; the book was announced as forthcoming in Appleton's *International Scientific Series*; and there were to be French and German translations of it. Neither the book nor any further 'Illustrations' appeared, so the published proof remained incomplete. But, as Peirce later said, the essential part of it, which opened the way to all the rest, consisted in showing, without mentioning him, that pragmatism was 'scarce more than a corollary' of Bain's definition of belief. What Peirce neglected to add was that it was only by applying his own three universal categories to the conceptions of belief, doubt, and inquiry, without notifying the reader that he was doing so, that pragmatism was shown to be such a corollary (CP 5.12; see Fisch 1954, 434–8).

Our fourth question is, what were the actual doubts that that first proof of pragmatism was intended to remove? Presumably the doubts that had been raised in the earlier 1870s in the Metaphysical Club in which pragmatism was born, and the doubts that had troubled Peirce since that time. As to what those doubts were, our evidence is still insufficient, and I therefore postpone the attempt to answer this question.

Our fifth question is, if Peirce had proved pragmatism in the later 1870s, and had got the greater part of the proof into print, why was he setting himself, more than a quarter of a century later, to prove it all over again, as 'the one contribution of value' that he had to make to philosophy? And our sixth question is, had he made any intervening attempts, and, if so, why, and in what respects had they failed?

To answer the sixth question first, there had been several intervening attempts, three of which must be mentioned here.

(1) In the spring of 1893, under the title *Search for a Method*, Peirce worked on a collection in book form of nineteen or more of his previously published essays, in chronological order, with the six 'Illustrations' in the middle. The 'method' of the *Search* was pragmatism, but a pragmatism purged of its initial errors by the addition of many new footnotes dated 1893, and with the original proof filled out by the other papers, both earlier and later. Before he

31 The 'Proof' of Pragmatism

had carried the notes far enough to remove his own intervening doubts, however, he had a falling out with the prospective publisher and the book never appeared (see Fisch 1974, 96–8).

(2) James arranged for Peirce to give a course of six lectures under the auspices of the Harvard department of philosophy in the spring of 1903, and Peirce chose to confine them 'to the single subject of pragmatism ... Its foundation, definition and limitation, and applications to philosophy, to the sciences, and to the conduct of life will make quite enough for six lectures' (Perry 1935 II 426–7). Peirce did not succeed in bringing the full proof as he then conceived it within the six, so he gave two additional lectures. The last was on mathematical conceptions of multitude and continuity. It included a brief exposition of Peirce's existential graphs. It was given under the auspices of the department of mathematics, but Peirce regarded it as an essential part of the proof (CP 5.201). Only in that lecture, for example, did it appear how a full proof of pragmatism 'would essentially involve the establishment of the truth of synechism' (CP 5.415). The editors of the *Collected Papers* found notebooks containing complete texts for the first seven lectures and published over a hundred pages of extracts from them (CP 5.14–212). They found no text for the last lecture, and the notebook later found is incomplete and lacks continuity (PP 316a(s); see Robin 1971, 41).

Peirce wanted to publish the lectures but James discouraged him. 'They need too much mediation, by more illustrations ... (nonmathematical ones if possible), and by a good deal of interstitial expansion and comparison with other modes of thought. What I wish myself is that you might revise these lectures for your Lowell course, possibly confining yourself to fewer points (such as the uses of the first-, second-, and third-ness distinction, the generality involved in perception, the nature of abduction – this last to me tremendously important) ...' (Perry 1935 II 427–8).

(3) So Peirce's Lowell Lectures in the fall of 1903, from which we took our start, became one more attempt at the proof of pragmatism, and the most successful so far. At the fourth lecture he distributed a printed *Syllabus*. In a draft of it Peirce had sketched the proof at the very beginning (PP 478 draft 2–23). This is one of his best statements of the case for pragmatism, but there wasn't room for it in the printed *Syllabus*, and it remains unpublished.[3] In the lectures themselves, pragmatism was omnipresent though seldom named.[4] The single last Harvard lecture on existential graphs, multitude and continuity was expanded into three ample lectures in the middle of the Lowell course. The key to the whole course was set in the opening lecture, 'What Makes a Reasoning Sound?', and carried through to the eighth, 'How To Theorize,' answering James's plea for more on 'the nature of abduction' (CP 1.606–11;

5.590–604; 7.182 n7). Looking back two years later, Peirce wrote: 'Considering how it stood in the mid-channel of pragmatistic thought to join ethics to logic, it seems to me strange that we had to wait until 1903 for any pragmatist to assert that logic ought to be based upon ethics. Perhaps some one of us had said it before; but I only know that it was then said in a course of lectures before the Lowell Institute in Boston, and was maintained on the ground that reasoning is thought subjected to self-control, and that the whole operation of logical self-control takes precisely the same quite complicated course which everybody ought to acknowledge is that of effective ethical self-control' (CP 5.533).

Peirce tried to interest George H. Putnam in publishing the lectures in book form, but Putnam thought the purchasers would be too few.[5] Extracts from the notebooks are scattered through seven of the eight volumes of the *Collected Papers*, but there is still no publication from which any sense can be got of the continuity of the argument within the series, or of its continuity with the argument of the Harvard lectures.

So when Peirce less than a year later promised a proof of pragmaticism as 'the one contribution of value' that he had to make to philosophy, it was still the case that the only exposition and proof he had got into print was in the unfinished 'Illustrations' of 1877–8.

We are now ready to answer our fifth question in summary fashion by saying that the reasons why he was now promising to prove pragmaticism all over again were: (1) The original proof was incomplete. (2) Within five years of its publication, he had come to have radical doubts about the tenability of the doctrine itself, as well as about his proof of it. (3) All his attempts to put something better in its place had failed of publication if not also of completion. (4) James, Schiller, Dewey, and others had recently been putting forth varieties of pragmatism which Peirce welcomed as signs of the vitality of the movement, but which seemed to him to carry the doctrine to such extremes as to put it beyond the possibility of proof. (5) It seemed to him therefore more urgent than ever to pull the doctrine back within provable limits, and then to prove it (PP 328, 1–2).

The *Monist* series in which this was to be done was at first to consist of three articles, 'What Pragmatism Is,' 'Consequences' or 'Issues of Pragmaticism,' and 'The Basis of Pragmaticism.' It was of course impossible that a three-article series should cover all the ground even of the eight Harvard lectures, to say nothing of the additional ground of the eight Lowell lectures. Much would have to be omitted. Peirce did not at first intend to bring in the existential graphs. But when the second article was already in print and he had produced several drafts of the third, he decided to devote the third to the

33 The 'Proof' of Pragmatism

graphs after all, and postpone the proof proper to a fourth. The third, on the graphs, appeared under the title 'Prolegomena to an Apology for Pragmaticism.' But as Peirce worked away on the fourth article, it also was divided into two, and eventually into three. The fourth was to begin the apology or proof, the fifth to contain the main argument, and the sixth the subsidiary arguments and illustrations. In a draft of the fourth, Peirce says he began the series 'under the clear conviction that no valid argument had ever been put forth for the truth of Pragmatism.' Nor, as of the spring of 1908, three and a half years later, had he seen any such argument. 'Of my own original promulgation of the doctrine,' he adds in a footnote, 'it is sufficient to say that it was published in a popular magazine, and therefore could hardly fail to be based on a begging of the question as it plainly is, in that it is entirely built upon the principle that that which a man believes is the proposition upon which he will be satisfied to act' (PP 296 draft A 1, 1–2). In a later draft, he called it a 'merely rhetorical defence' (PP 296, draft A_1, 2).

Meanwhile the publisher of the *Monist* had offered to reprint the *Popular Science Monthly* papers in book form and Peirce had begun editing them, with preface, introduction, and extensive revisions. Some confusion arose between that project and the continuation of the *Monist* series. Neither project was brought to completion. Extracts from a few of the numerous drafts appear here and there in the *Collected Papers*, but there is still no publication from which any sense can be got of the form the last revision of the proof would have taken in either of these two versions. The present paper cannot fill that need. It can only raise some relevant questions and urge some younger student of Peirce to take up the task.

This brings me to our last five questions. What were some of the not-yet-mentioned doubts about pragmatism and its proof as originally formulated that led Peirce to such reformulations as those of 1893, 1903, and 1904–10? What were some of the novel features of these reformulations? What role were the existential graphs to play in the final proof? How was it to 'involve the establishment of the truth of synechism' (CP 5.415)? And why did Peirce fail to complete the *Monist* series and the editing of the *Popular Science Monthly* series for reappearance in book form?

In the best draft of the fourth *Monist* paper Peirce wrote that he had 'passed through a doubt of pragmaticism lasting very nearly twenty years.' He asks if we happen to know what 'doubt' means. 'What "doubt" really denotes is to be insupportably discontent to dispose for oneself of the proposition that is said to be "doubted," in any suggestive way whatever, whether it be to affirm it to oneself, or to deny it, or yet to leave the question of its truth unsettled' (PP 300, 15–16).

What years did the 'nearly twenty' span? They ended not later than 1902, when the final proof 'first stood out clearly' in his mind (PP 300, 19). They began, therefore, not later than 1883, when he was lecturing on logic at The Johns Hopkins University. But they may have begun in 1878, when the 'Illustrations' ceased with the sixth. They ended, therefore, not earlier than 1896–7, the year in which he committed himself to the doctrine of real possibility and then invented his Existential Graphs. In either case, they included the year in which he projected the *Search for a Method*, and that may help account for his failure to finish annotating the 'Illustrations' for that volume.

And in either case I suggest that his doubts first became really acute in the year 1882–3 when he had but a single logic class of able graduate students at Hopkins meeting four times a week throughout the academic year. Most of them were mathematicians. The texts with which they began were 'The Fixation of Belief' and 'How To Make Our Ideas Clear.' The announcement of the course had said: 'Here, as everywhere throughout the course, the doctrine of the text will receive improvements, and the subject will be further illustrated by the aid of other works.' Probably neither before nor later were those two papers ever submitted to severer criticism. It would be good to know what improvements their doctrine received (Fisch and Cope 1952, 288, 370 notes 16 and 17).

The 'principal positive error' of the original exposition, Peirce later said, was its nominalism (CP 8.216). This was especially illustrated by his quoting the stanza from Gray's *Elegy* (CP 5.409). It was prominent also in the 'grievous error,' the 'damnable error,' of his application of the maxim to the idea 'hard' (CP 5.403; PP 289, 11, 16–17). He remained a nominalist in that sense until late in 1896, when he repudiated 'the nominalistic view of possibility, and explicitly' returned 'to the Aristotelian doctrine of a *real possibility*. This was the great step that was needed to render pragmaticism an intelligible doctrine' (PP 845, 28–29; cf CP 5.526–32).

In the first flush of pragmatism, he had perhaps receded even from the minimal realism to which he had committed himself in his Berkeley review of 1871 (Fisch 1967, 165). If so, that was not unnatural. As he put it in 1903, 'Berkeley is an extreme nominalist and Nominalism is itself of pragmatistic origin and its falsity is owing to its not fully planting itself upon pragmatistic ground ... [T]he faults of Berkeley's system ... arise simply from his deficient grasp of the pragmatistic principle' (PP 478, 21 of 2–23 draft sequence). Peirce's greatest single doubt about his original pragmatism was removed when he was able to see his own later and more thoroughgoing realism, not merely as a corrective of his earlier pragmatism, but as a consequence of his graduating into a more thoroughgoing pragmatism. It was a further development of pragmatism itself that led him to insist that 'a mere possibility may be quite real' and to abandon the 'strange rule' 'that every conditional proposition whose antecedent does not happen to be realized is true' – that is, to make room for strict implication without denying that there is a conditional *de inesse* (CP 4.580–1; Roberts 1973, 95–8).

One of Peirce's recurring sources of uneasiness about the original proof was its resting the case for pragmatism on psychological principles. Bain's theory of belief, of which pragmatism was 'scarce more than a corollary' (CP 5.12), was a psychological theory, presented in a psychological treatise and in a psychological manual, both of them used by James and others in psychological courses at Harvard (Fisch 1954, 432). Even the term 'idea' in 'How To Make Our Ideas Clear' was a term of psychology, not of logic. In his paper on 'The Algebra of Logic' in 1880 Peirce went so far as to begin with a physiological 'derivation of logic' (CP 3.154–61). But from the 1880s on he attached increasing importance to the classification of the sciences. In all his classifications, each science depends for principles on those above it and for data on those below it (CP 3.427; 1.189, 197). And in all of them psychology and physiology are down among the special sciences, below logic and metaphysics. In his earlier classifications, logic has only mathematics above it. Now pragmatism is a regulative principle of logic. Therefore it cannot be derived from physiology or from psychology. But on the other hand it can scarcely be derived from or based on mathematics. On what, then?

Peirce dates his serious study of ethics from 1883 (CP 5.111, 129; Fisch 1971, 194). As we have seen, his nearly twenty-years doubt of pragmatism began not later than that year. He soon began to entertain the possibility of basing logic on ethics. What gave him pause was that the only ethics discernible in the original exposition of pragmatism was an incipient hedonism. 'It is certainly best for us that our beliefs should be such as may truly guide our actions so as to satisfy our desires' (CP 5.375). It was only in conjunction with desires that the sensible 'effects' of the pragmatic maxim would have 'practical bearings'; that is, determine habits of action. Worse still, it was a vulgar and nominalistic hedonism. Desires were taken as ultimate given particulars, lacking in generality, not subject to dialectical or evolutionary development (CP 5.158). Moreover, all his efforts to base logic (including pragmatism) on ethics ran into the difficulty that ethics could be based only on esthetics, and that long seemed to him to lead once again into hedonism (CP 5.110). But by the early 1890s he had committed himself to a doctrine of agapism or evolutionary love, in opposition to hedonism. It was not until the Lowell Lectures of 1903 that he resolved this doubt by basing ethics on what he called 'a transfigured Esthetics' or 'esthetics in a transfiguration,' whose chief if not sole function is to determine the *summum bonum*.[6]

In the implicit ethics of the original exposition, what was not hedonism verged on stoicism, which Peirce found equally objectionable. As he put it in the article on pragmatism in Baldwin's *Dictionary* in 1902, 'The doctrine appears to assume that the end of man is action – a stoical maxim which, to the present writer at the age of sixty, does not recommend itself so forcibly as it did at thirty' (CP 5.3).

In later attempts at the proof, the language of desire, pleasure, and satisfaction gave way to that of rational purpose and of self-control; that of action to that of conduct; that of belief to that of assertion. That is, psychological language gave way to logical and ethical. But, as if partly to justify his previous use of desire and pleasure, at least once he showed even their generality, in his apple pie illustration (CP 1.341).

Peirce was a mathematician and one of his most serious and long-lasting doubts about pragmatism was whether it could be made acceptable to mathematicians and to himself as mathematician. As he put it in the Baldwin article, he 'subsequently saw that the principle might easily be misapplied, so as to sweep away the whole doctrine of incommensurables, and in fact, the whole Weierstrassian way of regarding the calculus' (CP 5.3; cf 32–3). He resolved the doubt in part by an extension of the conceptions of sensible effects and habits of action (CP 5.201–5, 539, 541). But he resolved it in part also by developing a theory of mathematics that made it an experimental and observational science, differing from the special sciences by experimenting on diagrams of our own construction. In mathematics as in the special sciences, a point is reached at which 'It is necessary that something should be DONE' (CP 4.233).

In a similar vein, one of the strengths he came to prize in his existential graphs is that, without loss of rigour and even with some gain, they lend themselves to a pragmatic understanding of logic itself. For example, even 'The definitions shall all be given in strictly pragmaticistic form; that is in the form of precepts of conduct, more definitely speaking, as *permissions* to do certain things under expressed general circumstances' (PP 280, 22).

But Peirce was not only a mathematician. He was a research scientist. And in the long history of classifications of the sciences, from Plato and Aristotle to the present, the nearest approach to a constant has been the distinction between theoretical and practical science. But pragmatism in all its forms, including Peirce's, seemed to blur if not to obliterate this distinction by turning theoretical into practical science. In 'How To Make Our Ideas Clear' Peirce had said that 'the whole function of thought is to produce habits of

37 The 'Proof' of Pragmatism

action' (CP 5.400). And in his Harvard lectures he said: 'Pragmatism is the principle that every theoretical judgment expressible in a sentence in the indicative mood is a confused form of thought whose only meaning, if it has any, lies in its tendency to enforce a corresponding practical maxim expressible as a conditional sentence having its apodosis in the imperative mood' (CP 5.18).

Nevertheless he was as insistent on retaining the distinction as Aristotle or any of his medieval interpreters, and nowhere more so than when he was making a more exhaustive study of the practical sciences than anybody before him or since (see especially PP 1343). In a draft of his third *Monist* article he listed among misapprehensions of pragmatism the notion that 'it involves a depreciation of pure science. In my case, I am confident it has worked in the opposite way; and I do not believe I should have been able to endure my life's exclusive devotion to pure thought if I had not been sustained by the pragmaticistic belief' (PP 284, 5). And in another draft he remarks that although research scientists 'look upon their work as purely theoretical ... they are nevertheless particularly given to thinking of their results as affording possible conditions for new experiments ... This shows that regarding a truth as purely theoretical does not prevent its being regarded as a possible determinant of conduct' (PP 283, 9–11).

One quite temporary solution of 1902 to the hedonism-or-stoicism objection was to admit a grade of clearness higher than the third or pragmatic, or even an indefinite series of higher grades. In the Baldwin article he says that the pragmatic maxim 'should always be put into practice with conscientious thoroughness, but that, when that has been done, and not before, a still higher grade of clearness of thought can be attained by remembering that the only ultimate good which the practical facts to which it directs attention can subserve is to further the development of concrete reasonableness; so that the meaning of the concept does not lie in any individual reactions at all, but in the manner in which those reactions contribute to that development' (CP 5.3). And in his Carnegie application of the same year he says: 'Moreover, my paper of 1878 was imperfect in tacitly leaving it to appear that the maxim of pragmatism led to the last stage of clearness. I wish now to show that this is not the case and to find a series of Categories of clearness' (CP 8.176n3; cf Eisele 1976 IV 30).

But in the following year, by firmly basing logic on ethics and ethics on esthetics, he reabsorbed into the third or pragmatic grade all that the fourth and higher grades had promised to yield (Kent 1975, 177). After that, his only revision in this respect was to 'subdivide the second grade into two, and the third into three' (PP 620, 18).

In the original proof, Peirce had not sufficiently guarded against such misunderstandings as that to attain the second grade of clearness is to have no further need of the first, and to attain the third is to leave the second behind; or that the first and second are transcended or *aufgehoben* in the third; or that all the first and second are good for is to make way for the third. In his later attempts at the proof, he emphasizes that the third grade presupposes the second, as the second presupposes the first; that it supplements them but does not supplant them, or even swing free of them; that they remain indispensable; and that they retain all the value they had without it.

By the beginning of 1909, if not earlier, Peirce had decided to devote the fourth article in the *Monist* series – the first of the three that were still to come, and that were to contain the proof proper, as distinguished from the prolegomena – to the second grade of clarity in relation on the one hand to the first and on the other to the third. This fourth article, he said, would present 'a theory of Logical Analysis, or Definition' which 'rests directly on Existential Graphs, and will be acknowledged, I am confident, to be the most *useful* piece of work I have ever done ... Now Logical Analysis is, of course, Definition; and this same method applied to Logical Analysis itself, – the definition of definition[, –] produces the rule of pragmaticism ... It is the very proof that I intended to bring out when I began this series of papers on Pragmaticism.'[7] In the following winter and the spring of 1910 he wrote at least six partial drafts of this fourth article. The sixth draft begins:

In my original essay on Pragmaticism I showed that there are three grades of attainment toward clearness of thought that are rendered distinct from one another by qualitative differences; the first and lowest imparting what may be more specifically called 'Clearness,' i.e. readiness in employing and in interpretatively applying the notion, idea, or other Sign to which it relates; the second imparting Distinctness, or analytic understanding of just what constitutes the essence of that meaning which the first grade has rendered Clear; and the third, or Pragmatistic, grade imparting what perhaps I may be allowed to call Pragmatistic 'Adequacy,' that is, not what has been, but what ought to be the substance, or Meaning, of the concept or other Symbol in question, in order that its true usefulness may be fulfilled.

I trust that, in that essay, I made my own opinion plain, that Pragmatistic Adequacy no more supersedes the need of Analytic Distinctness and of adherence to precise Definitions, than this latter, or Second Grade of clearness supersedes the need of intuitive or unintellectual Clearness in the specific sense just defined. It is evident that no abstract definition can possibly render needless the power of directly recognizing whether a given concept does or does not apply to a given image; and I believe I was sufficiently explicit as to my own opinion that no recognition of the utility of a

39 The 'Proof' of Pragmatism

concept, however just, could in the least affect the need of precisely defining it. (Please observe, by the way, that I speak of three distinct Grades of clearness, which I also call Kinds, but never Stages, as if one were done with before the next began; for the contrary will be found markedly their relation.)[8]

But I have since found reason to regret that I did not reinforce this opinion by arguments ... One of the purposes of the present paper is to supply that omission as well as I can.

Besides that, I think I may usefully formulate my notion of the proper way to perform the research requisite to the formulation of an accurate and precise Definition of a Concept which is already pretty Clear in the specific sense ... (PP 649, 1–4).

But if the 'principal positive error' of the original exposition was its nominalism, its principal negative error, its great sin of omission, was that, whereas Peirce had derived his pragmatism 'from a logical and non-psychological study of the essential nature of signs' (PP 137, 20; Eisele 1976 II 521), he said nothing whatever about signs in the 'Illustrations' of 1877–8. Pragmatism is a method for generating a particular kind of interpretant of a particular kind of sign, and a doctrine concerning that method; and it is only within the framework of the theory of signs that it is possible to give any precision to the method and the doctrine, so as to make their limitations fully evident and thus forestall needless objections and misunderstandings. Even in the Harvard and Lowell lectures of 1903 this was but inadequately done. As he worked on the *Monist* series, he was content at first to cover only so much of the theory of signs as was indispensable for relating pragmatism to it; but a more systematic and complete treatment of semeiotic came to seem at least a desideratum, and he began working on a new treatise to be entitled *A System of Logic Considered as Semeiotic*. This in turn called for a more elaborate development of his universal categories. He had already reserved a place in his classification of the sciences for a science that should have this as its task. He called it at first 'high philosophy,' then phenomenology, then phaneroscopy. Its place was between mathematics and the normative sciences. There are still several unpublished partial drafts of a treatise on phaneroscopy. The proof of pragmatism turned on the relations between the universal categories, the classes of signs, their objects and interpretants, the logical modalities, and especially the logic of vagueness. Whatever the solutions, Peirce was convinced that his existential graphs would yield the best possible representation of the modalities. Yet if he reached a graphical representation of them that completely satisfied him, it does not appear from anything so far published what it was. Closely connected with all this is the question just how the proof of pragmatism would essentially involve the establishment of synechism.

I may best approach my final conclusions by way of Peirce's division of semeiotic into speculative grammar, critic, and methodeutic (CP 1.191; CP 4.9). If he had finished and published his *System of Logic Considered as Semeiotic*, that division would have been its organizing principle. And if we were now about to examine the *System* for its contribution to the proof of pragmatism, we might well be asking ourselves such questions as the following. Does the pragmatism being proved belong to all three parts, and, if so, what elements or aspects of it belong to the first, to the second, and to the third? Or does it belong to only two of the three parts, and, if so, to which two, and why? Or to only one, and, if so, to which one, and why?

The best accounts of the 'proof' so far published are those by Manley Thompson (1953, 249–50, 259, 261) and John J. Fitzgerald (1966, 15, 126–32, 160–4). Both have these questions in mind, and both are aware that they are not explicitly answered in anything published by Peirce or included in the *Collected Papers* (Thompson 1953, 165). Both remark that in the pragmatism lectures of 1903 Peirce identifies pragmatism with 'the logic of abduction' and that this might seem to assign it to the first branch of critic (Thompson 1953, 286 n5; Fitzgerald 1966, 132). But Fitzgerald thinks the 1907 paper entitled by the editors 'A Survey of Pragmaticism' tacitly and rightly assigns it to speculative grammar (Fitzgerald 1966, 132, 136, 160, 165). And Thompson argues that Peirce's reason for identifying it with the logic of abduction is 'precisely because it completes the analysis of significant assertion given by speculative grammar' (Thompson 1953, 286 n5); that it 'begins the logic of abduction at the same time that it ends speculative grammar' (Thompson 1953, 263).

Neither Thompson nor Fitzgerald made use of the vast body of still unpublished Peirce papers. Had they done so, they would have found him, almost without exception, assigning pragmatism neither to speculative grammar nor to critic, but to methodeutic.[9]

So my final conclusions are two. First: The problem of the proof of pragmatism calls for further study of Peirce's still unpublished writings on phaneroscopy, semeiotic, existential graphs, the modalities, and the relations between pragmatism, tychism and synechism. Second: To facilitate such study we need a new edition of Peirce's writings, in a single chronological order, that will include extensive and expertly edited selections from these still unpublished writings.[10]

HANS G. HERZBERGER

Peirce's Remarkable Theorem*

I

Only once, late in his prodigious career, did Charles Peirce momentarily lose heart in defending his doctrine of categories. 'I am tired of arguing a truism,' he wrote in wavering hand, and he went on to launch a small polemic against those who still failed 'to see' his proposition.[1] What Peirce then held to be a truism seems extravagant to most of us today – the notion that monads, dyads, and triads comprise a complete metaphysical scheme. Yet for many years Peirce held this to be not only true but demonstrable, and also fundamental for philosophy:[2]

Three ideas are basic: those of something, other, and third ... In this mathematical proposition (for such it is shown to be), you have all logic and all metaphysics in a nut-shell.

The doctrine of categories, he stoutly maintained, could be logically justified. Toward this end he enunciated a *reduction thesis* to the effect that all higher polyadic relations were formally reducible to triads, and that no further reduction was possible.[3] Notwithstanding Peirce's long-sustained efforts to establish his thesis, no cogent demonstration has so far been found in any of his writings, and many scholars have long since written it off as one more uncollectable debt.[4]

But there is no statute of limitations on philosophical debts, and I submit that this one might eventually be settled through renewed effort and a dedica-

---

* This paper has benefited from discussions with Jacqueline Brunning on her own related research.

tion to fallibilism. Nothing more should be required than a willingness to go back to the texts keenly alert to the fallibility of all our beliefs concerning Peirce's logical philosophy.[5]

This paper re-examines Peirce's position in the historical context of his distinctive ideas on logic and definability. It undertakes to show why he found his reduction thesis so compelling, and also to some extent why present-day commentators find it so troublesome. In brief the explanation to be offered is that the logic Peirce invented is subtly different from, and yet deceptively close to, the logic we know today. In several respects Peirce's logic is close enough to ours to engender a comfortable sense of familiarity; and yet where it counts for present purposes I submit that a critical distance separates them.

Peirce built up a characteristic theory of definability from his chemical model of relational concepts, their 'valencies,' and the ways they 'bond' together. These unusual ideas have been almost wholly submerged under later developments.[6] Having lost contact with his presuppositions, we naturally find it difficult to follow his reasoning. Once these matters have been brought into awareness, exegetical problems can be turned to advantage. Peirce tells us that his doctrine of the categories helped him greatly in the study of logic,[7] and quite possibly our efforts to penetrate his philosophy now can help us to rediscover his logic. This is no small endeavour, and an honest mixture of approximation and conjecture is all I shall aim for on this first approach.

The historical stages in the joint evolution of Peirce's logic and philosophy are by no means clear. In retrospect one senses the gradual articulation of a vision through mutual accommodations, advances, and occasional retrenchments – not a tidy cumulative development. So I propose to work out what may be thought of as a retrospective treatment of the underlying vision. In pursuit of Peirce's working conception of definability, a 'bonding algebra' will be consolidated from his mature logical writings under the guidance of the reduction problem and other clues. Within the bonding algebra a qualified reduction theorem can be validated (Theorem T8), and a major counterargument to the reduction thesis can be headed off. That the counterargument holds within standard notions of definability (Theorem T11) sharply underscores the novelty of this kind of logical framework.

II

Peirce's commitment to the reduction thesis was steadfast almost to the point of tenacity. In varying terminology it appears at least two dozen times throughout his writings, from 1870 to 1909.[8] How central it was for him may be gathered from his pronouncement that 'nothing in philosophy is more

43 Peirce's Remarkable Theorem

important,'[9] and from the way he appraised his own contributions to pragmatism as 'entirely the fruit of this outgrowth from formal logic.'[10] And yet the proof was not forthcoming.

Often enough Peirce declared the reduction thesis to be provable; and even claimed to have published, somewhere, such a proof.[11] Later writings reveal some ambivalence on the matter of a proof. Once Peirce wrote that a proof would be extremely simple, but 'confusing,'[12] and on another occasion declared it 'too stiff for the infantile logic of our time.'[13] With characteristic resourcefulness he tried to make his thesis plausible through models, maps, and diagrams;[14] and finally appealed to his own considerable authority as a logician, informing the reader that thorough study of the logic of relatives 'confirms' the thesis[15] and renders it 'perfectly evident,'[16] and that analysis 'shows' it to hold.[17] Needless to say, none of these efforts counts as a proof.

However, the absence of a proof is not a refutation either. In the present case it could well be attributed to a complex of factors, including Peirce's temperamental distaste for algebraic methods.[18] Of his own algebra of dyadic relations he later wrote that 'it has too much formalism to greatly delight me.' Turning away from an algebraic treatment for his own logic of polyadic relations, he reproached Schröder as one who 'likes algebraic formalism better, or dislikes it less, than I.'[19] Whatever the merit of this explanation, algebraic logic has meanwhile matured into a highly effective instrument for the study – and quite possibly the resolution – of these problems.[20]

Although Peirce never proved a reduction theorem, his writings contain methods and examples around which a number of relevant proofs are readily assembled. In what follows two of his reduction methods will be used, and also the formula $[\mu + \nu - 2\lambda]$ in which Peirce quantitatively expressed his theory of conceptual valency.

III

One definitional process stands out as central to Peirce's reduction methods and to his whole thinking on the combination of concepts. This is the operation of *relative product*, which will now be used to study the reduction problem in miniature in order to proceed quickly to the heart of the matter. This definitional process will be given an extended formulation and its powers and limitations will be examined in the context of the reduction problem. Then, lessons derived from this study will be applied in a more comprehensive setting.

Relative product is familiar to most of us as an operation on dyadic relations. To use a favourite Peircean example, the relative product of [--is a lover of--] with [--is a servant of--] would be the relation [--is a lover of a servant of--]. Originally de Morgan's, this operation was taken over and extended by Peirce to apply to relations of any degree. So prominent did it become in Peirce's thought that some of his statements treat it as the very paradigm of conceptual combination. Around 1905 he wrote that 'the combination of concepts is always two at a time and consists in indefinitely identifying a subject of the one with a subject of the other.'[21] To frame this notion exactly, relations will be understood in extension, as sets of finite sequences.[22] Lower-case italics will be used for elements of the universe of discourse, and upper-case italics for sequences of those elements. The relative product of any two relations **R** and **S** then will be the set p(**R**, **S**) of exactly those sequences $X\frown Y$ such that for some 'linking' element w the sequence $X\frown w$ belongs to **R** while the sequence $w\frown Y$ belongs to **S**. In set-theoretic notation:

Relative product: p(**R**, **S**) = $\{X\frown Y: (\exists w)(X\frown w \in \mathbf{R}\ \&\ w\frown Y \in \mathbf{S})\}$.

The last subject of **R** here is 'indefinitely identified' with the first subject of **S**, as in the formation of the triad [--gives--to a friend of--] from the triad [--gives--to--] and the dyad [--is a friend of--]. The end result of this construction might be graphically represented by the diagram:

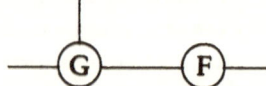

where the line joining **G** and **F** links the last subject of **G** with the first subject of **F**.

A miniature notion of definability may now be based on this single operation. Let any set of ordered n-tuplets be called a *relation of degree n*, or *n-ad*. Let a relation **R** be called p-*definable from the set* K of relations if and only if **R** can be obtained from members of K by repeated applications of the relative product operation. Following Peirce, reducibility will be understood as definability from relations of lower degree.[23] A relation **R** will be called p-*reducible within the domain* D if and only if **R** is p-definable from some set K of relations, each being a relation over D of degree less than that of **R**. The difference between definition and reduction is vital, for Peirce came to find triads underlying many common logical constructions. Within his framework these constructions count as definitions but not always as reductions.

Some negative results on reducibility are now very easy to obtain. The salient feature of relative product as a means of definition is the characteristic way it 'bonds' relations together. This is reflected in its 'algebraic type,' the

45 Peirce's Remarkable Theorem

TABLE 1
Degrees of simple relative products

r	s	r + s − 2
1	1	0
2	1	1
2	2	2
3	2	3
3	3	4

effect it has on the degrees of its operands. Each operand relinquishes one of its arguments, so that the relative product of any r-ad with any s-ad will have the degree $(r - 1) + (s - 1) = (r + s - 2)$, for any positive values of r and s. The significance of this calculation for Peirce's reduction thesis can be read off the table of values (table 1). Using relative product on a basis consisting of only monads and dyads, the table shows that nothing of higher degree than a dyad can be defined. This result corresponds to the negative part of the reduction thesis taken in miniature:

T1. *In the miniature setting, lower polyads ($n \leq 3$) are absolutely irreducible.*

This means that no triad is p-reducible within any domain, nor is any dyad, nor any monad.

The values in table 1 indicate that some tetrads will be p-reducible in suitable domains; it does not however show all tetrads to be so reducible. Nor is there any way to establish the positive part of the reduction thesis on the basis of algebraic types alone. To do so we must turn to Peirce's first reduction method, which shows how to reduce higher polyads under very general conditions. As a preliminary consideration it will be useful to record one essential limitation on p-reduction:

T2. *In the miniature setting, not all tetrads are reducible over their own domain.*

Proof: By counterexample, using a very small domain. Let $D = \{a, b, c\}$ and let **R** be that tetradic relation over D which consists of the four quadruplets:[24] *abba, baab, cbcb,* and *bcbc*. It will be shown that **R** is not the relative product of any two triadic relations over D. Let **G, H** be any two triadic relations over D, and let $\mathbf{P} = \mathsf{p}(\mathbf{G, H})$ be their relative product. Assume that $\mathbf{R} \subseteq \mathbf{P}$; it will be

TABLE 2
Schematic tabulation
of the triads G and H

G	H
abx	xba
bay	yab
cbw	wcb
bcz	zbc

shown that $P \not\subseteq R$. The assumption that $R \subseteq P$ constrains the triads G, H as indicated in table 2, where each of x, y, w, z is some so far undetermined member of the domain D. Now, if $x = y$, then $abab \in P$, so $P \not\subseteq R$; and if $x = w$, then $abcb \in P$, so $P \not\subseteq R$. By similar considerations, if any two of the elements x, y, w, z coincide, then $P \not\subseteq R$. But the domain D only has three distinct elements, so at least two of them are bound to coincide, and so $P \not\subseteq R$. Therefore $P \neq R$; there is no way of obtaining the given tetrad as the relative product of two triads over D.

In the above example, it would be quite possible to obtain R as the relative product of two triads over an expanded domain D+. For example, let D+ = $\{a, b, c, d\}$ and define G and H in accordance with table 2, with $x = a$, $y = b$, $w = c$, and $z = d$. In effect, this kind of expansion of the domain underlies Peirce's reduction methods. Here is one of his characterizations:[25]

I will show by example that a four can be analyzed into threes. Take the quadruple fact that *a* sells *b* to *c* for the price *d*. This is a compound of two facts: first, that *a* makes with *c* a certain transaction, which we may name *e*; and second, that this transaction *e* is a sale of *b* for the price *d*. Each of these two facts is a genuine triple fact, and their combination makes up [as] genuine [a] quadruple fact as can be found.

Taking R as the tetradic relation in question, what Peirce calls a 'quadruple fact' is one of its quadruplets *abcd*, which he reconstructs in terms of a particular 'transaction' *e*. The transaction *e* bears one triadic relation (call it G) to the first two members *a*, *b* of the quadruplet, and another triadic relation (call it H) to the last two members *c*, *d* of that quadruplet. Then the tetradic relation R will be the relative product p(G, H) of those two triadic relations G, H.

This method makes reference to certain auxiliary entities (the 'transactions') which play the role of the placeholders x, y, w, z of table 2. A limitation on the method is that these auxiliary entities must be thought of as belonging to the underlying domain, or as somehow constructed out of its elements by

47 Peirce's Remarkable Theorem

some so far tacit logical construction. As so far described, the method begins with some tetradic relation **R** over a domain D, and reduces **R** to a pair of triadic relations **G** and **H**, each defined over a possibly expanded domain D+. In favourable cases sufficient auxiliary elements will be available from D, so that D+ will coincide with D, and **R** will be reducible over its own domain.

From a formal standpoint it is inessential to this method of reduction that the auxiliary entities be abstract; they could as well be any members of the given universe of discourse.[26] The main requirement, as can be learned from studying the counterexample of T2, is that there be a sufficient quantity of auxiliary entities to permit their being correlated one-to-one with the quadruplets making up the tetradic relation **R**.

T3. *In the miniature setting, all tetrads are reducible within sufficiently large domains.*

Present purposes do not require a perfectly sharp criterion for 'sufficiently large' domains. A workable lower bound on the notion results from simple cardinality considerations:

A domain is 'sufficiently large' for purposes of reduction if its cardinality is no less than that of the relation to be reduced.

This condition holds whenever there is a one-to-one mapping from the n-tuplets of the relation into the members of the domain. The proof of T3 now is by construction, following Peirce's first reduction method. Let **R** be any tetradic relation over a domain D. Assume that D has a cardinality at least as great as **R**, so that there exists some one-to-one mapping of **R** into D. For any quadruplet $r = r_1 \frown r_2 \frown r_3 \frown r_4$ in **R**, let r^* be the corresponding element of D under this mapping. The element r^* assigned to the quadruplet r will be our formal counterpart for the 'transaction e' in Peirce's method described above. Now two functions **g** and **h** will be described, each of which maps members of **R** into triplets over D,

$$\mathbf{g}(r) = r_1 \frown r_2 \frown r^*, \qquad \mathbf{h}(r) = r^* \frown r_3 \frown r_4.$$

and two corresponding triadic relations **G** and **H** will be described, by the equations

$$\mathbf{G} = \{\mathbf{g}(r): r \in \mathbf{R}\}, \qquad \mathbf{H} = \{\mathbf{h}(r): r \in \mathbf{R}\}.$$

The first consists of all triplets correlated by **g** with members of **R**, and the second consists of all triplets correlated by **h** with members of **R**. Let $\mathbf{P} = \mathsf{p}(\mathbf{G}, \mathbf{H})$ be their relative product; it will be shown that $\mathbf{P} = \mathbf{R}$. Let $x = x_1 \frown x_2 \frown x_3 \frown x_4$ be any quadruplet belonging to **P**. By the definition of relative product, for some element w from the given domain, $x_1 \frown x_2 \frown w \in \mathbf{G}$ and

$w\frown x_3\frown x_4 \in \mathbf{H}$. But then by the construction of the triadic relations $\mathbf{G}$ and $\mathbf{H}$, w is uniquely determined as x^*, so it follows that the quadruplet x belongs to the original relation $\mathbf{R}$. Conversely, let $y = y_1\frown y_2\frown y_3\frown y_4$ be any quadruplet belonging to $\mathbf{R}$. By the construction of the triadic relations $\mathbf{G}$ and $\mathbf{H}$ then the triplet $y_1\frown y_2\frown y^* \in \mathbf{G}$ and the triplet $y^*\frown y_3\frown y_4 \in \mathbf{H}$, so the quadruplet y belongs to $\mathbf{P}$. This shows that $\mathbf{P} = \mathbf{R}$, and completes the p-reduction of $\mathbf{R}$ within the given domain.

There is no difficulty in extending this construction to all higher polyadic relations, as follows:

T4. *In the miniature setting, all higher polyads ($\mathbf{n} > 3$) are reducible within sufficiently large domains.*

Indeed, any n-adic relation over a domain D ($\mathbf{n} > 3$) is p-definable from a set of $\mathbf{n} - 2$ triads over the same domain, provided the cardinality of the domain is no less than the cardinality of the n-adic relation to be reduced; so 'sufficiently large' retains its original lower bound. To illustrate the reduction for the case $\mathbf{n} = 5$, each quintuplet $r = r_1\frown r_2\frown r_3\frown r_4\frown r_5$ belonging to a given pentadic relation $\mathbf{R}$ is to be correlated with the three triplets $g(r) = r_1\frown r_2\frown r^*$, $h(r) = r^*\frown r_3\frown r^*$, and $j(r) = r^*\frown r_4\frown r_5$. And so on, in each case iterating substantially the construction used in proving T3.

Now we have a precise version of the reduction thesis in miniature; a proof (from algebraic types) for the negative part of the thesis; and a proof (in T3) for the crucial tetradic case of the positive part of the thesis. A consequence of these results is:

T5. *In the miniature setting, Peirce's reduction thesis holds within any infinite domain.*

This is a corollary of T3 together with a theorem of Cantor's to the effect that any infinite set has the same cardinality as the set of all its n-tuplets. From this it follows that any infinite domain is 'sufficiently large' for purposes of reduction.

These results complete the exercise of evaluating the reduction thesis in miniature, by mapping out the powers and limitations of relative product as a means of reduction. The outcome so far appears to be consonant with Peirce's claims. The next part of the paper will undertake to show how the same results can be carried over to the full bonding algebra.

IV

Aiming for a minimal logical apparatus, Peirce gradually extended and elaborated relative product into a kind of omnibus definitional process for joining

49 Peirce's Remarkable Theorem

concepts through 'bonds of union.' He thought of his existential graphs as perfectly realizing this vision, carrying logical analysis 'to the fullest point possible in the nature of things,' and he claimed to limit that system to a single fundamental mode of logical combination.[27] Whether or not he achieved this aim depends on how 'modes of combination' are counted. Algebraic treatments tend to be slightly less economical; but even algebraically Peirce's mature definitional resources turn out to be strikingly compact.

The first generalization of relative product allows for corrections in a feature Peirce regarded as a defect of algebraic symbolism: the linear arrangement of 'correlates.'[28] Relative product conventionally bonds together the last argument of **R** with the first argument of **S**. Peirce came to use a definitional process capable of bonding together any chosen argument of **R** with any chosen argument of **S**.[29] Such a process could be formulated as a family of variations on relative product. Schematically, p_{ij} would be that member of the family which bonds together the ith argument of **R** with the jth argument of **S**. Algebraically it is more convenient to retain relative product as given, while admitting supplementary operations for transposing argument places. The familiar converse operation has this effect in the case of dyadic relations. For triadic relations Peirce formulated five operations he called 'transpositions.'[30] For the general polyadic case, Quine's treatment is more economical and will be adopted here.[31] Quine shows that two operations jointly suffice to bring about any desired permutation of arguments:

Major permutation: $\quad m(\mathbf{R}) = \{X \frown w : w \frown X \in \mathbf{R}\}$

Minor permutation: $\quad n(\mathbf{R}) = \{z \frown w \frown X : w \frown z \frown X \in \mathbf{R}\}$

Major permutation shifts the first argument into final position. Its effect on a triad is a kind of 'voice change'; thus m[--gives--to--] might be read [--is given to--by--]. Minor permutation switches the first two arguments, so that n[--gives--to] might be read as another sort of 'voice change': [-- is given by-- to--]. Now the effect of the schematic process p_{ij} can always be obtained by permuting first and then applying relative product. Thus:

$$p_{ij}(\mathbf{R}, \mathbf{S}) = p(m^i(\mathbf{R}), m^{j-1}(\mathbf{S}))$$

where $m^i(\mathbf{R})$ stands for the result of i successive major permutations.

The next increase of flexibility in bonding allows for multiple bonding within a single process of combination.[32] Again this could be accomplished schematically through a family of operations, and again the more natural algebraic treatment is to admit one repeatable supplementary operation:

Bonding: $\quad g(\mathbf{R}) = \{X : (\exists w)(X \frown w \frown w \in \mathbf{R})\}$

This operation 'indefinitely identifies' the last two arguments of any relation to which it is applied. For example, the result of bonding the triad [--betrays--to--] would be the monad [--betrays someone to himself]. In quantificational notation bonding would naturally be expressed as a composite process through an identity clause together with a double existential quantification; thus the result of bonding a relation expressed by '**R**xyz' might be expressed by the formula '$(\exists y)(\exists z)(y = z\ \&\ \mathbf{R}xyz)$.' For Peirce, however, bonding is a simple operation, represented graphically by joining two argument places:[33]

$$\text{—}\overset{}{\text{(R)}}\text{—} \Rightarrow \text{—}\overset{}{\text{(R)}}\text{⏋}$$

Two relations **R** and **S** can now be multiply bonded by suitable applications of relative product, permutations, and bonding. For example, if **R** is the triad [--gives--to--] and **S** is the dyad [--is a servant of--] then the monad [someone gives--to his own servant] can be expressed as gmp(**R**, **S**), with two bonds of union.

One more operation completes the bonding algebra:

Relative complement: c(**R**) = {$X \in \mathsf{D}^r$: $X \in \mathbf{R}$}

Complement is taken relative to the degree of the operand **R**, which accounts for the restriction '$X \in \mathsf{D}^r$' in its characterization. If **R** is any relation of degree r, its complement consists of exactly those r-tuplets from the universe of discourse which do not belong to **R**.

These five basic operations algebraically describe a system of definitional resources which could with slight reservations be ascribed to Peirce. Practically all of his remarks on definition and the composition of concepts can be encompassed within this system or explained in terms of it, including a number of otherwise intractable passages. And where it falls short on particular details, it provides at least a starting position for closer approximations.

All results from our miniature study of the reduction problem can be extended to this very much more comprehensive framework. Let a relation **R** be called *bonding-algebraically definable from the set* K of relations if and only if **R** can be obtained from members of K by repeated application of any of the five basic operations (p, m, n, g, c) of bonding algebra. And let a relation **R** be called *bonding-algebraically reducible within the domain* D if and only if **R** is bonding-algebraically definable from some set K of relations, each being a relation over D of degree less than that of **R**.

Since relative product belongs to bonding algebra, the reducibility theorem still holds:

T6. *In bonding algebra, all higher polyads* (n > 3) *are reducible within sufficiently large domains.*

51 Peirce's Remarkable Theorem

And by consideration of algebraic types, the main irreducibility result carries over intact:

T7. *In bonding algebra, lower polyads ($n \leq 3$) are absolutely irreducible.*

T7 follows from the fact that no operation of bonding algebra has an algebraic type higher than that of relative product. Bonding is degree-reducing, having the algebraic type $[r - 2]$, and the other operations are degree-neutral, having the algebraic type $[r]$. So table 1 can be reinterpreted as a table of maximum values for bonding-algebraic compounds of monads, dyads, or triads. Combining these two results as before:

T8. *In bonding algebra, Peirce's reduction thesis holds within any sufficiently large domain.*

And as before any infinite domain is sufficiently large. So far, then, bonding algebra shows at least hypothetically how the reduction thesis could have been defensible. Moreover it has many points of contact with Peirce's logical writings, especially those dealing with definition, analysis, and the composition of concepts.

V

Bonding algebra accords closely with Peirce's chemical model of concept-formation and what he came to call the 'doctrine of valency.'[34] In the 1890s two accounts of his chemical model appeared in print. In 'The Critic of Arguments' (1892) he compared relations with 'multivalent radicles' on the theory that 'in chemistry, unsaturated bonds can only be saturated by joining two of them, which will usually, though not necessarily, belong to different radicles.'[35] A few years later, in 'The Logic of Relatives' (1897), the comparison was repeated, again with explicit allowance for non-compositional joining of unsaturated bonds, and with stress on the pairwise character of this joining.[36]

According to this model, as I understand it, each concept has an ordered system of 'loose ends' which determine its valency. Any two concepts can be combined by joining together two loose ends, one from each component – as in the formation of relative products. Furthermore, any single concept can be transformed by joining together a pair of loose ends – as in bonding. These are the only two definitional processes which can affect valency, and in Peirce's view they are each aspects of a single generic process, pairwise bonding of loose ends.

Bonding algebra also coheres with Peirce's formula $[\mu + \nu - 2\lambda]$ for reckoning the degree of conceptual compounds. Peirce apparently derived this formula from a theorem he discovered while working on Listing's 'census theory.'[37] In application to the logic of relations it took the form:[38]

Valency rule: The union of any μ-ad with any ν-ad gives $[\mu + \nu - 2\lambda]$-ad, where λ is the number of bonds of union.

Relative product falls under the valency rule for the case $\lambda = 1$, which accords with my construal of that operation as introducing a single 'bond of union.' Each time bonding is applied, the index λ increases by 1. Thus $p(R, S)$ has degree $[r + s - 2]$, $gp(R, S)$ has degree $[r + s - 4]$, $ggp(R, S)$ has degree $[r + s - 6]$, and so on. Since permutations and relative complement are degree-neutral operations, any compound formable in bonding algebra will have degree $[\Sigma - 2\lambda]$, where Σ is the sum of the degrees of its elementary components and λ is the total number of applications of bonding and relative product. So the valency rule is validated within bonding algebra.

More generally, the valency rule by itself entails the reduction thesis, so long as λ is limited to positive values. Let any definitional process be called *valency-regular* if it conforms to the valency rule under the stated condition. Then:

T9. *The reduction thesis holds within any system of valency-regular definitional processes rich enough to permit the formation of relative products.*

Proof: The availability of relative product guarantees the reducibility of all higher polyads ($n > 3$) within sufficiently large domains (by T3). And the irreducibility of lower polyads ($n \leq 3$) follows from the valency-regularity condition. With λ restricted to positive values, the quantity $[\mu + \nu - 2\lambda]$ assumes its maximum values for given μ and ν when $\lambda = 1$, that is when concepts are joined by single 'bonds of union.' For this maximal case the formula simplifies to $[\mu + \nu - 2]$ or the algebraic type of relative product. Therefore table 1 can be reinterpreted as a table of maximum values for valency-regular compounds of monads, dyads, or triads. And the table shows that $[\mu + \nu - 2] > 3$ only if $\mu \geq 3$ or $\nu \geq 3$.

These facts were known to Peirce by the 1890s, and no doubt they helped to confirm his attachment to the reduction thesis. But they could only do so against the background of a conviction that valency-nonregular definitional processes were somehow dispensable. Since many standard definitional processes fail to be valency-regular, this background assumption requires careful scrutiny.

Both regular and nonregular operations figured in Peirce's earlier logical works; but from the outset they were not treated equally. The nonregular operations eventually thinned out, and Peirce quite deliberately undertook to explain them away by what I will call the doctrine of surrogate constructions. This doctrine, which is large enough to warrant a separate study,[39] will be broached here through two historically noteworthy cases.

VI

The valency-regular operations discussed by Peirce can all be derived within bonding algebra in known ways. These include the operations he called involution, backwards involution, transaddition, and relative sum.[40] Prominent among the valency-nonregular definitional processes discussed by Peirce is his version of Boolean product – the operation he variously called nonrelative multiplication, absolute product, or logical composition.[41] Its effect can be seen in the construction of [--is a French servant] from [--is French] and [--is a servant]; or in the construction of [--is a lover and a servant of--] from [--is a lover of--] and [--is a servant of--]. Applied to any two relations of equal degree this operation always results in another relation of the same degree. By this consideration, nonrelative product is clearly at odds with the valency rule. For example, valency-regular compounds of two triads are confined to degrees $[6 - 2\lambda]$, which are bound to be even; but the nonrelative product of two triads has an odd degree.[42] It is noteworthy therefore that Peirce made several ingenious efforts to explain away nonrelative product as somehow a special case of relative product together with auxiliary devices. These efforts began with his first major logical work, whose 'comma' notation can be understood as a first experiment in this direction.[43] And by 1882 Peirce wrote to O.H. Mitchell that he had found methods for 'altogether discarding' nonrelative product.[44]

Peirce's sophisticated handling of valency-nonregular processes can also be seen at work in his approach to the method of triple junction which he discovered on reading A.B. Kempe (1886). Kempe's work had a marked impact on Peirce's thought and prompted him, he tells us, 'somewhat to modify my position, but not to surrender it.'[45] There is much to ponder in the curious interaction between Kempe and Peirce, and much remains to be sorted out. For now I will attempt no more than to describe the triple-junction method algebraically and to place it in logical perspective vis-à-vis the theory of valency and bonding algebra.

According to the chemical model, combination of concepts is always two at a time. Upon studying Kempe's memoir it struck Peirce that triple combinations had a very special character, and could even be used to bring about a wholesale reduction of triads.[46] Following the pattern of the first reduction theorem (T3), the Kempe-Peirce method can be reconstructed with the help of a three-way analogue to relative product:

Triple junction: $t(Q, R, S) = \{X\frown Y\frown Z: (\exists w)(X\frown w \in Q \,\&\, Y\frown w \in R \,\&\, Z\frown w \in S)\}$.

This operation, algebraically unlike any so far encountered, has the algebraic type $[q + r + s - 3]$. It is not derivable within bonding algebra, nor does it conform to the valency rule. Let us see what could be done with it. Applied to three dyads it results in a triad, according to the diagram:

and it permits a wholesale reduction of triads along lines analogous to the first reduction method:

T10. *Using triple junction, all triads are reducible within sufficiently large domains.*

Proof: Let **R** be any triadic relation over a domain D, and assume that D has a cardinality at least as great as **R**, so that there exists some one-to-one mapping of **R** into D. For any triplet $r = r_1\frown r_2\frown r_3$ belonging to **R**, let r^* be the corresponding element of D under this mapping. Construct three functions g, h, j, each of which maps members of **R** into couplets of elements from D:

$$g(r) = r_1\frown r^*, \quad h(r) = r_2\frown r^*, \quad j(r) = r_3\frown r^*.$$

Now three corresponding dyadic relations **G**, **H**, **J** will be described, by the equations

$$G = \{g(r): r \in R\}, \quad H = \{h(r): r \in R\}, \quad J = \{j(r): r \in R\}.$$

Then by an argument similar to that in the proof of T3, the triple junction t(**G**, **H**, **J**) of these three dyads exactly coincides with the original triad **R**.

The effect of triple junction can be obtained in quantification theory for any relations of given degrees. For example, if **Q**, **R**, and **S** are any three dyads, their triple junction t(**Q**, **R**, **S**) can be expressed by the quantificational formula '$(\exists w)(Qxw \,\&\, Ryw \,\&\, Szw)$.' Therefore:

T11. *In quantification theory, all triads are reducible within sufficiently large domains.*

55 Peirce's Remarkable Theorem

And so the reduction thesis collapses under standard notions of definability.

Arthur Skidmore finds in these facts a refutation of Peirce's thesis, on the grounds that 'Peirce has given no reason to prevent the combination of plural predicates out of dyadic predicates which would not equally prevent the combination of plural predicates out of triadic ones.'[47] This would be a correct finding if Peirce's definitional resources exactly matched those of standard quantification theory. It would be undercut within the framework of bonding algebra, as can be seen by drawing out the consequences of theorems T6, T7, and T10. The result T6 shows that the first reduction method is available within bonding algebra; and the results T7 and T10 together show that the second reduction method cannot be implemented through bonding-algebraic operations. Furthermore, the operations underlying the two reduction methods contrast sharply with respect to Peirce's valency rule: relative product is a valency-regular operation but triple junction is not. These are principled distinctions which can be drawn naturally within the Peircean framework.

Triple junction is only one of numerous operations contrary to the reduction principle that are freely available within quantification theory. In every such case the doctrine of surrogate constructions applies: the operation in question can always be replaced by some valency-regular definitional process within bonding algebra.[48] Peirce's application of this doctrine became more systematic and rigorous following his encounter with Kempe's work; and in the manuscript 'On Logical Graphs' it was adopted as an explicit regulative principle. Having urged that 'we are bound to carry our logical analysis to the furthest point, when the analysis of thought is the very business we have in hand,' Peirce endorsed the proposition that 'combination other than in pairs can be expressed by means of combinations by pairs,' and argued in favour of the more restrictive mode of combination.[49]

VII

We now have before us a logical framework congenial to the reduction thesis and capable of marking Pericean distinctions. Having touched upon strikingly close connections between bonding algebra and Peirce's logical philosophy, it seems only fair to conclude by expressing certain reservations. Bonding algebra affords a resolution of the reduction problem which may after all be too neat to be true. Had Peirce's definitional resources exactly matched those of bonding algebra, it would not be easy to account for the absence of a reduction theorem in his works. Suppose, however, that he occupied a position close to the bonding algebra but involving marginal departures which very much complicated the reduction problem.[50] Or suppose that his views hovered round and about the bonding algebra without ever settling decisively upon it. In either case the bonding algebra could well serve its stated purpose as a base for closer approaches to his thought.[51]

APPENDIX
SOME FORMULATIONS OF PEIRCE'S REDUCTION THESIS

1870

'A relative term cannot possibly be reduced to any combination of absolute terms, nor can a conjugative term be reduced to any combination of simple relatives; but a conjugative having more than two correlates can always be reduced to a combination of conjugatives of two correlates.' ('Notation for the Logic of Relatives' CP 3.144)

1882

'Every quadruple or higher relative may be conceived as a product of triple relatives.' (Brief Description of a Notation for the Logic of Relatives' CP 3.317)

c 1885

'Thirdly, there are plural characters, which can all be reduced to triple characters but not to dual characters.' ('One, Two, Three: Fundamental Categories of Thought and Nature' CP 1.370)

c 1890

'Why not go on to find a new conception in four, five and so on indefinitely? The reason is that while it is impossible to form a genuine three by modification of the pair, four, five and every higher number can be formed by mere complications of threes.' (*A Guess at the Riddle* CP 1.363)

1892

'Non-relative and dual rhemata only produce rhemata of the same kind, so long as the junctions are by twos; but junctions of triple rhemata (or junctions of dual rhemata by threes) will produce all higher orders.' ('The Critic of Arguments' CP 3.421)

c 1894

'A thorough study of the logic of relatives ... shows that logical terms are either monads, dyads, or polyads, and that these latter do not introduce any radically different elements from those that are found in triads.' ('The List of Categories: A Second Essay' CP 1.293)

57 Peirce's Remarkable Theorem

c 1896
'The triad is the lowest form of relative from which all others can be derived.' ('On Logical Graphs' PP 482)

1897
'That out of triads all polyads can be constructed is made plain by Figure 17.' ('The Logic of Relatives' CP 3.483)

1898
'Now I call your attention to a remarkable theorem. Every polyad higher than a triad can be analyzed into triads, though not every triad can be analyzed into dyads.' ('Detached Ideas Continued' PP 439)

c 1899
'In 1867 ... I had a proof, duly published, that there was only a third category of characters besides nonrelative characters and dual relations ...' ('That Categorical and Hypothetical Propositions Are One in Essence' CP 1.565)

1903
'analysis will show that every relation which is tetradic, pentadic, or of any greater number of correlates is nothing but a compound of triadic relations.' ('Lowell Lectures' CP 1.345–7)

1904
'Mr. Russell's idea that there is a fourthness, etc. is natural; but I prove absolutely that all systems of more than three elements are reducible to compounds of triads; and he will see that this is so on reflection.' (Letter to Lady Welby 2 December 1904 in Lieb 1953 14)

1905
'It is a priori impossible that there should be an indecomposable element which is what it is relatively to a second, a third, and a fourth.' ('The Basis of Pragmatism' CP 1.298–9)

1905
'[Royce] attacks my one-two-three doctrine in the very field where it is most obviously defensible, that of formal logic ... The points are (1) that no dyadic relation can be composed of non-relative factors alone, and (2) no triadic relation can be composed of dyadic and monadic factors alone, although (3) every tetradic and higher relation can be composed of monadic, dyadic, and triadic factors.' (Letter to William James, August 1905 PP L224)

c 1906

'Indeed definitions being scrupulously observed, it will be seen to be a truism to assert that no compound of univalent and bivalent concepts alone can be trivalent ... Less obvious, yet demonstrable is the fact that no indecomposable concept can have a higher valency [than three].' ('Pragmatism (Editor [3])' CP 5.469)

1908

'I adopt this trichotomy because no logically indecomposable relative has a valency > 3, [tetrads] being demonstrably invariably capable of decomposition into [pairs of triads].' (Draft of a Letter to P.E.B. Jourdain 5 December 1908 PP L 230a)

c 1908

'It can further be said in advance ... that there must be an elementary triad. For were every element of the phaneron a monad or a dyad, without the relative of teridentity (which is, of course, a triad), it is evident that no triad could ever be built up.' ('Prolegomena to an Apology for Pragmaticism' CP 1.292)

c 1909

'If you choose to say that there are more than three correlates, that is a matter of indifference; for every relationship of more than three correlates is equivalent to a logical composite of two or more triadic relationships; but a triadic relationship cannot be built up from dyadic relationships.' ('Some Amazing Mazes, Fourth Curiosity' CP 6.323)

UNDATED

'The idea of four, or any higher number, is but a complication of three (which of course involves those of two and one). Thus, no combination of bivalent atoms can make a trivalent radicle; but with trivalent atoms radicles of every valency can be constructed.' ('The Three Categories and the Reduction of Fourthness' PP 915)

UNDATED

'It is impossible to analyze a triadic relation, or a fact about three objects, into a dyadic relation; for the very idea of a compound supposes two parts, at least, and a whole, or three objects, at least in all. On the other hand, every tetradic relation or fact about four objects can be analyzed into a compound of triadic relations.' ('Consciousness' CP 7.537)

Further variations will be found in the following paragraphs in CP: 1.371, 1.421, 1.476, 3.144, 3.639, 4.309, 4.445, 5.88.

EDWARD H. MADDEN

Scientific Inference: Peirce and the Humean Tradition

I

Peirce's whole philosophy is a complex mixture of characteristic nineteenth-century themes and original contributions of the highest order. In this paper we will be interested primarily in his philosophy of science, and most particularly in his analysis of 'induction' and 'abduction' and in his view that the scientific enterprise is self-corrective and yields cumulative knowledge.

Hans Reichenbach believed that Peirce prefigured his own more elaborate view that the repeated use of induction will lead to success if success is possible (Reichenbach 1939). Max Black also interprets Peirce as offering a self-corrective vindication of induction and then effectively criticizes the whole strategy of vindicating induction (Black 1954). Though he realizes that Peirce was not responding to Hume, Laurens Laudan in a recent very significant article nevertheless criticizes Peirce's view of self-correction as a degenerate form of Reichenbach's argument (Laudan 1973). It seems to me, however, that it is very misleading to link Peirce and Reichenbach in this way. Peirce never accepted the framework which makes sense of the concept of vindication, and he explicitly rejected Reichenbach's strategy. If we become clear about the differences between the two figures, we will be better able to appreciate what one of the giants of the Golden Age genuinely had to say about scientific inference. My procedure will be as follows: I will (1) examine the concept of a self-corrective vindication and present the views of Reichenbach and Laudan; (2) examine certain aspects of Peirce's views about belief, doubt, truth, and inquiry that throw light on his interpretation of scientific inference and its success; (3) characterize Peirce's mature view of the nature of scientific inference and his reasons for thinking well of it; and (4) specify in detail why linking Peirce and Reichenbach is misleading and why the whole concept of vindicating induction is foreign to Peirce's thought.

II

If the overall thrust of his view is to be properly appreciated, Reichenbach's 'pragmatic vindication' must be seen in the framework of the traditional problem of induction. What is the justification for extrapolating from an evidential base to a universal conclusion? If properties are conceived to be not wholly independent, they can, in various ways, be justifiably linked in such conclusions; if they are wholly independent, they cannot so easily be justifiably linked. The essence of the Humean tradition is the view that properties are wholly independent, and Reichenbach was a staunch Humean in this sense. What, in this event, can be said by way of justifying inductive generalization? 'Vindicators' like Reichenbach and Feigl conclude that such generalizations *cannot* be justified. All uniformity and limited variety principles fail to supply an adequate reason for relying on such extrapolations, and any inductive justification is bound to be circular. However, even though we cannot prove that induction is an adequate way of knowing the future, or the unobserved, we can show that *if* there is an adequate way of knowing the future, induction is the only method that can be shown beforehand to be adequate to the task. While we cannot prove induction to be a rational enterprise, we can prove that *the use of inductive techniques* is the only rational procedure. It may not do the job, but if it doesn't nothing else will. To the criticism that this advance over skepticism is miniscule, the reply has been that philosophers must learn to be grateful for small mercies.

More specifically, Reichenbach vindicated 'the rule of induction' by demonstrating that *if* there is a point, as a series grows longer and longer, when a relative frequency remains approximately constant, then the *repeated* use of this rule is capable of discovering that relative frequency (cf Reichenbach 1949, 1952; Lenz 1960; Madden 1960b). The use of this rule of induction (the straight rule) is self-corrective because its *repeated* use will lead to success *if success is possible* – and this can be proven beforehand, before any inductions are made. Reichenbach thought that Peirce anticipated his own view that the repeated uses of induction will lead to success if success is possible, and supported his view by citing one reference to Peirce (cf Reichenbach 1939, 188). Abduction, he thought, consists in studying facts and devising a theory to explain them. Its only justification is that 'if we are ever to understand things at all, it must be in that way.'

Laudan agrees with Reichenbach and Black that Peirce held a self-corrective vindication of induction and is critical of his formulation of this view. The trouble is, he writes, that Peirce showed quantitative induction to be self-corrective and then simply asserted without argument or justification that

61 Scientific Inference: Peirce and the Humean Tradition

qualitative induction or abduction is also self-corrective. 'Having shown (at least to his own satisfaction) that quantitative induction is self-corrective, Peirce then, without even the hint of a compelling argument, makes the crucially serious slide ... [and] acts as if his arguments about quantitative induction show all the other species of induction to be self-corrective as well' (Laudan 1973, 293). Reichenbach is to be admired over Peirce because he did not 'slide' but genuinely tried to justify the step. In addition to being critical of Peirce, Laudan is also critical of Reichenbach's students for the same reason. They spend all their time trying to justify the straight rule against a host of alternate rules that are also self-corrective in Reichenbach's sense (claiming, e.g., that only the straight rule is linguistically invariant), but neglect the question, which Reichenbach to his credit saw so clearly, whether there is any significant connection between the straight rule and the grounds for accepting scientific abductions (Laudan 1973, 296–8).

III

In order to understand Peirce's view of scientific inquiry, we need first to investigate his concepts of doubt, belief, knowledge, and inquiry in general.

As is well known, Peirce held a dispositional notion of 'belief.' For Peirce, 'A believes p' means that A has a capacity to act in a certain way and will act that way in certain circumstances. As Peirce says, a belief is 'a rule of action, or, say for short, a *habit*.'[1] Though he often speaks in psychological terms, his point is an epistemic or logical one. Doubt is the awareness that two beliefs are incompatible, and hence inhibits the exercise of a habit that effectively deals with reality. Legitimate doubt may arise in several ways. The incompatibility may be between beliefs based on perceptual grounds alone or between two non-perceptual propositions. If I am aware, for example, that the sherry tastes sweet yet believe that I am drinking Dry Sack, either because of the colour, bouquet, or label, then I see an inconsistency between two perceptual contents and thus legitimately doubt that I am drinking Dry Sack. Or I may believe that the earth is round and also believe that a ship will disappear on the horizon without the hull disappearing first and the mast tip last. In such a case the awareness of the inconsistency is between two non-perceptual propositions, and again the inconsistency is the basis of legitimate doubt.

Peirce contrasted such legitimate doubt with what he called sham doubt, the sort of doubt exhibited by sceptics and by philosophers like Descartes who discard ordinary knowledge claims on the way to some indubitable foundation for all knowledge claims. According to Peirce, these philosophers argue that since it is always logically possible for every empirical proposition to be false, empirical propositions cannot at any given time be said to be *known* to be true. Since it is logically possible that every empirical claim is mistaken, since they are always open to correction by future experience or can be shown to be intrinsically unreliable by the dreaming, evil demon, etc., arguments, it is impossible to say that we know an ordinary empirical proposition to be true. Their arguments do not purport to show that there is no ordinary, perceptual knowledge but only that if an ordinary proposition is true we cannot know it to be so on empirical grounds.

Peirce replies, in essence, that the possibility of being mistaken is no good reason for thinking one is mistaken – hence to doubt on such grounds is what he calls 'sham doubt' and to reject a knowledge claim on such grounds is unreasonable. It makes no more sense than to say that the possibility of being right is a good reason for thinking one is. To call something illusory, biased, or accidental, or to call it veridical, fair, or lawful, all require positive reasons. A perceptual statement is called in doubt when it conflicts with other perceptual and/or conceptual judgments, and accepted when there is no inconsistency encountered between this assertion and others one believes to be true. To extrapolate Peirce's argument, a general statement is called into question when it either conflicts with a system of scientific propositions or finds no explanation there, and accepted when it provides a *ratio essendi*, or is given a *ratio essendi*, within such a system. Some positive reason is presupposed for saying that something either is illusory or veridical, or is a fair or biased sample, or is a lawful or accidental universal. Simply pointing out that it is logically possible to be mistaken, Peirce thought, leaves wholly open the question whether one is or not, just as simply pointing out that it is logically possible that one is correct would leave wholly open the question whether one is or not. One fears that one has been shown to lack knowledge in one's ordinary beliefs by the sceptic's or Descartes' arguments, but one has only been shown by these arguments that it is impossible to *demonstrate* their correctness. But common sense does not, according to Peirce, require their demonstration. We are reassured when we remember that a positive reason is presupposed for saying either that a statement is mistaken or that it is correct – and hence that the sceptic presupposes all that we care about in his argumentative demand for a demonstration. This insight constitutes what might be called the common-sense strain in Peirce's philosophy.

Peirce discusses in great detail the way in which scientific inquiry specifically establishes its propositions.[2] Thomas A. Goudge's analyses of these discussions in TP part one, section VI remain unrivalled and are canonical in Peirce scholarship. So I can be brief at this point and only mention points that are crucial for my theme. Only qualitative induction concerns us now, though later quantitative induction will have to be considered very carefully.

63 Scientific Inference: Peirce and the Humean Tradition

By 'qualitative induction' Peirce meant the use of hypotheses in explaining events, a procedure he sometimes referred to as abduction or retroduction. Abductions are freely invented explanatory schemes arrived at by imaginative insight.[3] It is often a new way of looking at things made possible by using metaphors and analogies. If we conceive of the atmospheric blanket along the lines of a liquid, a sea of air, we can explain how a suction pump works and so on. After the passion to learn, the most indispensable capacity for the successful prosecution of science is the imaginative use of analogies. The pragmatic view of science allows any flight of imagination provided only that this imagination ultimately has some experienceable consequences. Moreover, hypotheses must have *explanatory* power; they show conceptual connections between events and not simply regularity among them. The crucial point is that imagination can produce a million hypotheses most of which would be silly; any imagined hypothesis is a good one only if it is a *reasonable* one in the sense that it has some *explanatory value* which then needs to be checked out in further detail. Peirce believed that Kepler's work was the best example of abduction at work. He condemned Mill for his simple-minded view that Kepler somehow simply generalized observational data. Rather Kepler was confronted with an anomaly requiring the formulation of an adequate explanation that did not fly in the face of established science and mathematics. Kepler went through agonies trying to find the proper explanation until his breakthrough insight that instead of looking at 'librations' vs 'ellipse,' it is crucial to see 'librations' as being equal to 'ellipse.' Kepler had finally seen the point; now the hypothesis and the data fit like hand in glove (Hanson 1958b 70–92).

There are a number of requirements for *testing* the consequences of an abduction to which Peirce returned again and again. For example, the consequences of a hypothesis must be *predesignated*. It works differently in the various kinds of induction. When the class 'sampled' consists of units, and the question at issue is the ratio between counts of occurrences, induction is numerical and a comparatively simple affair. The case of abduction is more complicated, however, for explanatory characters are not units, nor do they consist of units, nor can they be counted in the sense that one count is right and another wrong. Characters have to be estimated according to their *significance*.

Since there are many explanatory hypotheses and innumerable testable consequences in all areas of science, the crucial question is which hypotheses to test first and in what subsequent order. Peirce returned often to this question of what he called the Economy of Research Programs. This matter is important, for, as he well knew from his own experience, there is just so much money to support research, and a strategy is needed for getting the best results from limited resources. The value of each investigation is the amount of money it will pay us to spend on it. Hypotheses the consequences of which are most readily deduced and tested should be tested first, since if they are wrong they can be eliminated at least expense. The best hypothesis to test early is the one that can be most readily refuted if it is false. And would not such a one be the kind that is prolific in consequences? Peirce clearly emphasized the fundamental importance of disconfirmation in the logic of science. For fullness and precision of knowledge money is well invested up to a point. But the cost of the information also increases with fullness and accuracy; and it does not pay to push the investigation beyond a certain point in fullness or precision. And so on.

For Peirce, the result of structuring and testing abductions in these and many other ways is *scientific knowledge*. His analysis of scientific knowledge is part and parcel of his general analysis of 'believe,' 'doubt,' and 'inquiry.' As we have seen, good abductions, first, must be able to explain the facts and, second, pass the tests of experimentation, always being confirmed and never disconfirmed. Meeting these requirements leads to the avoidance of all surprise 'and to the establishment of a habit of positive expectation that shall not be disappointed.' Again, this point can be put into the non-psychological form we saw earlier. They explain what we experience, do not conflict with other explanatory abductions or with perceptual experience, and hence automatically preclude any reasonable doubt. Now in the light of these facts and in the absence of any legitimate positive reason for their falsity, these abductions can be said to be established conclusively. It would be unreasonable to doubt them. To say we ought to doubt them because it is always logically possible in spite of all appearances that they are false was, for Peirce, sham doubt and not a positive and hence legitimate ground for doubt. It should be ignored. Even where a high degree of exactitude and confirmation is not attainable, there is still no reason for saying that we have no knowledge at all. Simply because we cannot reach great certainty, say, about the life and teachings of Pythagoras, it does not follow that we should dismiss the subject as Zeller does as if we knew nothing about the matter.

While many abductions are conclusive, they are all, Peirce thought, nonetheless corrigible, and all scientists who hold them fallible. Though the logical possibility that an empirical proposition is false is no good reason for doubting it, this reason is a good one for being a fallibilist and holding that all empirical propositions are corrigible. Indeed, the scientific spirit requires that a man 'be at all times ready to dump his whole cartload of beliefs, the moment experience is against them.' Positive science can only rest on experience; and experience can never result in absolute certainty, exactitude, necessity, or

65 Scientific Inference: Peirce and the Humean Tradition

universality. 'That we can be sure of nothing in science is an ancient truth' and follows from the fact that ampliative inference is not deductive inference. But from the fact that ampliative statements are never incorrigible it does not follow that they are never established beyond reasonable doubt. Conclusiveness and incorrigibility must not be confused. Propositions are conclusively established when all the above inductive requirements are met. They are not incorrigible because it is always logically possible that they are false – a possibility that follows from the fact that they are not logically true propositions. All scientific propositions are empirical, and to say they are incorrigible would be to block further scientific inquiry – a view just as bad in its own way as to think that no scientific conclusions are ever conclusively established.

The use of inductive techniques, Peirce continued, is self-corrective. 'The true guarantee of the validity of induction is that it is a method of reaching conclusions which, if it be persisted in long enough, will assuredly correct any error concerning future experience into which it may temporarily lead us' (CP 2.769). Scientific inquiry, fully carried out, 'has the vital power of self-correction and of growth.' 'This is a property so deeply saturating its inmost nature that it may truly be said that there is but one thing needful for learning the truth, and that is a hearty and active desire to learn what is true.' Indeed, 'If you really want to learn the truth, you will, by however a devious path, be surely led into the way of truth at last' (CP 5.582).

Peirce stated his self-correction thesis in 1881, in 1883 three times, in 1896, in 1898 four times, in 1902 twice, in 1903, and in 1908. The formulations do not appear to have changed noticeably through the years and are fairly repetitious.[4] His formulations are quite undeveloped, and sometimes it is difficult to tell whether they refer to quantitative or qualitative induction or are meant to be inclusive. There are two prevailing models. (i) The scientist frames hypotheses and tests their consequences in public and repeatable experience and thereby keeps the hypotheses always falsifiable. If the hypothesis is false, then there is some such experience deducible from it that will show it to be false. And the repeated use of the scientific method will find that consequence in the long run. A disconfirming consequence that allegedly might exist and yet *never* be discovered in the long run is a 'metaphysical' consequence that can safely be ignored. Moreover, as an hypothesis is studied more fully it will be sure gradually to take on another colour, 'little by little to receive modifications, corrections, amplifications, even in case no catastrophe befalls it.' (ii) The correct relative frequency must ultimately reveal itself, if scientific inquiry continues, since induction is based on samples drawn at random and each sample is thus free to turn up with the same relative frequency. Judging of the statistical composition of a whole lot from a sample is judging by a method 'which will be right on the average in the long run, and, by the reasoning of the doctrine of chances, will be nearly right oftener than it will be far from the right.'

In one place Peirce gives examples of how methods of mathematical computation correct themselves when reapplied and give the correct answer in the end. This fact, he says, 'calls to mind' the *general fact* that scientific reasoning tends to correct itself 'and the more so, the more wisely its plan is laid.' He makes no effort to apply the mathematical model to abductive reasoning. He apparently uses it as a metaphor for that 'wise plan' which is presumably the strategy of research and the other elements of good inductive reasoning we have already examined.

IV

If the above account of Peirce's view of belief and scientific inference is accurate, it is clear that the whole concept of vindicating induction is foreign to his thought and that linking his name with recent and contemporary self-corrective views is misleading and unhelpful, to say the least.

(1) The concept of vindicating induction, as distinct from justifying it, makes sense only by interpreting Hume sceptically. 'Though vindications are only a shade short of Humean scepticism, philosophers should be grateful for small mercies.' The question is not whether the sceptical interpretation of the historical Hume is correct but what version of the latter the vindicators held. By their very characterization of 'vindication' Reichenbach and Feigl accepted the sceptical interpretation. Peirce, however, ignored the sceptical Hume for the same reason he rejected Descartes' strategy. His response (or lack of it) to the sceptical Hume is closely tied to his views on doubt and legitimate belief which we have already examined. But it will be clearer if we now bring in Peirce's explicit response to Descartes.

According to Descartes, since it is logically possible that I am mistaken in saying that I see the fire on the grate (I could be dreaming that I see the fire), it is incorrect to say that I know it to be the case. Peirce understood Descartes to mean that this argument entails that I can *legitimately* doubt it to be the case whether or not I actually do. But this doubt is sham doubt; Descartes knew perfectly well there was a fire in the grate. For genuine doubt to arise, there must be some inconsistency in our corpus of established beliefs or between such beliefs and a present perception. If the fire never warmed me no matter how close I got, there would be room for genuine doubt. All Descartes' argument does is to show that empirical propositions, etc., are not incorrigible, but Peirce not only was happy to agree with this view but insisted upon it.

67 Scientific Inference: Peirce and the Humean Tradition

He only insisted that the lack of incorrigible propositions does not entail the lack of warranted knowledge claims. By extension the same response would hold for the sceptical Hume. Since it is logically possible that I am mistaken in inferring 'All A are B' from 'All A are B at $t_1 \ldots t_n$' (nature could change its course), it is incorrect to say that I know that 'All A are B,' whatever the A and B might be. I can legitimately doubt any and all 'All A are B' assertions. But for Peirce this doubt would again be sham doubt; we know perfectly well that numerous scientific laws are conclusively established and explained. For genuine doubt to arise, there must be some inconsistency in our corpus of established beliefs or between such beliefs and a present perception. Such results are grounds for real doubt about the original 'All A are B' assertions. All the sceptical Hume's argument does is to show that scientific propositions are not incorrigible, but Peirce – and so on as above.

I am not claiming that Peirce's response either to Descartes or, by extension, to the sceptical Hume is adequate, though the response, it seems to me, is interestingly suggestive of recent discussions of a strong and weak sense of 'to know' and the uselessness of insisting that if we do not know in a strong sense then we do not know at all. It should be clear that I am showing only why Peirce thought he could ignore the sceptical Hume, or why he did in fact ignore him and why he did not think that scientific explanation needed any *vindication* in the contemporary sense. It does seem to me that his ground for rejecting sham doubt is much more interesting and important than his undeveloped self-corrective argument aimed against the arguments he did think engendered legitimate concern.

(2) 'Incorrigibility' is often confused with 'conclusiveness' and 'corrigibility' with 'inconclusiveness' and, given the conceptual connection between 'corrigibility' and 'fallibilism,' 'fallibilism' is confused with 'inconclusiveness.' This web of confusion plays its role when Peirce is said to have offered a pragmatic vindication of induction. Peirce is clearly a fallibilist; hence he believed that scientific explanations are inconclusive and require his self-corrective vindications. Peirce, however, insisted upon both fallibilism and conclusive explanations and so felt no need for a *vindication*. Let us examine this point more fully.

What is the relation in Peirce's thought between 'infallible' and 'incorrigible' and between 'fallible' and 'corrigible'? (cf Greenlee 1971). Presumably this: A *person* is infallible if the *proposition* he believes to be true is incorrigible – that is, if the proposition is not only exempt from error but exempt from the very possibility of error. Peirce held that a person is always fallible because no proposition is exempt from the *possibility* of error. Any proposition is corrigible because it is always logically possible that it is false. For Peirce, however, such corrigibility did not constitute evidence for rejecting it. To do so is to make the sceptic's and Descartes' mistake. Incorrigibility must not be confused with *conclusiveness*. For a proposition to be conclusively established it was sufficient for Peirce that there be no positive reason against it, and much in its favour. Incorrigibility is the stronger claim that there is no possible circumstance in which it could be false. Peirce believed that many empirical propositions have been established conclusively though none of them, by their empirical nature, are incorrigible. To say that they were incorrigible would be to block further scientific inquiry, as much a cardinal sin as to think there is insufficient reason for accepting any scientific statements as conclusively established.

It might seem that Peirce's self-corrective thesis leads directly to his ultimate community view of truth, and that these views flatly contradict his claim that scientific propositions can be conclusively established. It is crucial to see that this is not the case. To be sure, Peirce's self-corrective view and his analysis of truth are related. For Peirce true propositions are those which will always be confirmed and never disconfirmed, and this is the ideal that the scientific community ultimately would exhibit. It would be the ultimate achievement of the self-corrective process of science in contrast to the unreliable and endlessly conflicting other modes of making knowledge claims. So Peirce, too, ultimately allows incorrigibility into his inductive philosophy. For him, to say that the ultimate scientific community knows p to be true but p is possibly false *is* a self-inconsistent statement. However, for Peirce the community view of truth (and knowledge) is a definition of a concept that could never be known to be exemplified. It is the ideal that the scientific community seeks on every question, but it is never a *criterion* of knowledge (cf Meyers 1971). These criteria are the scientific techniques and methods that we have been discussing, and the proper use of them gives us those conclusively established scientific knowledge claims that forever fall short of the incorrigible ideal claims.

(3) That Peirce saw no need to *vindicate* scientific inference follows clearly from the fact that he thought the real problem is to explain why scientists had established conclusively so *many* hypotheses in such a short time, why science in fact has progressed more rapidly than might well be expected.

The success of abductive inference, Peirce thought, depends upon an affinity between the reasoner's mind and the structure of nature. The *hope* that such an affinity exists is the only thing that gives inquiry its point in the beginning. Possible abductions, if not strictly innumerable, 'exceed the third power of a million,' and therefore the chances are overwhelmingly against coming to correct hypotheses by trial and error or chance. But we do have a large number of conclusively established scientific theories. Hence what began as a hope, 'even a desperate one,' turns out to be itself well founded. Galileo, Kepler, Gilbert, and Harvey, he thought, all appealed at the most critical stages of their reasoning to instinctive judgments. Evolutionary theory, in fact, makes this way of interpreting the matter essential: a newborn chick does not rummage through all possible theories until it hits upon the good idea of picking up something and eating it. The chicken pecks by instinct. But if we are going to allow every poor chicken to have an innate tendency toward a positive truth, why should we think that man alone lacks this gift? No, we must acknowledge that man's mind has a natural adaptation to imagining correct theories of some kind, 'and in particular to correct theories about forces, without some glimmer of which he could not form social ties and consequently could not reproduce his kind.' 'Intelligent guessing,' then, is what comes naturally or instinctively to mind, and such intelligent guessing may be expected to lead us to a hypothesis that is sustained by experience and thus to leave the vast majority of possible hypotheses unexamined (Tomas 1957, 237–52).

The success of abductive inference means that science grows; it yields cumulative results. The fit between men's hypotheses and the structure of the world grows better and better. How does this scientific growth and progress toward the truth come about? Peirce believed that its growth is evolutionary and must be understood in terms of variations in abductive hypotheses and their natural selection by the facts of the world that confront each in the testing procedure. The variations of a hypothesis may be minute and unintentional à la Darwin or minute and the result of effort à la Lamarck, or the variations may be large and drastic, what had come to be called mutations. Peirce particularly contrasted the 'Lamarckian' and mutational views. The former is the model for ordinary scientific progress, the latter for dramatic and epochal progress. The physical journals each month, Peirce said, publish the results of a great number of new researches. It is well-done work, but this sort of research does not affect seriously what we already know. It simply extends it bit by bit. Even though such researches do not yield splendid discoveries of great magnitude, they constitute scientific progress. Such was the case in the classification of the chemical elements in the lapse of time from Berzelius to Mendeleev. Science, however, mainly progresses by great leaps, by decisive conquests, by mutations in theories. Peirce cites Pasteur's overthrow of Claude Bernard's view as a good example. For long the medical world had been dominated by Bernard's dictum that a disease is not an entity but merely a set of symptoms. Pasteur, however, was unmoved by this 'nominalism' and 'positivism.' He began with the phylloxera. He found it influenced the 'optical activity' of the sugar and thus pointed to a ferment and therefore to an entity. He began to extend the doctrine to other diseases. The method of cultures and inoculations carried the day 'and here we see new ideas connected with new observational methods and a fine example of the usual process of scientific evolution' (Tomas 1957, 225; cf 220–5).

(4) If Peirce was not *vindicating* induction, however, what *did* he intend when he argued that scientific inference is self-corrective? After all, he was defending science in some sense. The answer is clear-cut and twofold. First, he was *justifying* science against competitive claims. He makes a great deal of this point in his writings on scientific inference. Science has competitors and denigrators in religious authority, a priori philosophy, and intuitive reasoning. Indeed, even the soothsayer thinks he has a better way of discovering truth than the scientist. Against all of these opponents Peirce directed his self-corrective justification of scientific inference. Moreover, the self-corrective nature of science, as Peirce viewed it, is straightforward and uninvolved and follows analytically, so to speak, from the very concepts of scientific inquiry and scientific method. It involves no assumption about the uniformity of nature, or any other sort of assumptions. This self-corrective feature is not something to be proved but is that very feature which distinguishes it from the other traditional modes of making knowledge claims. The point is that the scientific method of framing hypotheses and testing their consequences in public and repeatable experiences keeps the hypotheses always falsifiable. If the hypothesis *is* false, then there is some such experience deducible from it that will show it to be false. And the repeated use of the method will find that consequence in the long run. For Peirce, a disconfirming consequence that allegedly might exist and yet *never* be discovered in the long run is a 'metaphysical' consequence that can safely be ignored. It is, to be sure, a logical possibility, but, again, to doubt any well-established scientific principle on such grounds is not permissible since such doubt would be sham and not genuine doubt.[5]

Second, Peirce was *justifying* scientific inference in response to the traditional affirming-the-consequent criticism. That he was influenced by his hero Kepler in this context I think is likely. Indeed, his self-corrective view is largely a generalization of Kepler's view on the subject. Kepler's *Apologia Tychonis contra Nicolaum Ursum* is a sustained piece of philosophy of science, and Peirce no doubt knew it as well as he did Kepler's purely scientific work. There are too many points in the *Apologia* that find an echo in Peirce's work to be a matter of chance.

Kepler detested the 'mathematical fiction' interpretation of astronomical hypotheses (Blake, Ducasse, and Madden 1960, 37–43). The old refrain was that such hypotheses cannot be interpreted realistically because the arguments to establish them commit the fallacy of affirming the consequent. Kepler was unimpressed with this criticism and believed that there is every reason in the world to accept the Copernican view as physically true and not only a better calculator. That a true conclusion should follow from false premises is due to chance and bound to be discovered eventually. 'So at last it comes about, through the interweaving of syllogisms in demonstrations, that, given one proposition not agreeing with the truth, an infinite number of such will follow' (Blake, Ducasse, and Madden 1960, 41). If we start with a false assumption, then, we will in the long run find negative results and discard it. That we can do, Kepler continues, unless we allow a person who wants to save a hypothesis at any cost to introduce ad hoc hypotheses indefinitely – or, as he puts it, 'unless you freely allow the arguer to assume an infinite number of further false propositions, and never to be consistent with himself in his arguings up and down' (Blake, Ducasse, and Madden 1960, 41). But the case is quite otherwise with the Copernican hypothesis. There predictions and postdictions and explanations never run into negative results that require ad hoc hypotheses. Consequently, since the Copernican hypothesis does not run into negative results but explains the data well, and the long-run argument takes care of the logical objection, it is simply unreasonable not to accord a realistic status to it.

The connection between Kepler and Peirce is striking and requires little explication. If you insist that hypotheses in physics and astronomy have the same logical form, as Peirce did, then it is a short step to generalize Kepler's long-run arguments in astronomy to the self-corrective argument for science in general – which is precisely what Peirce did in part of his own self-corrective argument. We may also surmise that Kepler's work in the philosophy of science also provided Peirce a model for his fallibilism. Kepler originally held that Copernican propositions were not only conclusively established but are 'certainly true.' Eventually, however, he admitted that such propositions are corrigible and that 'the reality of the Copernican hypothesis is highly probable' (Blake, Ducasse, and Madden 1960, 37–9).

To be sure, self-corrective justifications were not unfashionable in the nineteenth century and generally exhibited similarities with Kepler's views, so it is difficult to say unreservedly that Peirce was inspired by the *Apologia*. It seems likely given Peirce's eventual first-hand acquaintance with and admiration for Kepler's works, but in any case he in fact used the very sorts of argument advanced by Kepler and used them for the same purposes. That Peirce's or any other of the nineteenth-century self-corrective justifications did the required job, even in their own non-sceptical contexts, remains in doubt.

(5) Peirce's self-corrective view is significantly different from Reichenbach's, since the latter's vindication purported only to show that the repeated use of the straight rule will lead to success *if success is possible*. Peirce,

however, had no doubt about the success of scientific inferences. The qualifications 'if success is possible,' 'if there is such a point,' etc., all reflect Reichenbach's acceptance of the argument that we cannot *know* success is possible because it is always logically possible for nature to change its course. Peirce, on the other hand, is not trying to prove that *if* science is successful it must be by induction; as we have seen, he thought the unusual success of science itself requires explanation. Reichenbach cited only one quotation to show that Peirce had offered a pragmatic vindication (Reichenbach 1939, 188). However, this quotation is out of the mainstream of Peirce's discussion of the self-corrective ways of science. Indeed, if one examines the context in which it occurs it is doubtful that Peirce is even talking there about the self-corrective ways of science at all. I do not mean to reduce the question to an exegetical issue over one passage. The crucial point is that, given the whole of Peirce's view as I have presented it, the only qualification Peirce is making here as elsewhere is not the sceptical one but the usual one that if you want to know the truths of nature you must use the scientific method and not the a priori, intuitional, mystical, or authoritarian ways.

In an interesting paper, John W. Lenz (1964) agrees that the non-conditional interpretation of Peirce's self-corrective view is warranted by the texts. Nevertheless he also finds evidence for a conditional interpretation which, he thinks, is philosophically interesting also and very close to Reichenbach's view. On this view, Peirce is supposedly claiming that the repeated use of inductive procedure will discover the limit, or the truth, in the long run if there is any ascertainable limit or truth to be discovered. This view, Lenz argues, is an analytical consequence of Peirce's metaphysical theory of truth (Lenz 1964, 153ff). He further argues that the non-conditional version of Peirce's self-corrective argument also rests upon a metaphysical foundation, namely, his rejection of the absolutely incognizable. What cannot be known, Peirce alleged, simply cannot be. This doctrine thus rules out the possibility of a series without a limit. According to Lenz,

using the inductive method, we could not know, of course, what the limit was, but neither could we know *that there was no limit*. In this respect the true constitution of such a series would be unascertainable, and hence such a series could not be. In short... he leaves off the qualifying phrase: 'if there is any ascertainable truth,' because, in view of his rejection of the incognizable, there is no need to add it. (1964, 157)

Lenz's paper leaves us with this puzzle: why would Peirce advocate the incompatible non-conditional and conditional views in close proximity? This fact is incompatible with the notion that he sometimes held one view and sometimes another, or that he changed his mind on this issue. Moreover, if the two views are incompatible and each supported by a metaphysical principle of Peirce, it follows that these principles are incompatible. It is not clear, however, how or that the theory of truth and the unreality of incognizables are incompatible.

(6) Peirce did not hold a version of Reichenbach's vindication, since he explicitly rejected the strategy of reducing abductions to quantitative inductions while Reichenbach explicitly adopted it.

Both Peirce and Reichenbach held a frequency view of probability, or Peirce, in his later years, a dispositional frequency view. (Professor Goudge was the first to clarify and criticize in a significant way Peirce's views on probability, and my discussion at this point again would have been difficult without his analysis (TP part one, section IV A).) Reichenbach believed that a frequency interpretation, through the concept of 'weight,' can be given to propositions about single cases and scientific hypotheses. Peirce strongly rejected the view that a frequency interpretation can be given to scientific theories. He believed that probability concepts apply only to what he called quantitative induction and have no application to abductive inference. To speak of the probability of scientific hypotheses is meaningless, since probability judgments can refer only to arguments which are true and false a given number of times. Peirce makes his point in this colourful way: 'It is nonsense to talk about the probability of a law, as if we could pick universes out of a grab-bag and find in what proportion of them the law held good' (CP 2.780). Moreover, since probability judgments always involve a numerical ratio, it is again illegitimate to talk about the probability of scientific hypotheses since no numerical value is ascribable to them (CP 2.657). He preferred to say that abductions are 'strongly supported' rather than that they are 'probable.' However, it was not simply a quibble over words for Peirce: 'There are *other* quite proper senses in which the word "probable" can be used; but the sense I have defined [the frequency sense] is the only one that can properly be treated by the calculus of probabilities.'[6] Whatever view he might take at a given time about the 'proper' use of the word probable, it is clear that Peirce was a conceptual pluralist about probability judgments. He would have agreed with Carnap that a frequency analysis is inapplicable to scientific hypotheses and that there are two irreducible senses of the term. (He rejected in advance, however, Carnap's notion that the probability of scientific abductions could be given a numerical value.)

The crucial point in this discussion of Peirce's views on probability for our purposes is that it commits him to a rejection of the strategy behind Reichenbach's self-corrective vindication of the straight rule. The strategy was to vindicate the straight rule and through the concept of weight to reduce 'probability of scientific hypotheses' to induction by enumeration, and hence to have vindicated the former also by the self-corrective argument. From Peirce's viewpoint this whole approach is wrong, since he insisted that probability judgments apply to quantitative induction but not to abduction.

This issue is a difficult one, however, and requires a qualification. A good case can be made that Peirce had the ingredients of Reichenbach's concept of weight and used it in giving a frequency interpretation of an abduction. Peirce suggests in one place that some 'single cases' can be explained by the frequency theory. 'Thus, when an ordinary man says that it is highly probable that it will rain, he has reference to certain indications of rain – that is, to a certain kind of argument that it will rain – and means to say that there is an argument that it will rain, which is of a kind of which but a small proportion fail' (CP 3.19). Here, like Reichenbach, Peirce allows one to predict the occurrence of a single event – rain now – and this is certainly an abduction. Moreover, he justifies the prediction by referring to the frequency with which such arguments have been successful and failed in the past. Literally speaking, Peirce is inconsistent. He has both said that frequency interpretations cannot be given to abductions and yet shows how it can be done. It is, however, a literal inconsistency and not an essential one. Some abductions are about repeatable events; hence a frequency analysis in terms of 'weight' can be given to them. But the crucial abductions, those of scientific theories, be they model or non-model in nature, are not 'repeatable' abductions and resist any frequency analysis.

(7) As a final effort to dissociate Peirce from the sceptical Hume, let me warn against any falsificationist interpretation of Peirce's view of science. There are a number of places in my characterization of this view where such an interpretation does not seem implausible. Much of his discussion of consequence testing, as we have seen, emphasized the fundamental importance of disconfirmation in the logic of science. He was always interested in the most economic way of getting rid of a maximum number of competing abductions. The best hypothesis is the one that opens itself widest to disconfirmation. And his whole discussion of the Economy of Science in terms of research programs and how to choose and conduct them has a similar contemporary ring about it. That there are similarities I have no doubt, but they are a matter of detail and not basic orientation. The falsificationists, like the inductivists, are Humeans – one might say Humeans with a vengeance – and Peirce would be no more impressed with their view that the logical possibility of a change in the course of nature is a good reason for rejecting any positive instance as evidence for a generalization than he was with Reichenbach's conclusion from the argument. The Cartesian-Humean way of approaching philosophy was simply ruled out by Peirce's deep draughts at the Kantian springs and the wells of common sense.

DON D. ROBERTS

The Labeling Problem

Antony Flew has recently discussed the difficulty of 'supplying appropriate means of identification and criteria of identity for incorporeal personal substances' (Flew 1966, 33). Flew was concerned with the notions of God and the self, and raised such issues as the legitimacy of certain applications of the term 'God.' These issues are not new, we realize, and yet it was Flew's discussion, coupled with an interest in the perennial philosophical topic of the clearness of ideas, that suggested to me an approach to a slightly different or more general difficulty, namely: How is it that the names or classifications or descriptions that we use in philosophy generally come to be attached to or associated with a given element of our experience?

Names and classifications and descriptions certainly embody complicated uses of language, and the general difficulty I have just mentioned admits of no quick and easy solution. Furthermore, I am not prepared to present a theory about how names in general are applied or about how names in general ought to be applied. Rather, my interest now is in what might be preliminary to such a theory; that is, I am interested in examining how certain particular names or descriptions have been, and perhaps ought to be, applied. I have found this kind of examination useful in clarifying my own ideas, and to illustrate it in this paper I focus on two familiar applications of the concept or term 'knowledge,' one from Plato and one from Bergson. But first, a few remarks are necessary.

INTRODUCTION

To avoid adopting the terminology of others (such as Flew) inappropriately, and to facilitate my discussion, I introduce the term 'labeling' for any use of language to name or describe or classify the various elements of our experience, and I call any word or group of words used for this purpose a 'label.'[1]

I take for granted at the start that everyone uses labels, attaching them to or associating them with his experiences in the same quick and easy way in which ordinary conversation is carried on. How does a person come to attach a label to, or associate it with, a specific element of his experience? I think we might tentatively agree that the general answer is this: Either the person learns from other people how to use the label, or he invents the label and the association, or by reason of a kind of ingenuity or perversity he invents a new association for an old label.

What I call 'the labeling problem' is the problem of determining how *in practice* we come to apply the labels we use. This is a problem concerning the propriety of a particular use of a label. The question we must ask ourselves is this: What is it in a particular situation that determines our use of a label? or, What is it we recognize in the situation that makes our use of the label seem appropriate? In some cases, conscious interpretation of some sort or another seems to be involved, as when we try to label some activity legal or illegal. In other cases, recognition seems to involve no such interpretation, as when we identify a voice on the telephone, or a rose by its scent. But let me illustrate.

God tested Abraham and said to him 'Take your only son Isaac, whom you love, and go to the land of Moriah, and offer him there as a burnt offering.' Abraham obeyed, traveled to a designated mountain in Moriah, and took the knife in his hand – whereupon the angel of God called to him and told him not to harm Isaac, 'for now I know that you fear God' (Genesis 22: 1–2, 10–12, 15–17).

Now if one of us, today, heard a voice telling us to sacrifice one of our children, we might hesitate before labeling this as a message from God. The nature of the demand itself might strike us as unbecoming a God. But Abraham apparently did not hesitate, and is considered the Father of the Faithful partly because of his unquestioning obedience. But how did he know, how could he tell that this was the voice of God? According to Genesis, God had been talking to Abraham for more than twenty-five years by this time, so that Abraham presumably recognized the voice. Did Abraham subsequently reflect upon the content of the message? If he did, that apparently had no effect on his labeling it a message from God.

And then, being convinced that this command came from God, how could he later be sure that the second voice, commanding him to spare Isaac, was also a message from God? How could he know this? Could not this other voice be the voice of some tempter?

If only we could talk to Abraham, it would seem fair to ask him about his first conversation with God, over twenty-five years before the incident at Moriah. How could he tell that first time, before recognition was a possibility, that this was the voice of God?

In the third week of August 1976 the Associated Press ran a story about a

77 The Labeling Problem

Detroit man who got instructions from God to find and kill a 'fat, false prophet dressed in red.' The press release states that this man drove about the city until he found, on a street corner, a '239-pound self-styled minister' wearing a red robe. The man took a carbine out of his car trunk and pulled the trigger, but the gun misfired. Receiving no further instructions from God, he then used his gun as a club and beat the minister to death. Now we must ask, how did the man from Detroit know, how could he tell that his instructions came from God?

The great bulk of our labels were invented so long ago that it is probably not possible to fix their first usage. We come by them in the ordinary course of learning a language. And most of us must be taught by others, unlike Tarzan who, according to Edgar Rice Burroughs, taught himself to read – a mental feat that cannot be too much admired (Burroughs 1972, 41–2, 60). Parents, teachers, authors are among those who teach us how to use labels; scientists[2] and writers of fiction[3] are among those who invent new labels and their associations; poets,[4] psychiatrists, and philosophers are among those who invent new associations for old labels.

With respect to this last category, I think we may agree that poets and even psychiatrists invent new uses for old labels out of ingenuity; but it is tempting to say that philosophers sometimes do it out of perversity. Tempting, but probably not fair. And certainly not fair if a definition of philosophy given by Arthur Koestler is correct: Philosophy is the systematic abuse of a vocabulary specially designed for that purpose (Koestler 1975, 89).

Koestler does not put this forward as a completely serious definition of philosophy, but as a joke. It strikes us as funny, however, because there is some truth in it. I would like to see philosophy develop in such a way as to be less vulnerable to this kind of humour, and I think a concerted effort by the community of philosophers to establish a common terminology would have this effect and other beneficial effects as well.[5] One of the major benefits of such an effort would be the clarification of our ideas. As I indicated earlier, it is my determination to get my own ideas clear that led to my interest in labels and labeling. It is my purpose in the rest of this paper to present the labeling problem in such a way that 'fruitful reasoning can be made to turn upon it, and that it can be applied to the resolution of difficult practical problems' – which, the reader will recall, is one way Peirce described what he called the third grade of clearness (CP 3.457).

EUTHRYPHRO

One of the many interesting and useful things in Plato's dialogue *Euthyphro* is the description given of Euthyphro's progress through three stages of confidence.[6] Socrates and Euthyphro meet at the magistrate's office, and each is surprised to see the other there. Euthyphro discovers that Socrates is being prosecuted for impiety of one sort – inventing new gods and not believing in the old ones – and Socrates discovers that Euthyphro is prosecuting his own father for impiety of another sort – allowing a labourer to die through neglect.

To our modern way of thinking, Euthyphro does get off to a good start. When Socrates appears amazed that someone would prosecute his own father for the murder of a stranger, Euthyphro quite rightly asks 'What difference does it make whether the murdered man were a relative or a stranger? The only question that you have to ask is, Did the murderer kill justly or not? If justly, you must let him alone; if unjustly, you must indict him for murder, even though he share your hearth and sit at your table' (Plato 1956, 4b). Socrates does not deny this, but he thinks that the matter is sufficiently serious to require hard thought, and so he asks the question which sets the task for the dialogue: 'Do you mean to say, Euthyphro, that you think that you understand divine things and piety and impiety so accurately that, in such a case as you have stated, you can bring your father to justice without fear that you yourself may be doing something impious'? (Plato 1956, 4e).

In asking this question Socrates confronts Euthyphro with what I have just called the labeling problem; in the remainder of the dialogue Euthyphro will try four times to explain how he applies the label 'piety.' And Socrates will develop some of the consequences of these definitions in order to determine Euthyphro's meaning.

As Euthyphro offers his first definition of piety – 'piety means prosecuting the unjust individual who has committed murder or sacrilege, or any other such crime' – he allows no uncertainty as to his state of mind: 'Observe, Socrates, I will give you a clear proof ... that it is so (Plato 1956, 5e). Nevertheless, this first definition is inadequate; it tells us one or two of the actions that are pious, but is does not tell us what it is that makes them pious.

Euthyphro tries again: 'What is pleasing to the gods is pious, and what is not pleasing to them is impious' (Plato 1956, 6e–7a). This definition does no better, for Socrates notices that it requires that all the gods agree about what is pious; he is not certain that the gods agree about this, and he asks Euthyphro to make it clear to him that they do. We now observe the second stage in Euthyphro's progress: 'I could make that clear enough to you, Socrates; but I am afraid that it would be a long business' (Plato 1956, 9b).

Euthyphro tries a third definition, and Socrates shows him that it also will not do the job. Apparently realizing that he is running out of ideas, Euthyphro confesses 'Socrates, I really don't know how to explain to you what is in my mind' (Plato 1956, 11b). And here we have the third stage in Euthyphro's progress. He does attempt a fourth definition, but that too fails, and Euthyphro is in too much of a hurry to spend any more time reflecting upon his course of action.

At first Euthyphro is completely confident; he promises to give Socrates what he has 'already given to others,' a decisive proof that piety is what he says it is. Next, still confident that he can make himself clear to Socrates, he recognizes that it would be 'a long business.' Finally, he admits 'I simply don't know how to tell you what I think' (Plato 1961, 11b).

I suspect that we have all heard this remark, or echoes of it, outside the Platonic dialogues – in conversations, in classrooms, on radio or television broadcasts; perhaps we have read it somewhere; and it might not surprise us to learn that we have used it ourselves. But I think that the remark implies that there *is* something in the mind which – for some reason or another – cannot be put into words, and it simply does not seem obvious that this is true in any sense which would 'save' Euthyphro, or us.

At the start of his conversation with Socrates, Euthyphro claimed to have knowledge about a certain topic. Now we have learned from our teachers to associate the label 'knowledge' with the ability to put something into words, and make it understandable to others. This labeling association, which we shall call 'K_1,' for short, is reflected in many of our activities – in our participation in telephone conversations, in university life, and in publications such as this volume of essays honouring Thomas A. Goudge. Euthyphro apparently learned the same association for the label 'knowledge': early in the dialogue he claimed to have 'already given to others' the *proof* that piety was what he thought it was, and he promised to prove it to Socrates.

Euthyphro did try. In his conversation with Socrates he offered four definitions of piety, but all four turned out to be unsatisfactory. It is fair to grant that one or another of these definitions had convinced other people, but they do not convince Socrates and, thanks to Plato, they do not convince us. At the end of the dialogue there is *no* indication, there is no *evidence*, that Euthyphro did have anything better in mind.

Nevertheless, Euthyphro apparently thought he had something better in mind: 'Socrates,' he said, 'I really don't know how to explain to you what is in my mind.' That is, Euthyphro continued to make the same knowledge claim – what they were searching for was somehow 'in his mind' – but he admitted that he could not put it into words. With this admission, he abandoned the labeling association K_1.

It would seem fair to ask Euthyphro *why* he thought he had something better in mind. It would seem fair to ask him what labeling association he was using in place of K_1. My question is, how did he convince *himself* that he knew what piety was.

I believe that the answer to this is found at the beginning of the dialogue, when Socrates asked Euthyphro if he really thought that he understood divine things and piety and impiety. He replied, 'If I did not understand all these matters accurately, Socrates, I should not be worth much – Euthyphro would not be any better than other men' (Plato 1956, 4e–5a). This one-sentence reply abbreviates a valid argument in the form *modus tollens*, and Euthyphro's instance of this argument form, complete with the suppressed premiss and conclusion, is as follows: 'If I did not understand all these matters accurately, then I would not be any better than other men; I am better than other men. Therefore, I understand all these matters accurately.' So the answer to my question is this: Euthyphro accepted the Socratic dictum that knowledge is virtue (his first premiss is a consequence of this), and he thought that he was better than other men; and this is why he retained the label 'knowledge' even while admitting that it was not associated with something he could put into words. The labeling association he seems to have used in place of K_1 is this: associate the word 'knowledge' (or 'understanding') with anything whose name or description can be substituted, in his argument, for the phrase 'all these matters accurately.' Let us call this labeling association 'K_e,' in honour of Euthyphro.

It is tempting to dismiss Euthyphro's use of K_e on the ground of the falsity of the minor premiss, claiming that Euthyphro is not better than other men. But even though he is portrayed as somewhat vain in the dialogue, we can afford to be charitable to a man who takes justice so seriously. I am inclined to think that it is the major premiss that leads to real difficulties, or perhaps the combination of the two. My favorite published application of K_e may help clarify what I mean. In Looking-Glass House, Alice had just finished reading the poem 'Jabberwocky' – 'Twas brillig, and the slithy toves ... ' – and she said, to herself: ' "It seems very pretty ... but it's *rather* hard to understand!" (You see [remarks Lewis Carroll] she didn't like to confess, even to herself, that she couldn't make it out at all.) "Somehow it seems to fill my head with ideas – only I don't exactly know what they are!" '[7]

The chief value of K_e, and its major disadvantage, is that it enables one to avoid confessing a failure to understand. I think that both Euthyphro and Alice suffer from the belief that a failure to understand is somehow morally reprehensible. I think that many members of university communities suffer from this belief, and it has disastrous effects on our work. For instance, when this belief is coupled with what William James called an 'impatience of results,'[8] it sometimes encourages us to pretend to know what we do not know, and this pretence effectively blocks the way of inquiry for us.

The dictum that knowledge is virtue is part of our legacy from Socrates, and although it is susceptible to various interpretations I am loath to abandon it entirely. Perhaps one difficulty is that we tend to identify knowledge and virtue. If we stop short of that, holding instead that knowledge is a proper subclass (subset) of virtue, it does not necessarily follow that ignorance is morally reprehensible. Still, ignorance is embarrassing, and embarrassment can block the way to inquiry. Perhaps embarrassment would be less of a problem to us if we were to make our philosophy, more and more, an attempt to FIND THINGS OUT.[9]

Perhaps there is one more reason why Euthyphro's argument fails to persuade some of us. He seems to assume, contrary to all the evidence that Plato presents in the dialogue, that he had the correct idea of piety already, somehow, in his mind. This 'correct idea of piety' is, in my judgment, a figment of the imagination. It is natural that Euthyphro would make this assumption, for he is, in a sense, one of Plato's children; and traces of the theory of forms and the doctrine of recollection should not surprise us.

We too, in a sense, are Plato's children; but it no longer seems possible to most of us that learning is merely the uncovering of ideas that are already present in the mind. And perhaps because of this we may notice that Euthyphro's discussion with Socrates suggests a better model of learning: namely, we start with the best ideas we have, and derive from them their consequences; this shows us specific ways in which the ideas are inadequate for our purposes, specific ways in which they might be improved; we then improve them and try again. Notice that this model has no need for Euthyphro's figment. It seems to me a better model because it captures what we actually do when we learn, and it does not go beyond our experience.[10]

BERGSON

In 1903 Henri Bergson published an essay entitled 'An Introduction to Metaphysics.' The essay is remarkable for the felicity with which it expresses an important and perennial approach to philosophy.[11] Its first seven paragraphs, which constitute a kind of prologue, contain the elements of all that follows. In the first paragraph Bergson distinguishes two ways of knowing a thing:

The first [way of knowing] implies that we move round the object; the second, that we enter into it. The first depends on the point of view at which we are placed and on the symbols by which we express ourselves. The second neither depends on a point of view nor relies on any symbol. The first kind of knowledge may be said to stop at the *relative*; the second, in those cases where it is possible, to attain the *absolute*. (Bergson 1949, 21)

It might be convenient to set this out in a diagram, as in the figure below.

	FIRST WAY	SECOND WAY
IMPLIES THAT	We move round the object	We enter into the object
DEPENDS UPON	1. Our point of view 2. Our symbols of expression	1. No point of view 2. No symbols
ATTAINS	The relative	The absolute

The words are familiar to us, but some of the combinations may be new. For example, we know what it ordinarily means to move around an object of knowledge; we do it when we walk around a tree, or examine a slide under a microscope. But we expect Bergson to show us what it means to 'enter into' the object of knowledge; and we expect him to show us *that* it is possible to do so and *how* one is supposed to do it.

We know in general what a point of view is, and the term 'symbols of expression' suggests a language; but we expect Bergson to tell us *how* a way of knowing a thing *can* operate without either, and we expect him to tell us what it *does* depend upon.

The last line in the table is a bit less obvious; he will have to help us with it. And we should keep in mind that when he spoke of attaining the absolute, he added the qualification: 'in those cases where it is possible.'

In the remaining six paragraphs of his prologue, Bergson illustrates his two ways of knowing by applying each to the following seven examples:

1 the movement of an object in space;
2 the hero of a novel;
3 the set of photographs of a town, taken from all possible points of view;
4 the translation of a poem into all possible languages;
5 the 'extraordinarily simple impression' that a certain passage in Homer makes upon him;
6 some person, other than himself, raising an arm;
7 the province of the positive sciences in general and the life sciences in particular.

Bergson's treatment of some of these examples makes good reading indeed. To give the flavour of it, I quote from his fourth paragraph, which contains two of them.

83 The Labeling Problem

Were all the photographs of a town, taken from all possible points of view, to go on indefinitely completing one another, they would never be equivalent to the solid town in which we walk about. Were all the translations of a poem into all possible languages to add together their various shades of meaning and, correcting each other by a kind of mutual retouching, to give a more and more faithful image of the poem they translate, they would yet never succeed in rendering the inner meaning of the original. A representation taken from a certain point of view, a translation made with certain symbols, will always remain imperfect in comparison with the object of which a view has been taken, or which the symbols seek to express. But the absolute, which is the object and not its representation, the original and not its translation, is perfect, by being perfectly what it is. (Bergson 1949, 22–3)

In the case of the town, the instrument of knowledge[12] is the camera, and the symbols of expression, the photographs. To an indefinitely large set of photographs of the town, Bergson contrasts the 'solid town' itself. In the case of the poem, the symbols of expression used by the translators are their several languages. And to a series of better and better translations or 'images,' Bergson contrasts the 'inner meaning of the original.'

The five remaining examples likewise reveal contrasts. In the case of the movement of an object in space, the contrast is between sets of positions plotted in terms of various systems of co-ordinates, and the experience of moving. In the case of the hero of the novel, the contrast is between lists of character traits, reports of speech, descriptions of actions, and the 'simple and indivisible feeling' which one would presumably experience if he could identify himself with the hero.

To his own translation of a passage in Homer, supplemented by an endless string of explanations, Bergson contrasts the 'extraordinarily simple impression' that the passage makes on him. To an unending enumeration of the number of points through which an arm passes as it is raised, he contrasts the 'simple perception' which is available only to the person whose arm is moved. And to analyses and comparisons of the organs and other anatomical elements of living beings he contrasts 'possessing a reality' by the 'placing oneself within' that reality.

These examples confirm and enrich both our ordinary understanding of what it means to move round an object of knowledge, and our ordinary understanding of what sorts of things might function as symbols of expression. Thus, in the discussion of his first way of knowing, Bergson has improved upon the labeling association K_1. He continues to associate the label 'knowledge' with the ability to make something understandable to others by putting it into words, but he does not insist that the carrier of this knowledge be words. He adds to the list of acceptable symbols of expression such things as sets of geometrical co-ordinates, and photographs. Let us call this new labeling association, with the broader set of acceptable sumbols, 'K_2.'

With regard to the second way of knowing, which Bergson calls 'intuition,' the matter is somewhat more complicated. If we ask Bergson what labeling association for 'knowledge' he is using now, in place of K_1 and K_2, he might reply that the question is unfair: labeling associations have to do with the use of symbols, but his second way of knowing does not. Nevertheless, intuition is supposed to be a way of knowing, it is supposed to result in knowledge. So it would seem to be fair, provisionally at least, to associate the term 'knowledge' with whatever results can be obtained from Bergson's method of intuition. Let us call this association 'K_3,' for short. If we can get clear on the method and its results, we might be able to improve our statement of K_3.

(1) What then are the results or products of intuition? They are a 'solid town in which we walk about'; the 'inner meaning of the original' poem; the 'possessing' of a reality by 'placing oneself within' that reality; an 'experience' (of moving), a 'feeling' (of being the hero of a novel), an 'impression' (produced by reading Homer), a 'perception' (of the motion of one's own body).

According to K_1 and K_2, a 'solid town' may be an object of knowledge, but is not itself knowledge. According to Bergson, the town, the object of the representation, is 'the absolute'; but intuition, in those cases where it is possible, attains the absolute. Hence intuition, if it applies to Bergson's own example, attains the 'solid town,' and hence by K_3 the 'solid town' qualifies as knowledge.

According to K_1 and K_2, the meaning of a poem is found by translating it into words (say) whose meaning is already known, whether these words belong to the same language as that in which the poem was originally written, or a different language. According to Bergson, there is something which he calls 'the inner meaning' of a poem and which is, apparently, identified with the original and with no translation of it. This is 'the absolute'; but intuition, in those cases where it is possible, attains the absolute. Hence intuition, if it applies to Bergson's example, attains the 'inner meaning,' and hence by K_3 the 'inner meaning' qualifies as knowledge.

According to Bergson, the 'extraordinarily simple impression' that a passage in Homer makes upon him cannot adequately be communicated to someone who does not know Greek by the methods of translation and commentary and more of the same. Presumably intuition can do better.

Now there is no ordinary sense in which a solid town qualifies as knowledge, and if Bergson's sense for this – K_3 – is to be made clear, we must discover what it means to 'attain the absolute' in this case, that is, we must discover what it means to 'attain the solid town.' Bergson does not help us

85 The Labeling Problem

here. What is it exactly that intuition gives that the set of photographs, for instance, do not give? Bergson does not say, and because his metaphysics claims to dispense with symbols, perhaps he cannot say.

What is it to possess the inner meaning of a poem? Bergson grants that translations may give 'a more and more faithful image' of the poem they translate, so that symbols (the language of translation) would seem to have some application here; but he claims that these translations will always remain imperfect. Is this merely a matter of definition (for example: the original is perfect by definition, but a translation is not identical with the original; hence the translation is not perfect), or is it a reasoned judgment? If the latter, is it fair to ask Bergson how he arrives at this conclusion? What is it, exactly, to which he compares the translations (in order to show that they are imperfect), if the original cannot be put into words or symbols? How is such a comparison carried out? Bergson does not say.

(2) Consider now the method of Bergson's intuition. In connection with his examination of the movement of an object in space (example 1), he suggests a four-step procedure for practising his second way of knowing, a four-step procedure for intuiting:

1 attribute an interior to the object;
2 attribute states of mind to the object;
3 be in sympathy with those states of mind;
4 insert yourself into those states by an effort of imagination (Bergson 1949, 21).

The words seem clear enough; and if, as 'object,' we pick some person or other, the method seems at first to be practicable. But Bergson does not mean to restrict it this way, at least at first, as we have seen. It is, however, exceedingly difficult to know how to apply these steps to something like a comet, say, moving through space. How shall we attribute an interior, or states of mind, to a comet? Bergson does not give us adequate instructions. He tells us that this involves an 'effort of the imagination,' but every time I put forth such an effort I discovered that I was making use of the symbols of the imagination. Is there some other way to do it? Apparently, for according to Bergson intuition makes use of no symbols whatsoever.

Bergson also presents the following definition: 'By intuition is meant the kind of *intellectual sympathy* by which one places oneself within an object in order to coincide with what is unique in it and consequently inexpressible' (Bergson 1949, 23–4). Suppose, for example, that we are interested in the hero of some novel and want to discover 'what belongs to him and to him alone,' that is, we want to discover 'that which constitutes his essence' (Bergson 1949,

22). Since we are assured that the first way of knowing is inadequate for this purpose, we intend to employ intuition. This consists in placing ourselves within the object – the hero – 'in order to coincide with what is unique in [the hero] and consequently inexpressible.' What we must ask Bergson is this: How will we know when we have accomplished this? If what is unique in the hero is inexpressible, how will we ever recognize it when we are confronted by it? How will we know for sure that we have coincided with this hero and not with some other hero?

Let us imagine that Bergson is approached by a young student who wishes to become his disciple, and who desires above all things to master the method of intuition. How would Bergson teach it to him? If the student asks how Bergson himself learned to apply the label 'intuition,' what would be the answer? How would Bergson gauge the progress of his pupil? Ordinarily we judge the success of a method by its results; but the results of intuition are said to be inexpressible, so that ordinary communication cannot be used to establish success. Nevertheless, suppose that Bergson agrees to coach him in spite of all this, and suppose that after much work the student succeeds. But how does he know this? Perhaps Bergson tells him so. But how does Bergson know?

I do not see how we shall ever get over this difficulty. Thirty-five years before Bergson wrote this essay Peirce established that intuitions cannot reasonably be supposed to carry labels identifying themselves as intuitions (CP 5.213–24). Bergson does not suggest any other means of identification. A sceptic might say that the second way of knowing stands or falls with the filling in of such details.

K_3 was introduced as the association of the term 'knowledge' with whatever results can be obtained from Bergson's method of intuition. The examples considered above, however, fail to clarify the use of the method, and cast doubt on the existence of its purported products. For the present, then, I conclude that Bergson's special use of the label 'knowledge' is ill advised. (At the same time, I recognize that this part of my paper is no more than a beginning of a serious study of Bergson's philosophy.)

It is risky to propose simple explanations for complicated matters, but I believe that the development of this second way of knowing is based in part on a mistaken notion about the purpose and function of knowledge. Bergson thinks the purpose is to reproduce either the object of knowledge itself (example 3) or the experience one would have if confronted by the object (example 5); and he is critical of the first way of knowing because it fails to do this. This kind of reasoning is called 'the reproductive fallacy' by Richard S. Rudner, who thinks it is caused by a confusion between a description and what is described. As he points out, it certainly seems clear that 'to be a description of the taste of soup' is not 'to *be* the taste of soup' (Rudner 1966, 69, 70n).

For some reason or other, Bergson wants more from the first way of knowing than it can provide, more than it purports to provide. He imagines that the first way of knowing has an 'eternally unsatisfied desire' to embrace its object (Bergson 1949, 24), but it is Bergson who has the desire. I think it was in order to satisfy this desire that Bergson postulates the existence of his second way of knowing; and he apparently let himself believe that the existence of his desire is good evidence for the existence of the faculty necessary to satisfy that desire. I think there is no other evidence for it.

These considerations certainly support Thomas Goudge's judgment that Bergson's philosophy is the work of a mind which is 'mystical and unsympathetic to logic.' And I cannot improve on another of Professor Goudge's conclusions: 'Despite the felicitous language in which it is expressed and the wealth of empirical detail it surveys, the Bergsonian metaphysics contains many mysteries' (Bergson 1949, 20).

POSTSCRIPT

I came to my labeling problem gradually, after prolonged exposure to Peirce's writings on cognition and on the importance of getting one's ideas clear. It now appears that I might have come to it in any number of ways, say from a study of Plato or Wittgenstein[13] or Kierkegaard.[14] Its chief value, in my opinion, is that it can be used to help us get our own ideas clear; this is the use of it which I recommend to the reader.

PETER A. SCHOULS

Peirce and Descartes: Doubt and the Logic of Discovery[*]

For at least two reasons Peirce would not consider it very promising to go to the works of Descartes in order to be instructed about the logic of discovery. The first is that, in philosophy and science, discoveries are made in the context of genuine doubt. Descartes' doubt is interpreted by Peirce as 'make-believe.' Since 'make-believe' results in the re-adoption of beliefs held before such doubt was brought to bear, it does not lead to new discoveries. Second, Peirce holds that, to the extent Descartes attaches importance to 'intuition,' his position is one which is irrelevant to a discussion of the logic of discovery. For Peirce, all cognition is mediated or discursive. Basic to Descartes' system, however, is the conviction that through intuition truths can be known in isolation, that there are items known per se. Since hypotheses play a role only in discursive reasoning, we cannot look for anything relevant to the logic of discovery in the area of reason's intuitive functioning.

I would like to show that a discussion of the logic of discovery is as relevant within the framework of Descartes as it is within that of Peirce. That these two frameworks are substantially different is well known and therefore needs no elaboration. What I will argue is that Descartes' theory with respect to making discoveries in philosophy and the sciences shows considerable affinity with Peirce's statements about the logic of discovery. That there is such an affinity in spite of disparity in frameworks is, possibly, a surprising fact which may lead to further insight into the logic of discovery. Development of this theme, however, is beyond the scope of my study.

For the purposes of this study it will be assumed that there is a logic of discovery. This assumption involves the belief that there is room for philo-

---

[*] I am indebted to my colleagues Richard N. Bosley and John King-Farlow for discussion of several aspects of this paper.

89 Peirce and Descartes: Doubt and the Logic of Discovery

sophical analysis in the area of reasoning which constitutes the formation of hypotheses or which leads to the selection of hypotheses.

I will first show that Peirce misinterpreted the nature of Cartesian doubt and hence misjudged the extent to which Descartes applied doubt. After that I will argue that Cartesian doubt is not the futile procedure which Peirce claims it is. In both of these parts my discussion will be placed in the context of the relevance of doubt to the process of discovery. In the second part it will become clear that, within the Cartesian framework, there is a kind of 'intuition' which is inextricably bound up with discovery in both philosophy and the sciences. But since the nature and role of doubt will be the central theme of discussion, that of intuition will be left undeveloped.

I

Peirce holds that doubt stimulates us to inquiry, but that Cartesian doubt is not a genuine doubt and therefore leads only to idle discussion. The questions to be asked are: what does Peirce take doubt to be? what, in Cartesian doubt, leads Peirce to reject it as futile? and, what does Descartes take doubt to be?

For Peirce, the 'irritation of doubt' leads to inquiry which, in turn, leads to the adoption of new beliefs. Doubt involves 'irritation' because it comes to a person from 'outside': 'genuine doubt always has an external origin, usually from surprise.' For that reason 'it is as impossible for a man to create in himself a genuine doubt by ... an act of the will ... as it would be for him to give himself a genuine surprise by a simple act of the will' (CP 5.443). As he says elsewhere, 'belief came first, and the power of doubting long after. Doubt, usually, perhaps always, takes its rise from surprise, which supposes previous belief; and surprises come with novel environment' (CP 5.512). Since doubt comes from outside it disturbs the inner state of belief, which is the 'calm and satisfactory state which we do not wish to avoid' (CP 5.372). Once this calm has been disturbed it cannot be regained by simply rejecting the doubt; mere rejection of genuine irritation precludes the attainment of genuine calm and satisfaction. One might as well attempt to feel well fed by rejecting the feeling of irritation of hunger. In this respect doubt 'is like any other stimulus.' Nevertheless 'just as men may, for the sake of the pleasures of the table, like to be hungry and take means to make themselves so, although hunger always involves a desire to fill the stomach, so for the sake of the pleasures of inquiry, men may like to seek out doubts. Yet, for all that, doubt essentially involves a struggle to escape it' (CP 5.372n). Not only may one like to seek out doubt, one ought to seek it out. A philosopher therefore 'invents a plan for attaining doubt, elaborates it in detail, and then puts it into practice ... and it is only after having gone through such an examination that he will pronounce a belief to be indubitable. Moreover, he fully acknowledges that even then it may be that some of his indubitable beliefs may be proved false' (CP 5.451). For 'what has been indubitable one day has often been proved on the morrow to be false. He grants ... that it may be so with any of the beliefs he holds. He really cannot admit that it may be so with all of them.' If he does the latter 'he loses himself in vague unmeaning contradictions' (CP 5.514). The very existence of such contradictions. the very fact that 'earnest and industrious students' of philosophy have not been 'able to come to agreement upon scarce a single principle,' that philosophy is therefore still 'in its infancy' (CP 1.620) is, in the context of some of the other sciences, one of the surprising facts of the environment which, once noticed, disturbs the philosopher's inner state of satisfaction.

This much is sufficient as an indication of what Peirce takes doubt to be. It is within the context of these assertions that Peirce rejects Cartesian doubt. It is rejected for several reasons of which two are of greatest importance.

First, the 'Cartesian error' is that 'of supposing that one can doubt at will.' But only that can be doubted at will which is not believed to begin with, and hence doubting at will is not genuine doubt, for genuine doubt presupposes genuine belief. Furthermore, 'The breaking of a belief can only be due to some novel experience, whether external or internal,' and 'experience which could be summoned up at pleasure would not be experience' (CP 5.524). Thus Descartes' doubt is futile because, given that he doubts at will, it is either doubt of what he does not believe (which is futile because genuine doubt is of what one believes) or it is doubt of what one does believe but the doubt is not occasioned by experience (and this is futile because such doubt involves no grounds to change the beliefs to which it is applied). In neither case are there genuine *reasons* for doubt.

To believe that one can doubt at will is only one of the Cartesian errors. A second error is that Descartes 'teaches that philosophy must begin with universal doubt' (CP 5.264). But we 'cannot begin with complete doubt' for then there would not be any genuine beliefs which can be doubted. Instead, 'We must begin with all the prejudices we actually have when we enter upon the study of philosophy' (CP 5.265). 'A person may, it is true, in the course of his studies, find reason to doubt what he began by believing; but in that case he doubts because he has a positive reason for it, and not on account of the Cartesian-maxim' (ibid).

It is of interest to note that some commentators believe Peirce to have been inconsistent in his critique of Descartes. In this context Buchler (1939), for example, draws attention to a passage which Peirce wrote some ten years after the initial formulation of his critique of Descartes, namely: 'Leibnitz ... missed the most essential point of the Cartesian philosophy, which is, that to accept propositions which seem perfectly evident to us is a thing which, whether it be logical or illogical, we cannot help doing' (CP 5.392). Buchler missed Peirce's point here, for this passage is not a commendation of Descartes but a condemnation of Leibnitz. The context of this passage, the immediately preceding section, begins with the statement 'When Descartes set about the reconstruction in philosophy, his first step was to (theoretically) permit scepticism ...' (CP 5.391). Leibnitz is faulted on two counts here. (i) Leibnitz is wrong in criticizing Descartes for accepting a certain starting point in 'perfectly evident propositions,' for all philosophy must start from strongly held beliefs. (ii) Descartes, however, started with universal doubt, and Leibnitz should have shown the futility of such a starting point. In his critique of Descartes Peirce was eminently consistent. Some twenty-five years after its initial appearance he writes: 'Descartes marks the period when Philosophy ... began to be a conceited young man'; for 'to make believe one does not believe anything is an idle and self-deceptive pretence' (CP 4.71). It now remains to be seen whether Peirce's criticism was appropriate.

I will argue that, in view of the role which doubt plays in Descartes' position, Peirce's criticism of Descartes is quite invalid. Many critics of Descartes and, as far as I know, all expositors of Peirce, have accepted Peirce's criticism of Cartesian doubt as valid. And I am not aware of any critic who maintains that, in spite of Peirce's protestations, there is a considerable degree of resemblance between Descartes' and Peirce's statements on the role of doubt in the context of making discoveries. Nevertheless, I believe that a study of major works of both Descartes and Peirce will show that instead of there being a great gulf between the two positions on this point, they are in effect very similar to one another.

In the *Rules for the Direction of the Mind*[1] Descartes gives the following definition of his *method*:

Method consists entirely in the order and disposition of the objects towards which our mental vision must be directed if we would find out any truth. We shall comply with it exactly if we reduce involved and obscure propositions step by step to those that are simpler, and then starting with the intuitive apprehension of all those that are absolutely simple, attempt to ascend to the knowledge of all others by precisely similar steps. (HRI, 14; AT10, 379)[2]

According to Descartes, the kind of *reasoning* which gives us science involves, first, the establishing of an order in or among the problems which are considered to be relevant, an order which is obtained through reducing involved issues into simple or simpler ones; and, second, a (re)construction in terms of these simple(r) elements. Putting these statements about method and about reasoning side by side, we see that a description of the activity of reason coincides in essence with a statement which presents a definition of method (Schouls 1972a). As will become clear later on, it is the systematic use of *doubt* which pushes the 'reduction' involved in both of these statements. Thus doubt functions in both the definitional statements of reason and method. But this does not mean that one can speak of *universal* doubt, not even in the modified way Peirce speaks about this within the context of his own position ('what has been indubitable one day has often been proved on the morrow to be false ... it may be so with any of the beliefs he holds. He really cannot admit that it may be so with all of them'). For Descartes, any belief now indubitable (in a sense to be explicated) is always indubitable; and since there now are indubitable beliefs, it cannot ever be the case that all beliefs are dubitable.

'Universal' doubt is supposed to be present in the *Meditations*. However, Descartes states explicitly that all he has doubted in the *Meditations* are 'prejudices.' 'I have denied nothing but prejudices, and by no means notions ... which are known without any affirmation or denial' (HR2, 127; AT9, 206). What can withstand the test of doubt, even in the *Meditations*, is any concept (e.g. 'unity') or principle (e.g. 'something cannot proceed from nothing') known per se, whenever such a concept or principle is actually intuited. This implies also that the function of reason by which we know 'without any affirmation or denial,' namely, intuition, is held to be absolutely trustworthy. There is considerable evidence internal to the *Meditations* that this is indeed how its argument goes (see Schouls 1972b, 1973). What is considered dubitable in the *Meditations* is any belief or supposed item of knowledge which is based on sensation, any belief which is the result of the deductive function of reason, as well as, by implication, the trustworthiness of the deductive function of reason itself. For all of these, there are *genuine reasons* for doubt. However, even if within the two areas of sense-based and deductive knowledge there were to be universal doubt, such doubt would fall short of being a *genuine* universal doubt. For the intuitive function of reason, and its objects known, remain beyond doubt. Doubt of what is indubitable is nonsense, for Peirce as well as for Descartes. 'Universal' (or 'metaphysical') doubt is a doubt which makes us ask whether *all* we know clearly and distinctly can be trusted given that *some* of it, namely our perceptions or intuitions of items known per se, can be. This formulation clearly puts limits to 'universal' doubt.

Even universal doubt thus limited is further qualified. The statement 'two and three make five' is an item of deductive knowledge, and hence dubitable.

93 Peirce and Descartes: Doubt and the Logic of Discovery

But even after universal doubt is introduced in the form of the omnipotent deceiver, we find Descartes saying: 'always when I direct my attention to things which I believe myself to perceive very clearly, I am so persuaded of their truth that I let myself break out into words such as these: Let who will deceive me, He can never ... cause it to be true ... that two and three make more or less than five ... ' Since I am persuaded of the truth of items like these I am unable, psychologically, to doubt them. The universal doubt is, therefore, 'very slight, and so to speak metaphysical' (HRI, 158–9; AT7, 36). Nevertheless, this metaphysical doubt is 'not merely through want of thought or through levity, but for reasons which are very powerful and maturely considered' (HRI, 148; AT7, 21). There are *reasons* for it. Some of these reasons are presented in the *Regulae*, more than a decade before Descartes wrote the *Meditations*. Consider, 'scarce anything has been asserted by any one man the contrary of which has not been alleged by another' (HRI, 6; AT10, 363). Even in the most secure of sciences, mathematics and geometry, there has been little development since the ancients, and 'the order of their propositions alone makes us aware that they had no true method for discovering ...'[3] At least as important a reason for metaphysical doubt arises from the thinking behind the wholesale rejection of scholastic positions by many of Descartes' immediate predecessors as well as by many of his contemporaries. To the extent that there was such a rejection, the intellectual world had lost its moorings. And through François Sanchez' *Quod nihil scitur* and especially Montaigne's *Essais* ('The Apology for Raimond Sebond' (1576)), the sceptical positions of Sextus Empiricus and Pyrrho of Elis were current. This scepticism was sufficiently prevalent to make it obligatory for anyone 'who wanted to establish any firm and permanent structure in the sciences' (HRI, 144; AT7, 17) to present a method for the discovery of truth which would make 'it impossible for us ever to doubt those things which we have once discovered to be true' (HRI, 140; AT7, 12). The absolute scepticism of the Pyrrhonics was rejected outright by Descartes to the extent that he rejected only 'prejudices.' And to make clear that he rejected 'universal' doubt even within the realm of 'prejudice,' within the realm of sense-based and deductive knowledge, Descartes introduced his hypothesis of the omnipotent deceiver. For in order to remove the 'hyperbolical and ridiculous' (HRI, 198–9; AT7, 89) doubt here, to remove the metaphysical doubt from statements like 'two and three make five,' statements which we find psychologically indubitable, he shows that the only reason for such a doubt is not a good reason. For the only reason for such a doubt would be the existence of an omnipotent deceiver. Reason in its intuitive function tells him that the only omnipotent being is God, and that God cannot be a deceiver. Since it is intuition which tells him this,[4] the 'argument' is not circular. For there is no 'train of reasoning' in intuition, and hence there is nothing to which the charge of circularity can apply. Once intuition has shown that the hypothesis of an omnipotent deceiver is self-contradictory (cf HR1, 172; AT7, 53) there remains no good reason for universal doubt even within the areas of sense-based and deductive knowledge.

There are good grounds for taking the argument of the *Meditations* to involve a validation of reasoning. If Descartes had started from the position of Pyrrhonic scepticism, if his had been a position of genuine universal doubt, the charge made by Arnauld (HR2, 92; AT7, 214) that the argument is circular would be inescapable. Given the intellectual climate in which Descartes wrote, Arnauld's misinterpretation is understandable. It is ironic that Descartes could have defended himself with words used by Peirce:

I shall be arrested, at the outset, by a sweeping objection to my whole undertaking [of providing grounds of validity of the laws of logic]. It will be said that my deduction of logical principles, being itself an argument, depends for its whole virtue upon the truth of the very principles in question; so that whatever my proof may be, it must take for granted the very things to be proved. But to this I reply, that I am neither addressing absolute sceptics, nor men in any state of fictitous doubt whatever. (CP 5.319)

It is equally ironic that, had he thought of it, Descartes could have put his position in a form which might have pleased Peirce (cf CP 5.189):

The following surprising facts are observed: (i) 'scarce anything has been asserted by any one man the contrary of which has not been alleged by another,' and (ii) many believe that one can never be certain about anything (call these C_1 and C_2, respectively);

But if it were true that there is some omnipotent being who 'has employed his whole energies in deceiving me' (call this A_1), both C_1 and C_2 would be matters of course,

Hence, there is reason to suspect that A_1 is true.

However, A_1 is a self-contradictory hypothesis.

Given that the only justification for the belief of C_2 is the existence of an omnipotent deceiver (and the hypothesis asserting the existence of such a being is self-contradictory) C_2 is an unfounded belief.

For C_1 there are explanations in addition to that of A_1:

The surprising fact is observed that 'scarce anything has been asserted by any one man the contrary of which has not been alleged by another' (C_1);

But if it were true that people have been thinking in haphazard, unmethodical ways (call this A_2), C_1 would be a matter of course,

Hence, there is reason to suspect that A_2 is true.

95 Peirce and Descartes: Doubt and the Logic of Discovery

Moral: It is therefore necessary to proceed methodically if we are to come to agreement in philosophy and the sciences.

II

It now remains to be shown that Cartesian doubt plays an important role in the context of making discoveries, and it remains to be pointed out that doubt is bound up with some form of intuition which, in the same context, plays an equally important role. In order to elucidate this function of doubt, the context in which it does its work, namely, that of making discoveries, needs to be explicated. This explication will at the same time reveal that there is a considerable affinity between Descartes' and Peirce's theories about how discoveries are made. That there is indeed such an affinity will be apparent if we keep in mind some of the salient aspects of Peirce's logic of discovery. It will therefore be helpful first to present a brief sketch of the relevant aspects of Peirce's theory.

Discoveries, according to Peirce, cannot be made without 'guesses,' but if one tries to account for even a single hypothesis which turned out to be correct, *chance*, or uncontrolled use of the imagination, cannot account for it. 'If you ask an investigator why he does not try this or that wild theory, he will say, "It does not seem *reasonable*"' (CP 5.174).

When a man desires ardently to know the truth, his first effort will be to imagine what that truth can be. He cannot prosecute his pursuit long without finding that imagination unbridled is sure to carry him off the track. Yet nevertheless, it remains true that there is, after all, nothing but imagination that can ever supply him an inkling of the truth. He can stare stupidly at phenomena; but in the absence of imagination they will not connect themselves in a rational way. (CP 1.46)

Thus, since 'imagination unbridled' does not provide anything 'reasonable,' one may conclude that *controlled* imagination is necessary if one is to succeed in connecting 'phenomena ... in a rational way.' What is it that controls the imagination?

Peirce writes that abduction

furnishes the reasoner with the problematic theory which induction verifies. Upon finding himself confronted with a phenomenon unlike what he would have expected under the circumstances, he looks over its features and notices some remarkable character or relation among them, which he at once recognizes as being characteristic of some conception with which his mind is already stored, so that a theory is suggested which would *explain* (that is, render necessary) that which is surprising in the phenomena. (CP 2.776)[5]

This gives at least two of the elements which control the imagination: (i) the nature of the problem to be explained ('he looks over its features'), and (ii) prior knowledge ('some conception with which his mind is already stored'). But there are more. 'By hypothesis, we conclude the existence of a fact quite different from anything observed, from which, according to known laws, something observed would necessarily follow.' It is reasoning which 'explains' 'from effect to cause' (CP 2.636). Thus (iii) the imagination must present us with a new idea. As he writes elsewhere, 'Abduction is the process of forming an explanatory hypothesis. It is the only logical operation which introduces any new idea' (CP 5.171). Furthermore, (iv) the 'new idea,' the postulated 'fact,' must be such that it can be the 'cause' of the 'effect' observed, must be capable of being the explanation of the problem at hand. And (v) from the 'cause' thus postulated the 'effect' must follow 'necessarily.' One does not, of course, know at the moment the hypothesis is formulated whether from it the 'effect' will indeed follow with necessity; only subsequent reasoning and/or experimentation will reveal whether or not that is so. The postulated 'cause' must therefore be such that this necessity can be initially (i.e. prior to subsequent reasoning and/or experimentation) *presumed*; that is, it is a priori, *as it were*. This *presumed* necessity is stressed by Peirce in statements like 'Abduction merely suggests that something *may be*' (CP 5.171).

This brief sketch of the most salient relevant elements of Peirce's theory will suffice for purposes of comparison with Descartes. We can now turn to Descartes and indicate the relevance of both doubt and intuition for making discoveries. The fulfilment of this task will, at the same time, provide us with the elements of similarity between Peirce's and Descartes' positions.

After the *Meditations* were completed, but before they were published, Mersenne questioned the order of exposition in them. Descartes replied that 'in all my writing I do not follow the order of topics, but the order of arguments'; and 'in orderly reasoning from easier matters to more difficult matters I make what deductions I can, first on one topic, then on another. This is the right way, in my opinion, to find and explain the truth' (DPL 87; AT3, 264).[6] Thus the way to find systematic truth involves order and deduction. Deduction (as opposed to intuition) consists in a 'train of reasoning,' and 'in every train of reasoning it is by comparison merely that we attain to a precise knowledge of the truth.' For 'all knowledge whatsoever, other than that which consists in the simple and naked intuition of single independent objects, is a matter of the comparison of two things or more, with each other'

97 Peirce and Descartes: Doubt and the Logic of Discovery

(HR1, 55; AT10, 439). The question is: assuming that one knows one of the two (or more) 'things' to be compared with one another, how does one obtain the relevant 'other thing'? This is a question in the area of the logic of discovery. It is an issue which Descartes took to be of crucial importance. 'In fact practically the whole of the task set the human reason consists in preparing for this operation' of comparing two things or more with each other. When the comparison 'is open and simple' (as it is when one recognizes that there is an immediate logical implication between thought and existence, or between thinking and entertaining ideas) 'we need no aid from art, but are bound to rely upon the light of nature alone, in beholding the truth which comparison gives us' (ibid). Deductive thought is not that of recognizing immediate logical relations. It therefore involves 'art,' involves reliance upon 'aids' to 'the light of nature.' These aids are 'imagination' and 'sense.' Imagination functions in metaphysics, where it helps reason in proposing judgments as 'things' to be compared with whatever is already known, in order to continue the train of reasoning. It also functions in, for example, applied physics where, with the help of 'sense,' the imagination presents reason with models and experiments. The judgments, models, and experiments have to be presented in a certain order so as not to 'interrupt the flow and even destroy the force of my arguments' (DPL 87; AT3, 264). Thus questions about right order, and proper use of imagination and sense, are questions about how to 'find truth,' are questions about the logic of discovery.

There are, says Descartes, two 'orders' or 'procedures' of exposition, those of 'analysis' and 'synthesis.' Of these, only 'analysis' 'shows the true way by which a thing was methodically discovered and derived, as it were effect from cause' (HR2, 48; AT7, 155). Thus analysis is held to be not just a method of exposition; it is also taken to be the method of discovery. For analysis involves a double achievement. There must be a reduction of the unintelligible or only partially intelligible complexity which initially confronts the thinker, to the fully intelligible starting point(s) for his deduction. And there must be an actual deduction of systematic knowledge on the foundation of this starting point (or on the foundation of these starting points). If we look at some of the details which Descartes presents with respect to 'analysis' as the method of discovery, this will allow us to say more about both doubt and imagination. Most of these details are presented in the *Discourse* and in the *Regulae*. In the *Discourse* (HR1, 92; AT6, 18–19) we read that there are four precepts which, if one does not 'fail, even once, to observe them,' are quite sufficient as methodological principles for the discovery of truth. The third of these precepts concerns deduction or composition. In spelling it out Descartes makes an implicit reference to the role of the imagination. For there he states that one must direct one's 'thinking in an orderly way, by beginning with the objects that were simplest and easiest to understand, in order to climb ... to the knowledge of the most complex ... *assuming an order among those objects which do not naturally precede each other.*' The assumption of order involves the formation of hypotheses; and that is the work of the imagination. For the assumed order is an 'imaginary' order, and the 'things' that 'follow' once the order is assumed are 'conjectures' (*praejudicia*) (cf, e.g., HR1, 31; AT10, 404–5). Through intuition and doubt one determines whether the assumed order is the right order, and whether the conjectures formed with the help of the imagination constitute correct judgments. Before we turn to these roles of intuition and doubt let me support the material presented thus far by showing that it fits the picture we have of Descartes' procedure in the *Meditations*.

As we saw before, Descartes claims that all he has doubted in the *Meditations* are 'prejudices' and *not* things 'known without any affirmation or denial'; that is, prejudices are judgments. Thus we see that when Descartes attempts a continuation of his train of reasoning in the *Meditations* by means of introduction of judgment, he often warns us of its status of 'conjecture' or 'prejudice': 'I am ... a thing which thinks. And what more? I shall exercise my imagination.' (*Imaginabor; J' exciterai encore mon imagination.*') (HR1, 152; AT7, 27; AT9–1, 21); and: 'But among these ideas, some *appear to me* to be innate, some adventitious ... (*mihi videntur; me semblent*') (HR1, 160; AT7, 38; AT9–1, 29)). The judgments made are non-intuitive, and are to be confirmed or disconfirmed; that is, these 'conjectures' or 'prejudices' are hypotheses.

There is another way of making the latter point, a way which will immediately make clear the role Descartes assigns to doubt. The first methodological principle stated in the second part of the *Discourse* is restated in a somewhat truncated form in the fourth part, where Descartes writes that in his 'search after truth' he decided to 'reject as absolutely false everything as to which I could imagine the least ground of doubt' (HR1, 101; AT6, 31). The full statement of this principle in the second part of the *Discourse* gives grounds for the statement made earlier about the phrase 'universal doubt,' in that it gives us the extent of 'everything' in 'to reject as absolutely false everything ... ': 'to accept nothing as true which I did not clearly recognise to be so: that is to say, carefully to avoid precipitation and prejudice in judgments, and to accept in them nothing more than what was presented to my mind so clearly and distinctly that I would have no occasion to doubt it.' With respect to most judgments I introduce to further my train of reasoning I can, prima facie, imagine some ground of doubt. If I can conceive some ground of doubt then these judgments must be hypotheses. *Hence the methodological exercise of doubt shows up judgments as, prima facie, hypothetical.*

99 Peirce and Descartes: Doubt and the Logic of Discovery

We are now also in the position to answer the question, *why* is there some ground of doubt with respect to these judgments? There are grounds of doubt because these judgments, rather than being given by the understanding alone, are produced with the help of the imagination. It is for this reason that propositions which state a 'truth known *per se*' – e.g., 'there must at least be as much reality in the efficient and total cause as in its effect' – are not hypothetical and hence can withstand the test of doubt. They are not the result of the activity of the imagination; there was no 'comparison,' and there were no steps of reasoning which led to their apprehension. Instead, they are singly 'manifest by the natural light' (HR1, 162; AT7, 40).

It is not, in the first place, through doubt that Descartes attempts to make progress in philosophy and the sciences. Like Peirce, he holds that it is imagination which plays the more fundamental role in the development of knowledge. It is through the imagination that we generate our enumerations. Since what is enumerated must be prima facie relevant to the particular problem, the use of imagination does not introduce an element which leads to 'vague and blind inquiries ... relying more on good fortune than on skill' (HR1, 30; AT10, 403); it is not as if we proceed 'at random and unmethodically' (*casu et sine arte*) (HR1, 31; AT10, 405). Again, like Peirce, Descartes holds that what is called for is a controlled, disciplined use of the imagination. How Descartes relates reason, method, and imagination is indicated especially in the *Regulae*.

The opening paragraph of Rule IX states:

We have now indicated the two operations of our understanding, intuition and deduction, on which alone we have said we must rely in the acquisition of knowledge. Let us therefore in this and in the following proposition proceed to explain how we can render ourselves more skilful in employing them, and at the same time cultivate the two principal faculties of the mind, to wit perspicacity, by viewing single objects directly, and sagacity, by the skilful deduction of certain facts from others.

Rule IX then deals with 'perspicacity' which, as is stated also in Rule XI, constitutes a further discussion of intuition. Rule X deals with sagacity. Its references to 'play,' 'invention' and 'imaginary order' indicate that Descartes is here dealing with the use of imagination. Since 'sagacity' is said to be 'one of the two principal faculties of the mind,' and since a discussion of 'sagacity' is a discussion of the proper uses of the imagination, imagination must be taken to be subsumed under 'one of the two principal faculties of the mind.' How it is thus subsumed is indicated in Rule X. But in Rule XI Descartes writes that Rule X deals 'with enumeration alone.' The function of the imagination therefore is to provide 'enumerations.' The examples given in Rule X make clear that it is concerned with the kind of enumeration which can be called 'inductive enumeration,' with the 'enumeration or induction' which is an 'inventory (*perquisitio*) of all those matters that have a bearing on the problem raised' (HRI, 20, AT10, 388). That the imagination must be disciplined is stated many times in this Rule. It obtains its discipline especially from 'all play with numbers and everything that belongs to Arithmetic, and the like'; for 'It is wonderful how all these studies discipline our mental powers, provided that we do not know the solutions from others, but invent them ourselves.' Thus, in general, the relation between reason, method, and imagination can be put as follows: imagination is necessary to generate deductions; it can fulfil its role only if it is disciplined, and it is disciplined by the mathematical sciences which themselves are structured by method. We can say that inductive enumerations are generated through a methodic use of the imagination. Since the method of discovery is applied in the search for '*causes*' of certain 'effects,' what is presented by the imagination to the understanding will be possible 'causes' for experienced 'effects.' (In this respect Descartes' position is like that of Peirce as far as concerns what I have identified as Peirce's fourth element which controls the imagination.)

This gives us the general answer to the question raised earlier: given that the mind has before it one of the 'things' to be compared, how does one obtain the other 'thing' which is relevant? The general answer is that for any 'effect' the imagination will present as simple a 'cause' as possible or a number of alternative 'causes.' These 'causes' are presented as hypotheses, conjectures, *praejudicia*. That 'cause' which can be seen (i.e., intuited) to have a necessary connection with the effect is the 'cause' which can withstand the test of doubt. (In the 'necessity' of the 'connection' there is at least some affinity with Peirce's fifth element.) In that case the 'cause' or hypothesis in question becomes a true proposition; another link had been forged to the deductive chain. If we reformulate the question stated in the first sentence of this paragraph to read: (a) what, precisely, is the next step to be taken in the argument, and (b) how do you know that this is the right next step to take? it will be seen that the general answer provided gives us criteria for answering the second part of the question. However, more needs to be said about both the first and second parts.

The imagination allows the understanding to assume as 'causes' that for which it does not yet have proof. These conjectures are not presented without *some* justification. Just as 'chance circumstances' will 'never bring about' the requisite experiments in the sciences, they will not call to mind the relevant judgments in metaphysics either (cf HRI, 215; AT9-1, 20). Enumeration of possibly relevant 'causes' is in terms of the contents of the ideas of these

101 Peirce and Descartes: Doubt and the Logic of Discovery

'causes'; and which idea is a possibly relevant one is at least in part determined by the 'effect' to be explained. Thus each inductive enumeration calls for a suitable preparation. Such preparation, besides involving competence in the method, involves thorough familiarity with the details of the argument as far as it has been developed. For, as Descartes writes about 'analysis,' about 'the true way to which a thing was methodically discovered,' 'it contains nothing to incite belief in an inattentive ... reader; for if the very least thing brought forward escapes his notice, the necessity of the conclusion is lost' (HR2, 48-9; AT7, 156). (Compare this with Peirce's first element.) Once a complete enumeration of possible 'causes' has been made, the role of the imagination is over. What remains to be done is to find means for eliminating those conjectures which are false. In the process of eliminating the false conjectures the deduction is 'discovered.' The means for elimination is provided in the methodological principle of doubt. Suppose that a conjectured 'cause' cannot be seen or intuited to have a necessary relation to the relevant item or items of knowledge already possessed by the mind or to the relevant parts of the body of knowledge already developed. Suppose that it cannot even be seen to be related to some still only partially intelligible item which is thus only partially incorporated in that body of knowledge. Then such a 'cause' is considered to be dubitable and is rejected as false.

In the *Regulae* Descartes states that

if our method rightly explains how our mental vision (*mentis intuitu*) should be used, so as not to fall into the contrary error, and how deductions should be discovered (*deductiones inveniendae sint*) in order that we may arrive at the knowledge of all things, I do not see what else is needed to make it complete; for I have already said that no science is acquired except by mental intuition or deduction. (HR1, 9-10; AT10, 372)

The use of 'discovered' in 'how deductions should be discovered' involves another point of importance to be noticed. Descartes uses the same word in his comment on analysis as the method of discovery: 'Analysis shows the true way by which a thing was methodically discovered and derived, as it were effect from cause' (*Analysis veram viam ostendit per quam res methodice et tanquam a priori inventa est* (HR2, 48; AT7, 155)). The Haldane and Ross interpretation of *tanquam a priori* by means of the phrase 'as it were effect from cause' is justifiable in view of what Descartes says elsewhere about the method of discovery, especially in view of *Principles* I, XXIV ('in this way we shall obtain a perfect science, that is, a knowledge of the effects through their causes' (HR1, 229; AT9-2, 35)). *Invenire* does not mean 'discover' in the sense of 'invent.' Given the philosophical tradition in which this word plays an important role[7] as well as the context in which Descartes uses it, its meaning is 'to find,' 'to come upon,' that is, to discover something already in existence. Further, the use of 'as if' in *tanquam a priori* seems to indicate that Descartes is using 'a priori' in a somewhat metaphorical way. Finally, since in its usual sense *invenire* is an achievement-verb, it must be understood as doing duty here for itself and for some activity-verb, such as 'seeking,' 'looking for,' 'following.' For it is such a verb which would naturally receive the adverbial modifications *methodice* and *tanquam a priori.* Thus *methodice* pertains not to a mode of discovery but to a mode of proceeding in making the discovery possible; and *a priori* qualifies a description of the method and not of the result or achievement attained by following it. Hence a pedantic but properly clear translation of the whole sentence would be: 'analysis shows the true way through which a thing has been sought-and-discovered methodically and, as it were, a priori.' In the light of the qualifications just introduced, its meaning would be: Analysis shows us the true way which, if followed methodically and, as it were, a priori, leads the seeker to the discovery of a thing. 'Following methodically' would include the controlled use of imagination to give an inductive enumeration of all the relevant (i.e., *possibly* necessary) 'causes.' To this work of the imagination attaches an '*as it were, a priori*' element: *which* of the several 'causes' it proposes will be the correct one is to be settled through doubt and final intuition. Whichever of the 'causes' can withstand the test of doubt will be intuited as the correct one(s). In metaphysics there will be only one such 'cause.' In 'theoretical' physics there may be several causes. If we want to develop an 'applied' physics, we will have to ask which of these several possible 'causes' is the actual 'cause.' This question is to be settled through experimentation, that is, through the use of imagination and 'sense' combined. (Again, much of this shows considerable affinity with Peirce's fifth element.)

The remaining two of the five elements which were singled out in Peirce's position, (ii) and (iii), also play a role for Descartes. That *new* ideas must be introduced Descartes stresses repeatedly. One of the major distinctions between traditional deduction and the Cartesian method of analysis is that whereas the former never introduces any new ideas, whereas 'syllogisms' are of use only to 'give practice to the wits of youths' and the cause of progress in knowledge 'is profited but little by those constraining bonds by means of which the Dialecticians claim to control human reason' (HRI, 4–5; ATIO, 363–5), The Cartesian method allows both the attainment of the foundation for philosophy and science, and the construction of the edifice of philosophy and science upon it.

That prior knowledge ('some conception with which his mind is already stored') is an element also operative for Descartes can be seen most easily through an example. Once it has been established, in the *Meditations*, that 'I am, I exist, is necessarily true each time that I ... conceive it' (HR1, 150; AT5, 25) what follows immediately, necessarily, self-evidently, in a 'natural precedence' as I think 'in an orderly way' is that there must be a content to my thinking, that there must be 'thoughts,' 'ideas.' No hypothesis is needed for this step. The next step, 'without interrupting the order of meditation which I have proposed to myself, and which is little by little to pass from the notions which I find first of all in my mind to those which I shall later on discover in it' is that 'it is requisite that I should here divide my thoughts into certain kinds'; and 'Of my thoughts some are, so to speak, images of things, and to these alone is the title "idea" properly applied' (HR1, 159; AT7, 36–7). But *are* ideas 'images of things'? that is, is there an extra-ideational reality? This calls for a thought-experiment, complete with hypotheses; for since there is no 'natural precedence' here, we must 'assume an order.' Therefore: 'But among these ideas, some appear to me to be innate, some adventitious, and others to be formed by myself.' We are here asking whether *all* of the ideas I find in myself could be either innate, or adventitious, or fictitious; that is, by using the imagination we are proposing hypotheses which are prima facie plausible as 'causes' for known 'effects.' Any hypothesis advanced as explanation for all of our ideas in terms of an adventitious or fictitious theory is open to doubt, and hence rejected. Only the hypothesis which states the innate theory can withstand doubt. Through contemplation of the innate idea of God it is learned that God exists, that nothing less than God could have been the cause of the idea of God. The sought-for conclusion is found once, through intuition, there is discovered a necessary relation between an 'effect' (my idea of God) and a 'cause' (God). Whether other extra-mental realities exist, whether there are after all good reasons for the tripartite division of ideas in terms of their causes, are problems Descartes takes up later. Important to notice now is that the step from 'idea of God' to 'God' as the cause of this idea does not follow for Descartes unless we introduce 'prior knowledge.' For nothing 'follows from' the idea of God unless we introduce the causal principle. This principle 'is manifest by the natural light' (HR1, 162; AT7, 40), is known per se, is an innate principle. Qua innate, it is a 'conception with which the mind is already stored' par excellence.

CONCLUDING COMMENTS

It has been said that Peirce 'chose his influences where he found them, deeming all historical figures alike with regard to their proponence of doctrine, despite considerations of chronology. Thus if Kant was his Teacher, and Duns Scotus his Friend, Descartes became his Adversary' (Feibleman 1960, 69). As far as his stand on the logic of discovery is concerned there is far less of an adversary position between Peirce and Descartes than Peirce (or his expositors) believed there to be. Perhaps this is because Descartes was less of an adversary of Duns Scotus than his comments on the scholastic period would lead one to expect. Perhaps it is because although 'modern science and modern logic require us to stand upon a very different platform' from that of Descartes (CP 5.265), in the construction of his own platform Peirce has unwittingly salvaged some of the planks of Descartes' – perhaps because no other planks are available, so that all that could be done was to cut them to different size, arrange them in a different order, and cover them with a different coat of paint. Or (to use some of Gerd Buchdahl's words) perhaps it is because although Descartes was 'unaware of the vast importance of the method of hypothesis in the ages to follow, there can be no doubt that the concept itself was for him as much a matter of course as it was to be for his more empiricist successors. Indeed ... how could it have been otherwise? For is not that notion ... a mere commonplace?' (1963, 405).

T.L.S. SPRIGGE

James, Santayana, Tarski, and Pragmatism

I   INTRODUCTION

Santayana was one of pragmatism's more effective critics, as he was also of idealism, for he had more sense of the inwardness of these theories than did such other major critics as Russell and Moore, who argued from a somewhat similar realist point of view. There has always been a question, indeed, whether there are not substantial elements of pragmatism in Santayana's own philosophical positions. In this essay, I shall consider how far Santayana's criticisms of pragmatism are valid, and whether there is anything in the idea that he was, after all, in some respects a pragmatist himself.

This topic seems a suitable one in a volume in honour of a philosopher and scholar who has done so much to promote serious discussion of pragmatism, in particular by his *The Thought of C.S. Peirce*, a study which greatly forwarded the serious critical and scholarly study of Peirce's philosophy and which continues to be an invaluable guide thereto and essential reading for all those interested in this phase of American philosophy. In the nature of the case, however, when discussing Santayana's relation to pragmatism, one must dwell rather on James than on Peirce, since it was primarily the former's pragmatism which figured in Santayana's field of vision. In concentrating on James' pragmatism I shall also take the opportunity of suggesting certain clues as to its understanding which have often escaped James' readers. Read with these clues in mind, it may seem even more divergent from Peirce's pragmatism (or pragmaticism) than is commonly recognized, a fact which should be of interest to Peirce scholars, who are so often concerned to speak of James' misuse of Peirce. It would certainly have been interesting to consider what logical relevance, if any, Santayana's criticisms, which, historically speaking, were not directed at it at all, may have to Peirce's theory of meaning and truth, but that would be beyond the compass of this essay, as it would also be to analyse the precise relations between Peirce's and James' pragmatism. I must leave the reader to work out the implications of my discussion of James and Santayana for this himself, recommending to him especially Professor Goudge's suggestion that there was a much closer affinity between Royce and Peirce than James and Peirce (TP 333). What I do hope to show is that, however unclear James may have been as to his relation to Peirce, it is a mistake to think of his pragmatism as merely a botched version of Peirce's, with no independent interest, and I believe its 'deep structure' can be brought out particularly clearly by considering Santayana's, at one level, somewhat devastating criticisms of it. In any case, the words 'pragmatist,' 'pragmatism,' 'pragmatic,' and so forth, when they occur in this essay without qualification, relate to James' pragmatism and its offshoots.

The following sections are concerned respectively with Santayana's criticism of Jamesian pragmatism, with James' pragmatism itself, and with pragmatic elements in Santayana's own thought.

II SANTAYANA'S CRITICISM OF JAMES' PRAGMATISM

In criticizing James' pragmatism, Santayana treats it as purporting to be, above all, a theory of truth. Actually James professed a pragmatic position regarding the meaning of concepts generally, and presented his theory of truth simply as its application to one of our most basic concepts. There was, evidently, something rather odd here, for a pragmatic analysis of what is involved in asserting a proposition would seem to apply the pragmatic mode of analysis one time more than is necessary, if it is first applied to the concepts in the proposition and then to the question what is involved in regarding the proposition as true. On the other hand, James was surely right to think that pragmatism as a theory of meaning does imply a certain theory of truth, and that it is here that one of its chief interests must lie.

Santayana's most extensive discussion of the pragmatic theory of truth comes in *Character and Opinion in the United States* but in the chapter called 'Later Speculations,' not the one called 'William James.'

Let us remind ourselves of what James did, in fact, say in succinct exposition of what, on the pragmatist view, truth is.

True ideas are those that we can assimilate, validate, corroborate and verify. False ideas are those that we can not... Truth happens to an idea. It becomes true, is made true by events. Its verity is in fact an event, a process: the process namely of its verifying itself, its veri-fication. (James 1907, 201)

107 James, Santayana, Tarski, and Pragmatism

Then most notoriously:

The true, to put it very briefly, is only the expedient in the way of our thinking, just as 'the right' is only the expedient in the way of our behaving. Expedient in almost any fashion; and expedient in the long run and on the whole of course; for what meets expediently all the experience in sight won't necessarily meet all farther experiences equally satisfactory. (James 1907, 222)

Santayana's first, and perhaps from his point of view most important, criticism of this theory of truth is that it confuses truth with the correctness of ideas, conceived of as mental events. If truth consisted in correct ideas, then there could have been no truth if there had been no minds, not even the truth that there were no ideas. That there could have been truth, truth about the physical world, even if there had been no psychological phenomena, was one of Santayana's most cherished convictions. However, his objection to treating truth as a property of certain psychological phenomena rested on grounds logically deeper than his commitment to materialism, for he held that even if the sole constituents of the world had been various streams of sensation and opinion, the truth about the world would have concerned the definite relations holding between these sensations and opinions, something which none of these opinions might have regarded at all. (Santayana 1920, 153-4)

Let us pause briefly for some comment on the value of this first criticism.

Santayana has some tendency to think that what he calls the truth is a reality which the pragmatist has no real wish to deny, but which he simply overlooks, when he rather confusingly equates truth with the correctness of ideas. After all, the pragmatist implicitly acknowledges 'the actual and concrete truth ... that there are many states of mind, many labouring opinions more or less useful and good' (Santayana 1920, 154). The pragmatist does not really think that the truth that these are just the opinions held that there are held has to be a truth opined by someone in order to be so.

It might seem that the divergence between Santayana and the pragmatist, so far as this particular point goes, is simply over which of two realities, which they both acknowledge, is best called truth. Both think that certain things are so, and both think that ideas divide into the correct and the incorrect, but while Santayana regards 'truth' in its primary meaning as referring to the being so of what is so, the pragmatist thinks it is applied more properly to the correctness of ideas.

But I do not think this is an entirely just estimate of the situation. For Santayana thinks that as well, so to speak, as the aggregate of existence generally and all its constituents, or let us say more simply the world and all that therein is, there is something else which is the truth about the world.

James was clearer about this matter than I suspect Santayana gave him credit for, and claims, in effect, that one does not deny that there is a non-human world in denying that there is a non-human truth (James 1909 xiv ff). The world might exist, with particular constituents in particular relations, but world and truth about the world are not the same thing (as Santayana would acknowledge) and this something more than the world (and its constituents in relation) which is the truth about it must consist, according to James, in correct ideas about it which occur in minds.

Santayana may well be right in holding that truth is a reality other than the world and its constituents, and other than any actual opinions (past, present, or future) about it, but he tends, not always, but at times, to treat the pragmatist as odder than he is in denying this, by talking as though, taken strictly, he denied a reality with characteristics and constituents which may always stretch beyond any complete human grasp thereof.

Santayana had definite reasons for thinking that unopined truth was a reality in addition to the aggregate of physical and mental reality (past, present, and future) about which it was the truth, though these reasons are never developed in any very systematic way. I have said something about them elsewhere (Sprigge 1974, 170–2), and here I shall only remark that Santayana is rather unclear as to just what sort of reality these reasons lead him to identify under the heading of truth. Sometimes, as in *Character and Opinion*, he talks as though what he calls *the* truth consisted of a system of individual true propositions (conceived of as items non-linguistic and non-mental), yet this is not really the sort of theory which emerges from *The Realm of Truth*.

In any case, the issue as to whether truth is to be equated with correct opinions as to what is (is not, etc.), or whether there could be truth quite unopined, is not a simple one, and simply to identify it with the question whether there are definite realities uncognized by human beings would be a mistake. We must leave it open whether Santayana or James is in the right here. I would, however, urge that though Santayana sometimes oversimplifies by tending to make such an identification, he is abundantly right in his insistence that there is (in some sense) a way things are, whether anyone knows this or not, and that pragmatist language does tend to blur, if not to deny, this simple point.

Three further comments may serve to conclude discussion of this, Santayana's first, criticism.

(1) There is a common compromise between the view that truth consists in correct judgments or opinions which have actually occurred, and the view that it is something quite other which correct opinions somehow participate in

or grasp. On this compromise view, a truth is a possible judgment, which would be correct if actual, having a status like a possible move in a game of chess which would have brought victory. I believe that both James and Santayana would reject this view, as exploiting the concept of possibility illegitimately (though Santayana's language occasionally inadvertently suggests it).

(2) Santayana, somewhat confusingly, runs the objection, which we have been discussing, to the view that truths are correct ideas in people's minds, with a really quite different objection to what pragmatists say about truth. This second objection is to the use of the word 'truth' to refer to the abstract quality in virtue of which truths are true, rather than to that which has this abstract quality. On this I would say that though it is certainly right to distinguish questions as to what sort of items or item have this quality (and thus are truths or the truth) from questions as to the correct analysis of this quality, his suggestion that the word 'truth' applies properly only to what has this quality, and not to the quality itself, is surely a mistake.

(3) If James' accounts of truth do, whatever his real position, rather suggest that there is not a way things are which is independent of what may ever be discovered to be so, the same would seem to apply to some of Peirce's formulations, as when truth is equated with 'the satisfaction which *would* ultimately be found if the enquiry were pushed to its ultimate and indefeasible issue' (TP 25; CP 6.485). As against this, Santayana might claim that there might well be truths of detail about the forgotten dead which no amount of historical research will or would lead men to formulate.

The second criticism which Santayana advances of the pragmatic account of truth is that, even as an account of what makes an idea *true* or *correct*, it confuses the idea as a psychological fact with natural relations, to be studied by psychology, with other phenomena, with the idea as an opinion. Truth belongs to an idea qua something with a certain logical purport, not qua a mere experienced datum.

There is certainly truth in this characterization of pragmatism in James' version. However, James treated the truth of ideas in this fashion, not through inadvertence, but because he thought a logical force which was not explicable in terms of what we actually experience was a myth and that, therefore, logical significance was ultimately to be explained in terms drawn from psychology (as James conceived this). Thus Santayana might seem to be helping to formulate pragmatism rather than showing ground for rejecting it.

He does, however, offer a ground for criticizing pragmatism when he gives his own formulation of what it is for an idea, qua opinion, to be correct or true, for he is pointing out what looks very much the standard use of 'true' in this context and noting that this sense seems to be missing from the pragmatist's account. 'An opinion is true' he reminds us, 'if what it is talking about is constituted as the opinion asserts it to be constituted' (Santayana 1920, 155). He goes on to insist that truth, in this sense, simply cannot be reduced to the natural relations of the opinion qua mere event. 'Psychologists, however [and his whole point is that James' pragmatism in essentially a psychologist's theory], are not concerned with what an opinion asserts logically, but only with what it is existentially; they are asking what existential relations surround an idea when it is called true, which are absent when it is called false. Their problem is frankly insoluble; for it requires us to discover what makes up the indicative force of an idea which by hypothesis is a passive datum; as if a grammarian should inquire how a noun in the accusative case could be a verb in the indicative mood' (Santayana 1920, 156).

I think we might formulate this point which Santayana and others such as Russell and Moore made against James' pragmatism rather more sharply, by borrowing a leaf from Tarski's book (Tarski 1944). It should be noted, however, that though the hint comes from Tarski, I use it to clarify Santayana's point, and therefore in a most un-Tarskian way. For Tarski, truth was a property of *sentences*, and he gave a highly artificial definition of it as such (or rather of 'true in L') with which I sympathize no more than would have Santayana, or James. So having acknowledged where the hint came from, I shall not further point out that what I say in its basis is not true to Tarski.

Suppose someone holds the opinion that snow is white. Under what conditions can it be a true opinion? Purely and simply under the condition that snow is white. In general, an opinion that p is true if and only p.

There are problems in treating the last sentence as a definition of 'true.' Put more formally it would have to run:

(T) For all p, *an opinion that p* is true if and only if p.

Note, incidentally, that I say '*an* opinion,' because we are concerned with opinions qua mental events, so that your opinion to a certain effect is a different individual from mine to the same effect. An opinion is here an act of opining.

Now suppose that we give a value to 'p' in what follows the quantifier ('For all p'). We might get: 'An opinion that birds sing is true if and only if birds sing.' The trouble is that 'birds sing' has a different significance in each case. In its first occurrence it is a noun phrase, a mere part of the antecedent clause of the biconditional, while in its second occurrence it is itself a whole clause or subordinate sentence. We do not, therefore, have a properly uniform replacement of the variable, or, if we do, we have nonsense, for what follows 'if and only if' is a noun phrase hanging in the air.

111 James, Santayana, Tarski, and Pragmatism

A certain looseness in the above can be reduced most briefly by revising (T) as follows:

(T, revised) For all *that p*, an opinion *that p* is true if and only if *that p*.

Here '*that p*' is the variable and its substituends are noun clauses commencing with 'that'. It is obvious that the second subordinate clause is incomplete, and could only be completed, unhelpfully, by adding 'is true' after the last 'that p.'

But though T (and its revision) is not properly formed, and cannot be a definition, we may still say that a proper definition of 'true' must be such as explains why we are logically obliged to assent to all propositions whose verbal formulation is like the following in a rather obvious way:

(E) If a man is of the opinion that birds sing, his opinion is true if and only if birds sing.

Yet if we consider the various pragmatic definitions of 'truth' and 'true,' they do not explain why we are under such a logical obligation, and suggest rather that we are not. There is no time to show this at length, with regard to these various definitions. The simplest examples derive from the definition of 'the true' as 'the expedient in the way of our thinking.' Surely a man's opinion (especially if this means 'his opining') that he is purposely sent to this earth with a mission to effect certain reforms might be highly expedient, even in the long run, without his actually having been sent here purposely by anyone.

This is a very real difficulty for the pragmatic theory of truth. I shall suggest later why James may have been less impressed by objections in the same spirit as this one than he might have been.

There is one last comment by Santayana on the pragmatic theory of truth which I wish to consider.

Having made the point that the pragmatic theory of truth fails because it tries to give a psychological explanation of something which cannot be treated psychologically, he elaborates on this theme by suggesting that pragmatism confuses the problem of the truth or correctness of ideas, which is not soluble in psychological terms, with a problem which is so soluble, namely the question of the relation between a sign and the thing signified. He then says: 'Of this relation a genuinely empirical account can be given; both terms are objects of experience, present or eventual, and the passage between them is made in time by an experienced transition' (Santayana 1920, 159).

This is very remarkable, and that for two reasons. First, this account of the relation between a sign and the thing signified is quite incompatible with the mature philosophical position of *Scepticism and Animal Faith* and *Realms of Being*; yet the former was published (1923) only a few years after *Character and Opinion* (1920) and is anticipated in earlier writings. The Jamesian account of signification here evidently accepted would mean that a sign present to my mind could not signify anything lying wholly in the past or anything incapable of being immediately present to any consciousness (as is true of all material things for Santayana).[1] Santayana's concession to pragmatism on this point must, surely, have been either carelessness or a vast and misleading simplification for the sake of brevity. (It is incompatible with what he says immediately afterwards about knowledge of the past.)

This comment of Santayana's is also remarkable, in a more general way, because one would expect signification (in the sense, really, of reference) to be just as much a logical notion, in a broad sense, as truth.

A partial explanation of both these points is this. Santayana tended to think that a mental act of thinking was about that part of reality which it was about, in virtue of its natural relations to certain features of the environment, but that what it opined about whatever nature settled that it was about was a matter of the *essence* intuited, something to which its relation is hardly psychological.

What I think is true in Santayana's comment is that a main motive behind James' pragmatism is precisely to explain how a present mental content can refer to something now out of consciousness, and that he was not at all clear as to the relation between successful reference and truth.

So much for Santayana's criticisms of pragmatism. I shall now make some comments of my own on James' theory.

III SOME MOTIVATIONS FOR JAMES' PRAGMATISM

There are, I believe, at least three distinct trains of thought in James' philosophical development which culminate in what purports to be a single pragmatic theory in which they are no longer distinguished.

First, there is the thesis which brings him nearest to Peirce on the one hand, and to logical positivists on the other, that the meaning of a concept is to be identified with the difference its being applicable makes to the observable circumstances and results of various possible actions. This is the aspect of pragmatism which has been most influential and which many would think its most important component.

Second, there is James' view, elaborated most carefully in 'The Will to Believe,' that under circumstances in which a choice between effective acceptance of one or other of several positive hypotheses is compulsory, and where empirical evidence (and other sorts of cognitive consideration) is neutral between them in its upshot, it is legitimate, and even desirable, that their varying satisfactoriness to our emotional nature and as spurs to useful activity should settle which is accepted.

Third, there is a view as to what it is for the mind to be thinking about something which is not a present content of consciousness. The view, roughly speaking, is that a mental content is the idea of an absent object X, if it serves to put one in effective practical relations with that object.

Each of these trains of thought gives a reason for insisting that truth has something to do with practice. On the first, a theory whose truth (i.e. the applicability of whose concepts to the relevant phenomena) made no difference to the experiences practice will produce is a meaningless theory; thus a true theory necessarily has practical implications. On the second, certain theories have as their main commendation their effects on our affecto-conative nature, and hence ultimately on our behaviour. It is an easy step from this to saying that their truth may lie in such effect. On the third theory, an idea is essentially a means whereby we adjust ourselves practically to our surroundings, and it is an easy step from this to saying that the success of such adjustment determines the truth of the idea (or of its application).

On the other hand, the theories point in somewhat different directions in some important respects. The first theory, in a pure state, implies that hypotheses the same in their observable implications have the same meaning, while the second is applied most typically to choice between theories which, on the first score alone, would not even be distinguishable alternatives. Again, the second theory has a tendency to produce the view that the truth that an object exists may lie in the satisfactoriness of believing that it does, while the third theory would lead rather to an emphasis on the independent *thereness* of the object as what our true ideas adjust us to. I think that some of the peculiarities of James' pragmatism, and in particular his sense that his critics always misunderstood him, are reduced if one realizes its triple basis. Whether a consistent theory of meaning and truth could be made out of these three ingredients, and how far James can fairly be credited with such a theory, are matters I cannot pretend to settle here.

What I wish to dwell on rather is the particular contribution made by the third theory to James' pragmatism, and the way a reference back to it explains certain peculiarities thereof. This third factor in pragmatism is also especially relevant when one considers Santayana's relation to it, and that for two reasons. First, it is this factor, if any, which justifies Santayana's claim that the so-called pragmatic conception of truth is really concerned less with the question 'What is truth?' than with the question 'What determines the significance (the reference) of a sign?' (One might think Santayana was making the more usual point that pragmatism is concerned with 'how to make our ideas clear,' in Peirce's phrase, rather than with truth, but this would be a mistake.) Second, there is an element in Santayana's own final theory of reference or intending which is akin to this component in pragmatism.

Among the philosophical problems which most perplexed James from time to time throughout his career was that of escaping Royce's main early proof of the existence of the Absolute, of a unitary cosmic mind of which all finite minds are constituents. At one time, Royce's argument seemed logically compulsive to James, but the conclusion was always repellent to him, partly for moral reasons (since all evil would then be willed by the mind we have to call divine), partly because the notion of the compounding of consciousness seemed dubious (though *after* he had won his alternative to the Absolute he gave up this objection to it). Therefore he was keen to find a way out.

Put with absurd sketchiness, Royce's argument is this (Royce 1885 XI). If all a person ever thought about were the contents of his own consciousness, he could never be in error, for they are simply what he makes them by his thought. However, we know that error exists, and therefore people must think of things other than contents of their own consciousness. This is only possible, however, if these contents are intended to symbolize the character of certain definite other things, which character, in fact, they misrepresent. Now X can only symbolize Y if a mind to whom both are present has set up such a symbolizing relation. This cannot be the individual mind, as its escape from its own mental contents to a reference to something beyond is precisely what as yet seems inexplicable. The only solution is a comprehensive mind, comprising my mind and everything I could possibly think of (thus the whole world), who uses, through me, my mental contents as symbols, adequate or inadequate, of other things. Since, in a sense, I am (a particular incarnation of) that absolute mind, what it means by these symbols in my consciousness is what I (really) mean.

James found this argument very compelling, as an examination of it, more thorough than is possible here, would show he was not unreasonable in doing. It was, therefore, imperative that he escape it by finding an alternative account of the reference of an idea.

The account he eventually came up with is adumbrated in 'The Function of Cognition' (1885) and 'The Tigers in India' (1895),[2] though he took some time to be satisfied that it was really an adequate alternative to Royce from a metaphysical as well as from a purely psychological point of view. The thesis is that an idea or concept, considered as a content in my consciousness now, is about something outside my consciousness if it could lead me to an experience in which that object is perceptually present. Thus an idea in my mind may be the idea of a certain distant building if it can serve as an instrument for getting me into the actual presence of the building.

There is a good deal of vagueness in James' statement of this view. He speaks of an idea as leading one to its object. Sometimes one rather has the impression that if only one lets the idea dominate the mind one will be led up to the environs of the object, willy nilly, as if in a hypnotic trance. But I think the key idea is really as I have said, that the idea is an instrument I can somehow use for getting in appropriate touch with the object (or at least with things causally related to it, as in the case of objects in the past). James insists that this relation between the idea and its object is something wholly within the experience of the individual mind, for one experiences the whole series of transitions which lead from the idea, via various intermediate experiences, to an experience in which the signified object is an element. As to whether the actual physical thing perceived is an element in the perceptual experience, or whether only a percept representing it is, James moved steadily towards the former view, but so long as he saw any need to distinguish between percept and thing represented thereby, he saw the first as an instrument for acting on the second.

To say that an idea is an instrument for bringing one into contact with its referent is not quite the same as saying that it is an instrument for evoking appropriate behaviour towards it, but the two are closely related, and James moves quickly between them. It is hard to see the value of a mere confrontation which is not also a useful adjustment. To see ideas as bringing one into relations with their referents, and not to value ideas according to the satisfactoriness of these relations, would seem odd. A pragmatic theory of truth naturally follows. If ideas are essentially instruments of adjustment, surely the ones we praise as truths must be those instruments which work well. Thus the pragmatic theory of truth seems to derive partly from a need to explain how ideas have reference.

With this factor in the development of James' pragmatism in mind, I believe that light can be thrown on three puzzles about it.

(1) Does pragmatism acknowledge a reality whose existence is logically prior to the utility of postulating it?

Attacks on pragmatism often assume that the answer is negative. In particular, pragmatism (in James' version) is accused of making the independent existence of God unnecessary to the truth of believing in it.

Upon the whole, I think one can say that, for James, when an idea is true, it is so in virtue of the existence of a specific reality outside thought to which it relates. Certainly this is what one would expect in the light of that thread in pragmatism which consists essentially in a theory of the reference of ideas. Although James scarcely makes it clear, his position surely is that reference and truth are not sharply distinct. To be about something at all an idea must put us into practical relations with it, must help our behaviour take account of it. The idea is true to the extent that these relations are good in their upshot, so that we and others are better off than we would have been if the idea had not helped us thus to adjust to the independent object.

Of course James is concerned to deny that true ideas somehow merely reproduce in their own medium the character and form of reality as it is apart from them. Such a denial is one of the main points of his pragmatism. He insists, in particular, that thought carves reality up so as to create divisions and junctions which would scarcely be there apart from it. But this does not mean that pragmatism is an attempt to reduce the independent *thereness* of things to the utility of our belief in it. Rather is it an attempt to explain thought as an adjustment to their logically prior *thereness*.

It is true that James holds that the realities to which thought is an adjustment are all in some sense experiential (and that even if they have a noumenal source it is not to it but to them that our ideas relate us). But that is a feature rather of his radical empiricism than of his pragmatism.

(2) Does not pragmatism fail to explain why we are logically obliged to accept all propositions formulable in sentences resembling E above?

It might seem to do so, for surely one might grant that some man's belief that A is F put him into the best practical relations possible with some specific reality, best for others also and in the long run, while denying or doubting that A was F. Also vice versa.

The solution to this may be found I think in realizing that a main inspiration for James' pragmatism is the need to explain the self-transcendence of certain mental occurrences. Now it one tries to solve this problem, one is thinking of mental occurrences as describable in non-intentional terms, and considering how an occurrence thus describable can refer to a reality beyond. Moreover, since, as I have suggested, James' explanation of this reference beyond of mental occurrences is also an explanation of their truth (this being completely successful reference), it follows that truth also needs to be explained as belonging to something not initially specifiable in an intentional manner. In short, James is concerned to explain why a particular mental event having certain sensory and quasi-sensory properties refers and is true, not why a belief that something is the case is true. The most obvious examples of such occurrences are image complexes or strings of words, considered in their immediate sensible character together with the fringe of other sensible items which vaguely surround them. For simplicity, let us concentrate on the reference and truth of an image, hoping that the reader will not forget the phenomenological subtlety of James' description of images.

If one asks why an image, in a man's mind at a certain time, or perhaps present over a period (not complicating the discussion by troubling too much over its trans-temporal identity) is *true*, one cannot answer it in the simple way one does if one asks why a belief that my dog is hungry is true. To say that my belief that my dog is hungry is true because my dog is hungry is to utter a mere substitution instance of a formula for producing trivialities. To say that an image of a certain shape and a certain uneasy felt quality is true because my dog is hungry is on a different level. James' view is that an image is true because it relates one fruitfully to a certain reality, and thus the truth of this particular image may be that it relates me fruitfully to my dog's hunger, but its referent and its truth condition is a matter of the particular work it does for me, not something to be read off a formula.

Many will think that if this is James' view, then so much the worse for James, since it is definitely beliefs which are true, not images and other quasi-sensory particulars.

I now move to a possible defence of James' position, true I believe to what he was essentially getting at, though I claim no textual support for such a rejoinder in detail.

James could reply that, of course, it is a belief that is true, but that a belief is, nonetheless, a mental event with its own inherent character, perhaps that of an image. Having established that an image puts a man into practical relations to a certain reality and that these relations will be good ones if and only if that reality is his dog's hunger, one can then call it his belief that his dog is hungry. In short, it is a belief that p, and a true one, because the reality specified as p is that to which it usefully relates him. It is not in virtue of being a true belief that p that it relates to this fact, but in virtue of relating him successfully to this fact that it is a true belief that p. (Had it related him to a reality in a way which failed to be useful, precisely and only because the reality was not one specifiable as p, then it would have been a false belief to the same effect.)

The same account could be given of verbal beliefs, but one must think of the words as just sounds which relate a man to a certain reality. We have no right to describe a man as recognizing the fact that p because he makes the noise p to himself in a straightforward unqualified way unless we personally use the phrase 'p' to describe a reality, whatever it may be, to which he was thereby usefully related.

It would be false to James to think of the specific reality to which a piece of knowledge, qua mental event, successfully relates us as itself a discrete fact in some non-human stack of facts. Rather, the reality to which I refer is a continuum to some element in which my thought (itself only a pulse in a continuous mental life) points, so that in virtue of that pointing one may use the same expression X in describing both the pointer and the pointed at, the arrow and its target. The arrow is called the knowledge that X, the target X; which of the two was first called X may vary from case to case.

One might put the matter roughly thus. James is describing what he sometimes called 'the truth function' as a relation between a mental occurrence and a quasi-external reality at a logical level before either of these is 'propositionalised.' The former is not called 'the knowledge that p,' and the latter is not called 'p.' Propositional descriptions of them as the knowledge or true belief that p and a state of affairs p really presuppose that the problems of reference and truth have been solved. One may then use E-type sentences, of the sort Tarski made famous, in order to explain what truth is, but this bypasses the serious problem. James' thought on these matters does not emerge altogether clearly because he does not make the distinction between descriptions of the truth-relation (and the reference relation) which explain why propositional language of a certain sort applies to the phenomena, and descriptions which trade on the fact that such language has already been applied.[3]

(3) Why is verification, not merely verifiability, said to be essential for genuine truth?

In Lecture VI of *Pragmatism* James tells us that we include among truths notions which are verifiable as well as those which are verified. (Verifiability meaning such as would be verified, if tested, not, as in the verification principle of meaning, verifiability as mere testability.) He makes it quite clear, however, that verifiable truths are only true in a secondary sense. This seems far more extreme a view than one would expect, if one thought only of those strains in James' thought which bring the theory near to logical positivism or to Peirce's pragmatic maxim. At least part of the basis for the strange claim that only notions which have passed empirical testing processes, as opposed to those which would pass them, are true in the strongest sense, seems to be this. A true idea must have a reference, and reference of an idea is essentially constituted by the working contact in which it puts us with its referent. Thus a completely untested idea scarcely has a referent, and can only be true in a secondary sense of cohering well with ideas which do have a referent, and assisting these more fully true ideas in their effective operation. Why James should hold this view becomes clearer, I hope, in the light of our previous discussion.

Although I have contrasted the strain in James' pragmatism which derives from his concern with the problem of intending, of the reference of ideas to things beyond, with the strain associated with Peirce's account of the meaning of concepts, I am not denying some overlap with Peirce's theory of signs. I am doubtful, however, whether he owes much to Peirce in this connection.

119 James, Santayana, Tarski, and Pragmatism

IV ELEMENTS OF PRAGMATISM IN SANTAYANA'S PHILOSOPHY

There is no space for more than a few remarks on this topic.

One may say, first, that Santayana stands clearly opposed both to the 'Will to Believe' element in James' pragmatism and to that part of it closest to Peirce's account of 'How to Make our Ideas Clear.' He does, however, have views which suggest the influence of the strain in pragmatism which we have mainly discussed.

We may note, first, that Santayana tends to see *given essences* as predicates of external realities just in so far as the intuition of them expresses my behavioural adjustment to them as a result of physical stimuli from them or their associates. Thus the self-transcendence of 'ideas' is for him as for James a matter of their relation to action which takes account of what lies outside consciousness. His epiphenomenalism, however, and his view that physical reality is, in some strong sense, external to consciousness, prevent him allowing that the adjustment is strictly caused by the intuition or could lead to the actual presence of the things adjusted to within my experience (cf Santayana 1923 XVIII).

James and Santayana, then, give partially similar accounts of the sense in which an 'idea' in my mind can refer to something beyond. They both invoke certain essentially dynamic and behavioural relations as being what link the two, rather than connections of a purely logical, ideal, or semantic type on the one hand, or connection via an absolute mind on the other.

It might seem, however, that, whereas truth for James was simply a fully successful reference, for Santayana it was something different. Santayana (1920) says as much himself in the passage quoted above (a passage inconsistent, as I noted, with his mature view of the physical).

It is true that, from Santayana's point of view (or one of them), someone who knew the literal truth about something would be *referring to that thing* in virtue of natural or dynamic relations in which he stood to it, but would *know the truth about it* in virtue of the identify between the essence intuited and that independently exemplified in the thing. Certainly with respect to both literal knowledge and absolute truth Santayana stands in complete opposition to pragmatism.

In fact, however, Santayana thought most human knowledge was symbolic in character. The essence attributed to the intended reality is an apt symbol for it rather than true of it, as what it exemplifies. If one asks in what the aptness of such a symbol lies, Santayana sees many different sorts of aptness as possible. Two, however, may be mentioned as of especial importance. First, the symbol may serve as an adequate guide to what needs to be done if I am to survive and obtain what I desire. (Strictly it is not a guide, but only expressive of the fact that my psyche is physically thus guided.) Second, it may arouse, or be associated with, emotions which it is appropriate I have to the reality in question. Scientific ideas, in so far as they fail to lay bare the literal truth, are often symbolically true in the former sense; religious ideas in the latter, e.g. the concept of God the Father may evoke (to speak loosely) appropriate feelings of dependence on a reality, nature, in relation to which I must be humble.

The first of these sorts of symbolic aptness is quite close to features of the truth-function which we have found James emphasizing. The second is, in a broad sense, a pragmatic function. Indeed, in his whole account of symbolic knowledge, and of conventional and dramatic truth, Santayana describes them in a way which for James would count as pragmatic. The difference is that Santayana always sees them in contrast to a literal truth (see Santayana 1938, passim).

The pragmatic elements in Santayana's account of religious belief raise the question why James and Santayana clashed so strongly on the status of religion. It is true that each valued religion for quite opposite effects on conduct and mood, one as a sweetener and a quieter, the other as a goad, but more fundamental is the fact that, amusingly enough, it was really James who was upset here at the other's unconcern with literal truth.

In what I have said on James' pragmatism, and on Santayana's relation thereto, I have, I hope, brought out two points which should be of interest to Peirce scholars. The one is that, indeed, James' position is sufficiently dissimilar to Peirce's to have justified the latter distinguishing it from his own 'pragmaticism'; the other is that, for that very reason, it should not be regarded as a mere botched version of pragmaticism, but as a theory of interest on its own account.

H.S. THAYER

Peirce on Truth

The essence of truth lies in its resistance to being ignored. (CP 2.140)

When Thomas A. Goudge's *The Thought of C.S. Peirce* appeared in 1950, it argued that a basic source of problems, and therefore of our understanding of Peirce's thought, was a conflict of loyalties and predilections between naturalism and transcendentalism. It is not necessary to comment on the merit of this informed and critical study, for it continues to be consulted by all students of Peirce's philosophy and has become a classic in the literature on the subject.

My subject is Peirce's theory of truth. As the discussion will show, certain developments of his theory reflect naturalistic and transcendentalist interests, thus to some extent bearing out Professor Goudge's thesis.

I

Peirce's writings contain many references to and brief discussions of truth. What he has to say on these diverse occasions is not always very clear; nor is there an obvious unity and coherence in the views expressed. The reader is faced with the recurrent difficulty in understanding Peirce's intentions as to whether we are encountering thematically distinct and historically separate phases of thought, or whether there is to be discovered beneath the fragmentary state of the writings an underlying consistency of procedures and a common objective.

Peirce's views on truth appear to have had a circuitous formation, reflecting and proliferating among aspects of a coherence theory, a theory of truth as correspondence of propositions with reality, and a pragmatic theory of true belief as the resolution of doubt by means of inquiry. It is not my purpose to discuss the question of whether one or several theories of truth are to be attributed to Peirce. I incline to think, with some hesitation, that despite a number of anomalous formulations, certain changes in emphasis and construction – effects, no doubt, of refractory difficulties and critical revision – there is one generally coherent and comprehensive sense in which Peirce understood the meaning of truth. The differences among Peirce's various statements on truth are due primarily, I believe, not to major changes of mind – and accordingly of theories – but to what seems to have been an attempt to explore several different ways of capturing and articulating one comprehensive theory of truth. The problem was how to arrive at a statement that was precise enough to be useful and informative while also retaining a sufficient generality so as not to exclude or fail to account for the characteristically varied uses the notion of truth has enjoyed in the sciences, the arts, philosophy, and the common language.[1] The effort culminated in the formulation of the meaning of truth within the general theory of signs and sign interpretation.

It is Peirce's general theory of truth and some of its suggestive features, rather than the particular history of its development, that I propose to discuss.

The most familiar of his statements on truth are the following:

(A) The opinion which is fated to be ultimately agreed to by all who investigate, is what we mean by the truth, and the object represented in this opinion is the real. (CP 5.407)

(B) Truth is that concordance of an abstract statement with the ideal limit towards which endless investigation would tend to bring scientific belief, which concordance the abstract statement may possess by virtue of the confession of its inaccuracy and one-sidedness, and this is an essential ingredient of truth. (CP 5.565)

Of the two (B) is a more careful and complete definition. As for (A), which is from 'How to Make Our Ideas Clear' – it occurs in a discussion devoted to the meaning of 'reality,' not truth. And in this respect, the notion of an opinion 'fated' or 'destined' to be agreed upon by all who investigate it, while having a role in the analysis of truth, is particularly important; for it formulates the criterion of what is meant by a 'real' object. In speaking of an opinion 'fated' to be agreed on, Peirce explained that he meant only what is sure to come about and cannot be avoided. He intended nothing more mysterious or superstitious than to say that a coin tossed often enough is 'fated' to land heads one-half the number of trials. It meant 'that which is sure to come true' (CP 5.407n). And he added that even if by some perversity an 'arbitrary proposition ... be universally accepted' by the whole human race, still if investigation should be carried 'sufficiently far,' even by another race of investigators, the 'true opinion must be the one which they would ultimately come to' (CP 5.408).

As will be seen below, this does not commit Peirce to requiring of a true opinion that it is, or must be, one that will be ultimately agreed to by all who investigate. For it is not necessary to our understanding of what we mean by some opinion being true that it *will* be agreed upon by those who investigate it; rather, for Peirce, the true opinion is one that *would* be so agreed upon. And there is an important difference between stating that something *will* be the case and that something *would* be the case.[2] Furthermore it is possible to believe truly that a coin would land heads one-half of a large number of trials, and to know what I *mean* by this belief being true (viz, that in the long run the ratio of heads to tails would approximate to one-half) without having to venture on the heroic but futile task of dedicating a lifetime to realizing this 'fated' result.

We may remark, in passing, that the propriety of regarding (A) and (B) as definitive, or fundamentally representative of Peirce's conception of truth is, of course, open to question. We do not have Peirce's authority for this supposition; nor, on the other hand, do we have his disapproval. It has been in the scholarly writing on Peirce that these statements have come to receive special attention. With the growing interest in Peirce's thought that commenced in the 1920s, and perhaps centred in M.R. Cohen's *Chance, Love, and Logic* (1923), which made some of the important material accessible then to readers, these statements gradually acquired the status of 'received doctrine.'[3] In 1939, in *Logic: The Theory of Inquiry*, Dewey gave these same statements a conspicuous place, recognizing their importance and expressing his endorsement of them (345). But it is worth observing that while William James made generous acknowledgments to Peirce as the founder of pragmatism, it is Peirce's views on *meaning*, not *truth*, that interested him. He has very little to say about the latter (indeed nothing about truth in his most notable discussions of Peirce's ideas). And in his own accounts of pragmatic truth, although James frequently cites Schiller and Dewey, he at best makes only remote allusions to Peirce's doctrine of truth (James 1975a 106–7 or James 1975b 88–9). Thus while Peirce undoubtedly did hold a theory of truth in his own lifetime, this was not what gained him notice. His well-known renunciation of pragmatism for pragmaticism (about 1905, see CP 5.414) was, I suspect, among other reasons partly motivated by his consciousness of the disparity between his own conception of truth and the theories expounded by James and Schiller.

It is therefore with some diffidence that I will suppose one generally consistent theory of truth can be attributed to Peirce and concerning which (A) and (B) are, while sketchy, representative. We can derive, however, some comfort from the fact that the unpublished manuscripts exhibit developments of ideas that are frequently in harmony with (A) and (B), and sometimes closely similar in language and substance. In any case, the theory stated in (A) and (B) is original and interesting on its own account and has come to be regarded and discussed – whether altogether justly or not – as Peirce's theory. So we may pass from merely identifying the statements in question to an examination of them.

II

Three eminent philosphers, Russell, Ayer, and Quine, have registered severe criticisms of (A) and (B). Russell (1939, 135–56) had several objections. Concerning (A), which so far as I know he had come upon in Dewey's quotation in the *Logic* and not in Peirce's original article, he expressed a perplexity shared by many readers over the notion of an opinion 'fated' for agreement. (But we recognized earlier the innocuous sense in which, for Peirce, an opinion may be fated for agreement.) He was equally troubled about the word 'ultimately' in Peirce's saying that there will be an opinion 'ultimately agreed to by all who investigate.' He remarks that if 'ultimate' is intended chronologically, truth will depend on the opinion of the last man alive in his last minutes as the earth becomes too cold to support life. But Russell sees that it is more likely that what Peirce had in mind is 'a series of opinions, analogous to a series of numbers such as $\frac{1}{2}, \frac{3}{4}, \frac{7}{8}, \ldots$ tending to a limit, and each differing less from its predecessor than any earlier member of the series does' (Russell 1939, 145). (This may be questioned as we will see.) Russell also found strange the notion of a 'confession of inaccuracy' as essential to truth. Finally he wondered why Peirce believed there was ever an ideal limit towards which investigation would bring belief. He queried, Is this an empirical generalization or unfounded optimism?

Quine (1960, 23) regards (A) to be mistaken for several reasons, but for two especially. He questions the assumption of there being one 'ideal limit' or unique resulting theory as the outcome of continuously applied scientific method to belief.[4] He also comments: 'There is a faulty use of numerical analogy in speaking of a limit of theories, since the notion of limit depends on that of "nearer than," which is defined for numbers and not for theories.'

Ayer restates another of Russell's objections (Ayer 1968, 27). There are many particular events which, while candidates for true or false beliefs, are likely never to be recorded. Such, for example, would be beliefs as to the

125 Peirce on Truth

number of grains of sand on a beach, or, as Russell instanced, whether I had bacon for breakfast. Concerning such questions of belief Peirce's definition of truth appears to be inapplicable for 'it can hardly be supposed that even in the scientific millennium a complete historical record will have been kept of every particular event.' He, too, questions the apparent assumption of there being a continuous investigation of opinion, of an 'endless investigation.' A further difficulty raised by Ayer can be put this way: it is trivial to hold that there are no real things or laws that will escape being known by science, if it is assumed that investigation will continue into infinity. For if something *can* be known, it will 'eventually' become known if you allow the eventuality to come under a process of discovery that can continue forever (Ayer 1968, 26).

These various objections raise two kinds of questions which, especially in dealing with Peirce's ideas on truth, are inseparable although distinguishable. There is first a question of accurately discerning Peirce's intentions; distinguishing if possible certain philosophic assumptions and convictions that he may have held and which colour and affect what he says. And there is the problem of determining precisely what he does say and means to say. Before turning to what I think is the most serious error underlying the above criticisms, two observations which take these questions into consideration are worth making.

(1) Both Russell and Quine view (B) as suggesting that in the scientific investigation of belief, opinions or theories become ordered in a series analogous to a series of numbers tending to a limit. While this is a natural interpretation of what Peirce says, it is, I think, doubtful. Peirce does not, so far as I can find, expatiate on the point and so settle the matter for us. But I think it is quite likely that what he had in mind was not that a series of opinions (or theories) tends like numbers to a limit. His intention was rather to suggest that a continuous critical activity of investigation and testing would 'like a series of numbers tending to a limit' render some asserted opinion ever more precise and accurate (and ever more definitely true or false). The persistent and 'endless investigation' he says (in (B)), would 'tend to bring scientific belief' to an 'ideal limit.' And in the same passage he says that it is an *'abstract statement'* (not a series of opinions or theories) that is in concordance with the ideal limit. It is the 'endless investigation' of the belief represented by the abstract statement (i.e., the belief as stated) that tends towards the 'ideal limit.' It is, in short, *inquiry,* a complex of critical operations and activities shared by those who investigate, that is the continuous activity – a remark he makes shortly before stating (B) supports this interpretation. When we assert an abstract proposition, he says, the truth of it depends on 'that proposition's not professing to be exactly true.' We hope that in the progress of science, 'error

will indefinitely diminish, just as the error of 3.14159, the value given for π, will indefinitely diminish as the calculation is carried to more and more places of decimals.' But he adds, 'what we call π is an ideal limit to which no numerical expression can be perfectly true.' I understand this to suggest that continued 'calculation,' that is, continuous inquiry concerning some initially stated proposition, would indefinitely diminish its error. But this is not to claim that single opinions or whole theories can be arranged in a series like the decimal expansion of π, and made to approach a limit.

(2) Russell thought the main question to be raised concerning (B) was the assumption that an investigation would continue indefinitely and would result in ultimate agreement. The question is: Is this belief well founded? We could press even further and ask: Is the belief that investigation will continue indefinitely and produce an ultimate agreement itself subject to endless investigation and ultimate agreement? Is (B), Pierce's theory of truth, true on its own terms? Here I will say dogmatically, for brevity, that it is usually a mistake to demand of any theory or definition of truth that it must itself satisfy the conditions it legislates, that it should be included in its own scope. For a theory or definition of truth is not an expression of belief or a statement in the same sense as the beliefs or statements whose truth (or falsity) it is designed to formulate. But to return to Russell's objection.

It is clear that we cannot *know* (i) that an investigation will continue forever; or (ii) that an endless investigation will tend to an ultimate agreement. Peirce, however, *believed* both (i) and (ii) and gave reasons. He says there is no ground for disbelieving (i). He was, we may recall, a child of the nineteenth century, the century of 'progress.' We, living in the age of the nuclear bomb, may be somewhat less confident than was Peirce about the irresistable evolution of concrete reasonableness. He thought that intelligent life on earth would probably not continue indefinitely, but even if it disappeared, there is a likelihood of intelligent life existing in other parts of the universe. The community of intelligent inquiry is a cosmic disposition! His most empirical reason for affirming (ii) is the history of science. He thought that the history of science exhibited clearly a gradual tendency to increasingly systematic agreement over the acceptance and rejection of beliefs. We speak of the growth of scientific thought, and the word 'growth' reflects these characteristics of the cumulative and logical unification of beliefs. Crises that might occur in scientific theory he would have explained, I think, as taking place within a wider context of settled beliefs and practices. Revolutions in science occur within frameworks and with reference to a settled background of prior and collateral information that makes their advent, detection, and our appreciation of what is at stake possible.

127 Peirce on Truth

There is an assumption which Peirce acknowledges: the number of questions will increase with the increase in knowledge; but the number of answers will also increase. This, he says, is because the method of providing answers will also increase in efficacy and precision. And in apparent inconsistency with (A) and (B) he was to comment that some questions may never be answered and some opinions are already 'true'; they will have, thus, acquired the status of ultimate agreement (CP 8.43).

For his beliefs in (i) and (ii) Peirce was guided by metaphysical convictions concerning cosmic evolution and a progressive reasonableness persistent in the universe and effecting the realization of law and rational order. But, in addition, there was a form of reasoning, most readily exemplified in mathematics (especially in the theory of errors), which profoundly inspired Peirce and is essential to understanding his conception of endless investigation and the ideal limit (cf. Thayer 1968, 105–20). The idea undoubtedly also had metaphysical significance for him in relation to what he regarded as cosmic reason. This paradigm of reasoning he once referred to as the 'marvelous self-correcting property of Reason' which 'belongs to every science' (CP 5.579). The self-corrective process is the essence of rationality. He regarded it as 'one of the most wonderful features of reasoning and one of the most important philosophemes in the doctrine of science ... namely, that reasoning tends to correct itself' (CP 5.575).

III

The above are some of the empirical and philosophical considerations that Peirce adduces in behalf of (A) and (B) and other similar accounts of truth. All this is well known to readers of Peirce and requires no elaboration here. Still, I think it must be conceded that, excepting (1) some pages earlier, what we have just reviewed can scarcely be regarded as answering the objections raised by Russell, Quine, and Ayer.

What I now wish to suggest is that the foregoing are not *arguments* for Peirce's theory at all. His belief that ours is a rationally evolving universe, and that there is 'fated' to be an enlightened ultimate opinion agreed to by those who investigate, are no doubt intimately related to the doctrine of truth as he conceived it. Indeed these ideas help explain why Peirce formulated his statements on truth in the manner of (A) and (B); they explain the idiom. But it is a serious mistake to suppose, as Russell, Quine, and Ayer have done (and in this they are not alone) that (A) and (B) state conditions that must be fulfilled by any one opinion (or belief) if it is to be true. What (A) and (B) say is not that to be true an opinion must necessarily be 'ultimately agreed to by all who investigate' and be the 'ideal limit' of an 'endless investigation.' Rather, (A) and (B) specify what it *means* to say that an opinion is true; the definitions state what conditions are entailed and *would* be fulfilled if an opinion is true. And this is to explain what is meant by ascribing truth to an opinion.

There is an important difference to be observed. If it is snowing, and I see that it is and form the opinion 'It is snowing,' my opinion is true. I do not mean that it is true because (or only if) it represents the ideal limit of endless investigation. For since no such investigation has occurred, the opinion will then be either false or neither true nor false – and this is absurd. Nor do I mean that the opinion will *in fact* ever be the subject of endless investigation. And in this respect Russell and Ayer are correct; there are many particular events and beliefs that in fact will never be recorded nor ever come under investigation to result in an 'ultimate opinion.' The point, for Peirce, is that what we mean by the truth of the opinion (e.g., that it is snowing) is that what it represents and asserts *would* be affirmed *if* it were subject to endless investigation. We also mean that an endless number of expectations of a certain description and numerous confirming instances of it would be entailed by the truth of the opinion.

In developing the meaning of truth Peirce was, I think, following the same procedure he recommended for the explication of any general predicate (notably in 'How to Make Our Ideas Clear'). If we mean by 'hard' 'would not be scratched by many substances,' we can form the translation and explication of the predication 'x is hard' as follows: We specify a certain experimental situation, E (the apparatus and conditions for scratch-testing on x); the operation of testing, O; and the general result, R (x is not scratched by many substances in repeated tests). We may then say: 'x is hard' *means*: if E and O, then R (x would not be scratched). Peirce's accounts of the meaning of truth exhibit analogous distinctions. To say x is true (where x is an opinion or statement) will mean: if E (the conditions for investigating x) and O (the operation of endless investigation of x), then R (x is fated to be agreed on by all who engage in O). We can state this last part of the description in another way suggested by Peirce, namely: if E and O, then the number of assentors to x approaches unity while the number of dissenters approaches zero (cf. Thayer 1968, 90–2; 118; 129–32).

It is to be noticed that we are not here saying that if x is true, it is *because* an endless investigation results in an ever-increasing assent and diminishing dissent to x among investigators. For no such investigation need occur; and even if it could occur, the investigation does not *make* x true (for in the same way, the operation of scratch-testing on x does not make x hard). Thus, all we are permitted to say is that the above conditional description makes explicit what we *mean* when we ascribe truth to some x.

The above sketch of Peirce's method of explicating the meaning of truth should also indicate how the idea of probability and the notion of meaning containing a reference to the future have roles in Peirce's definition of truth. For R represents a statistical fact: the distribution of assent and dissent in investigations tending to a limit (and that distribution, for Peirce, is like the tendency of a coin when tossed to come up heads one-half the number of tosses, as we noticed previously in II). Thus the meaning of 'x is true' contains a reference to this future result (to innumerable confirming instances of 'is fated to be ultimately agreed on by all who investigate.'

It is thus necessary to distinguish the following.

(a) Why we think a belief is true. As to this, Peirce says we think a belief is true if we have no reason to doubt it, or it satisfies us, or it appears 'unassailable by doubt' (CP 5.416). A further reason would be that the belief, as a premise or a habit of reasoning, always (so far as we know) leads to satisfactory consequences, no dissatisfactory consequences having been encountered to follow from it.

(b) How we know a belief is true. Here we instigate tests and develop inferences. And while fallibilism advises us that we never know fully that any of our beliefs is true (CP 1.141), we may have good reasons for thinking that some and indeed most of them are true.

(c) What we mean by ascribing truth to a belief.

In criticizing Peirce's definitions of truth, Russell, Quine, and Ayer treat (b) and (c) as comprising one question. They proceed, without any justification that I can discover, to confuse (b) and (c); to regard Peirce's definitions of the meaning of truth (that is (c)) as involving and depending fundamentally on questions of (b), of how truth so defined could ever be known, or how we could ever come to know a belief is true. And because Peirce defines truth (for purposes of (c)) as a statement represented by an ideal limit of endless investigation, they object that this is not acceptable because *in fact* no actual investigation will be endless nor will it be likely to arrive at an ideal limit. This question of fact, of how such truths can be known is, if anything, possibly germane to (b); but it does not present a fatal difficulty for what Peirce discusses under (c). Peirce himself kept this distinction clear. He wrote:

I do not say it is infallibly true that there is any belief to which [a] person would come if he were to carry his inquiries far enough. I only say that that alone is what I call, Truth. I cannot infallibly know that there *is* any Truth.[5]

It is curious that his critics should fail to distinguish issues of (b) and of (c), for they are the very philosophers who have effectively pointed out on other occasions that the meaning of 'truth' and the discovery of truth are not at all the same.

IV

Dewey was right in noticing the importance of (A) and (B) as fundamental statements of Peirce's theory of truth. But there are other formulations which, while not free of obscurity, are equally important. Many of these have a place – as perhaps all things were envisioned by Peirce to have a place – within the theory of signs. And he later expressed regret that he had omitted from the famous articles of 1877–8 a discussion of signs and of pragmatism as the method of interpreting signs.[6] We find a change, not so much in the meaning of truth, but in the development of specialized procedures of analysis and of concepts for expressing that meaning. As an alternative locution to the set of ideal conditions and operations forming the conditional explication by means of E, O, and R (as we represented these above), Peirce seems to have become more interested in the clarification of truth as a function of a certain kind of signs and the conditions they must satisfy through a procedure of interpretation.

When Peirce treats the meaning of truth in his theory of signs, he emphasizes the role of propositions. Truth, he says, belongs 'exclusively to propositions' (CP 5.553). Propositions are Dicisigns, 'the kind of sign that *conveys information*' (CP 2.309; 2.320), and 'professes to refer' or relate 'to something as having a real being' (CP 2.310).[7] Every sign is a *representamen*, 'something which stands to somebody for something in some respect or capacity,' It 'addresses' and acts in this capacity, and further 'creates in the mind of that person' another sign, which Peirce calls the *interpretant* (CP 2.228).[8] A proposition is thus indicative and is equivalent to a sentence in the indicative mood (CP 2.315). The indicative function is explained by noting that a proposition contains a subject and predicate, and 'the subject is a sign; the predicate is a sign; and the proposition is a sign that the predicate is a sign of that of which the subject is a sign. If it be so, it is true' (CP 5.553).[9] The proposition as a whole is 'a sign which separately indicates its object' (CP 5.569). The *interpretant* of the proposition is the mental representation thus determined, 'it represents the proposition to be a genuine Index of a Real Object ... for an Index involves the existence of its Object' (CP 2.315).

A true proposition will be one whose *representamen*, as interpreted, is indicative (and an Index) of real objects as so represented. Peirce states the point with emphasis: 'Truth is the conformity of a *representamen* to its object, *its* object, ITS object, mind you' (CP 5.554).

Since signs are interpreted by signs, and the interpretant sign is another sign of the same object, the interpretant of a proposition will be another proposition. A true proposition is one for which 'every interpretation of it is true' (CP 5.569). Any necessary inference from a proposition is also an interpretant. It is by developing the interpretants of propositions and their issue in perceptual experience that acceptance or refutation is made possible and truth or falsity is discovered:

an interpretant of the proposition would, if believed, produce the expectation of a certain description of [a] percept on a certain occasion. The occasion arrives; the percept forced upon us is different. This constitutes the falsity of every proposition of which the disappointing prediction was the interpretant ... A true proposition is a proposition belief in which would never lead to such disappointment ... (CP 5.569)

The picture that emerges from this very crude and inadequate sketch of Peirce's view is of a system of knowledge and communication as a vast and intricate structure of propositions and sets of equivalently related propositions as partial interpretants. The distinctively pragmatic aspect has to do with the method of developing interpretations of propositions. And a prerequisite of the method is that it enables us to derive in a systematic fashion such interpretants of propositions as will eventuate in predictive judgments and percepts of a certain prescribed and definite character. For this method will be essential to determining the *meaning* of signs as well as to determining their *truth*. The derivations issue as 'precepts' and formulae 'prescribing what you are to *do* in order to gain a perceptual acquaintance with the object of the word' (CP 2.330) or proposition. While for valuable practical purposes the interpretations are limited, theoretically they are never exhaustive or complete; there is no last analysis or final interpretation. Or perhaps, if there is a final interpretation of propositional signs, it will not be found in scientific thought but, Peirce suggests, in the esthetically good, the ultimate achievement and highest translation of all thought (CP 5.594).

Thus while the theory of truth found in (A) and (B) was absorbed in a wider theory of signs, the theory of signs culminates in a transcendental esthetic. Moreover, it was from the esthetic that Peirce conceived his proof of the truth of pragmatism. The line of thought moved back and forth or took a cyclical course.

While I regard Peirce's outlook to be primarily a form of philosophic naturalism, there are certainly propensities to transcendentalism incurrent in and crossing through the naturalistic grain. Even (A) stands to (B) as truth conceived in the spirit of experimental naturalism to a *semiotic* (CP 2.227) with offshoots into idealism. Affinities with Royce could be cited, and Peirce once remarked on the close alliance of pragmaticism and Hegelian absolute idealism (CP 5.436). Still, the transcendentalism might be given a Kantian rendition, referring to interpretative principles whose application is not a departure from all experience, but whose relevance and function is not confined to any one limited portion or single aspect of experience, and which have the whole varied domain of experience for their object. In this sense the theory of signs appears to be a transcendental construction.

In inspecting Peirce's theorizing about truth we seem to have found an illustration of the main thesis of Professor Goudge's book. I am not certain that the conflict in this case is irresolvable. While Goudge's thesis may require certain refinements and modifications, it remains, nonetheless, a valuable guide to our understanding of Peirce's thought and the problems that affected its troubled and incomplete development. And we are enabled to appreciate all the more perceptively Peirce's remarkable acumen, imaginative power, and richness of philosophic vision – the rarest of capabilities but which never quite achieved complete focus or realization. That destined incompleteness which he believed to be a positive virtue and essential ingredient of truth in the unfolding of reason in communal and public forms was just as surely to be Peirce's private tragedy.

MANLEY THOMPSON

Peirce's Conception of an Individual*

In the *Thought of C.S. Peirce*, Thomas A. Goudge concluded that 'Peirce "declared for realism" in several different senses, but that only one of these falls properly within the orbit of his naturalism' (TP 98). The conception of an individual Peirce finally developed seems to me entirely compatible with the version of his realism that accords with his naturalism, and which Professor Goudge regarded as philosophically more tenable. As Goudge pointed out, it is one thing to maintain that generals have a 'different mode of being from that of singular existents,' and quite another thing to maintain, as Peirce also claimed, that they are physically efficient, having 'the kind of compulsion which a final cause exerts' (TP 101–2). I say nothing about the latter claim in this paper, as my concern is solely with Peirce's struggles to account for the existence of individuals after he had started with a position that denied it. For in its earliest form, his realism led him to the extravagant claim that 'being at all is being in general' (CP 5.349) and that nominalism in affirming the existence of individuals maintains a 'metaphysical figment' (CP 5.312). The incompatibility of naturalism with realism in this sense is obvious. But with the position Peirce later came to espouse, 'there is no difficulty about the truth that whatever exists is individual, since existence (not reality) and individuality are essentially the same thing' (CP 3.613). The recognition that individuals alone exist while 'laws and general types' have, in Professor Goudge's apt phrase, 'an ineluctable reality none the less' (TP 101–2) is thus the crucial step in Peirce's development of his realism to a point where it becomes compatible with his naturalism. To say that generals are physically efficient then marks a further step that makes the compatibility very difficult if not impossible to

---

*This paper was read in slightly different form under the title 'Names and Individuals in Peirce's Semiotic' at the meeting of the Charles S. Peirce Society in New York, 28 December 1975.

maintain. It is also a step, I believe, that is irrelevant to Peirce's account of individuals, although I do not argue for the irrelevancy here.

Within the limits of a single paper I have not been able to go into the various doctrines in Peirce's philosophy, notably his theory of categories, that are intimately related to his contention that existence is the mode of being of individuals.[1] But I do not believe that this limitation obscures the general philosophical points I want to urge with regard to his conception of an individual. Professor Goudge ends *The Thought of C.S. Peirce* by quoting William James' remark that Peirce 'is a gold mine of ideas for thinkers of the coming generation.' Peirce's account of individual existence as the other side of his realism, when the latter is taken in the sense that places it 'within the orbit of his naturalism,' seems to me a rich vein in this mine that still remains largely untapped.

I THE FIRST DEFINITION OF AN INDIVIDUAL

In his contributions to Baldwin (1901, 537–8) Peirce distinguished 'two closely related senses' of 'individual' as a technical term in logic. In the 'more formal' of these senses, 'an individual is an object (or term) not only actually determinate in respect to having or wanting each general character and not both having and wanting any, but is necessitated by its mode of being to be so determinate' (CP 3.611). He remarks that this definition does not imply 'Leibniz' principle of indiscernibles.' For it 'does not prevent two distinct individuals from being precisely similar, since they may be distinguished by their hecceities (or determinations not of a generalizable nature)' (CP 3.612). He adds: 'Although the principles of contradiction and excluded middle may be regarded as together constituting the definition of the relation expressed by "not," yet they also imply that whatever exists consists of individuals.'

Thus, an object X that is both P and not-P and neither P nor not-P is an imposssible and therefore nonexistent object (where 'P' represents *any* general character whatsoever). But if X is determinant with respect to P – is not both P and not-P but determinately one or the other – and therefore an existent object, it is also an individual by Peirce's first definition of 'individual.' Kant's law of specification,[2] which Peirce refers to in the next sentence, presents a difficulty for this definition. The law denies the possibility of an absolute determination of *infimae species* and prescribes that inquiry always proceed as if every species contains subspecies under it. As Peirce puts it, the law 'treats logical quantity as a continuum in Kant's sense, i.e., that every part of which is composed of parts' (CP 3.612). But if an object is always logically composite, it can never be determinate with respect to every general character.

135 Peirce's Conception of an Individual

Peirce does not illustrate the point in his article in Baldwin's *Dictionary* but he gives a clear illustration of it in a paragraph in an early paper (1870) on the logic of relatives. He begins by explaining the phrase 'logical atom.'

> The logical atom, or term not capable of logical division, must be one of which every predicate may be universally affirmed or denied. For, let A be such a term. Then, if it is neither true that all A is X nor that no A is X, it must be true that some A is X and some A is not X; and therefore A may be divided into A that is X and A that is not X, which is contrary to its nature as a logical atom. (CP 3.93)

Peirce thus thinks of 'logical division' in the Aristotelian sense of dividing genera into species. But like Kant he rejects the Aristotelian epistemology that traditionally accompanies this view of logical division. It is neither true, for example, that all animals are rational nor that none are, and it is true that some are and that some are not. The genus animal then divides into the species rational and irrational. However, while it is neither true that all men (all rational animals) are snub-nosed nor that none are, and it is true that some are and that some are not, it does not follow that the species man divides into the subspecies snub-nosed and non-snub-nosed. Being snub-nosed is an accidental property and its presence or absence does not mark a difference in essential nature. Since all differences between individual men are accidental, man is an *infima species* – a logical atom incapable of logical division. That rationality does and snub-nosedness does not mark a difference in essential nature is established in the Aristotelian tradition by the epistemological doctrine of an immediate grasp – an intellectual intuition – of real essence. Without this doctrine logical division would be infinite and scientific knowledge in the Aristotelian sense of *episteme* would be impossible.

Kant's law of specification is an inevitable consequence of his (and Peirce's) rejection of any form of intellectual intuition. With this law specification is terminated only with the complete determination of an individual rather than an *infima species*. But since the process is infinite the determination is never achieved. Peirce concludes, 'A logical atom, then, like a point in space, would involve for its precise determination an endless process' (CP 3.93). A name or other singular term is always capable of logical division. 'The second Philip of Macedon,' Peirce remarks, can be divided into 'Philip drunk' and 'Philip sober,' though 'we call it individual because that which is denoted by it is in only one place at one time. It is a term not *absolutely* indivisible, but indivisible as long as we neglect differences of time and the differences which accompany them. Such differences we habitually disregard in the logical division of substances' (CP 3.93). There is thus a distinction between 'the absolutely individual and that which is one in number from a particular point of view,' a distinction that is 'shadowed forth' in the two words *individual* and *singular*.

Peirce concludes the paragraph by remarking that 'as those who have used the word *individual* have not been aware that absolute individuality is merely ideal, it has come to be used in a more general sense.' He explains in a footnote that in this more general sense, whatever exists is individual. But this view of existence seems clearly false. 'For whatever lasts for any time, however short, is capable of logical division, because in that time it will undergo some change in its relations. But what does not exist for any time, however short, does not exist at all' (CP 3.93 n1). We thus reach the conclusion that whatever exists is general, i.e. is capable of logical division. 'So far,' Peirce says, 'there is truth in the doctrine of scholastic realism' (CP 3.93 n1). But, on the other hand, 'all that exists is infinitely determinate, and the infinitely determinate is the absolutely individual. This seems paradoxical, but the contradiction is easily resolved. That which exists is the object of a true conception. This conception may be made more determinate than any assignable conception; and therefore it is never so determinate that it is capable of no further determination' (CP 3.93 n1).

These remarks show that as early as 1870 Peirce was aware that the extreme form of scholastic realism he had espoused scarcely two years earlier, when he proclaimed 'that being at all is being in general' (CP 5.349), was in need of modification. There must be some sense in which individuals as well as generals are real. Why speak at all of being in general and not just of being if there is no contrasting sense of being – being in particular or the being of individuals? In his 1870 footnote Peirce's answer is that being in general contrasts with the ideal of being actually determinate as a logical atom, but not with the reality of being infinitely determinate. Being at all is general in the sense of never being actually determinate in all respects, and also individual in the sense of being infinitely determinate – of always being susceptible of further determination. While absolute individuality is merely ideal and as such does not exist, whatever exists is individual because infinitely determinate. In fact, it is just because being is thus individual that it is also general– never actually determinate in all respects.

Peirce's footnote, then, seems to provide an easy dialectical resolution of the nominalism/realism issue. Properly understood, each side is in its way true. Controversy arises only when realists take nominalists to be affirming the existence of absolute individuality, or nominalists take realists to be affirming the existence of a universal actually determinate in all respects. But this resolution does not fit the issue as Peirce had already drawn it in 1868. To the best of my knowledge, he never returned to this 1870 view of nominalism and realism. But the view points up some fundamental problems for his philosophy, problems with which he continued to struggle for the rest of his life, and it is instructive to see just where the resolution fails.

The resolution assumes that the cognitive process of absolutely determining an individual is the same as the process of discovering truth. But in 1868 Peirce had clearly distinguished the two processes. The absolutely determinate individual stands as an 'ideal first' at the beginning of the process of discovering truth or 'the real,' while truth stands at the end as 'that which, sooner or later, information and reasoning would finally result in' (CP 5.311). There are thus two points to be determined, the absolute beginning and the final end of inquiry. With intellectual intuition ruled out, both points must be conceived as limits of a process of inference.

In 1868 Peirce saw no difficulty in maintaining a fundamental difference between the two points even though both must be conceived as limits. It is one thing to seek an absolutely first premise and quite another to seek a final conclusion. The latter is simply that conclusion which the community of inquirers 'will always continue to re-affirm' (CP 5.311). It is thus a limit that may be actually attained while the process of seeking it continues. It has been attained when it will be re-affirmed by all further inquiry, which continues indefinitely. The process of seeking an absolutely first premise, on the other hand, is not a matter of communal inquiry. Each individual begins with his own private determination of what he takes to be real, and he can hope to achieve an absolute (a truly objective) determination only as he identifies his inquiry with that of the community – in other words, only as he seeks a final conclusion and not an absolutely first premise. But he may think of the information he has at any given time as deriving from his own 'previous cognitions which are less general, less distinct, and less vivid' than his present cognitions. He thus comes to the limiting notion of an 'ideal first, which is quite singular, and quite out of consciousness' (CP 5.311). While an absolute beginning, then, like a final conclusion, is conceived as a limit, it is unlike a final conclusion in being merely ideal and therefore unattainable. 'By an ideal,' Peirce explains, 'I mean the limit which the possible cannot attain' (CP 5.311 n).

Being in general thus contrasts not only with the ideal of absolute individuality, but also with being as infinitely determinate. 'There is nothing,' Peirce observes, 'to prevent our knowing outward things as they really are, and it is most likely that we do thus know them in numberless cases, although we can never be absolutely certain of doing so in any special case' (CP 5.311). Being in general is thus always susceptible of further determination only in the sense of always being subject to test by further inquiry. But all further testing may produce further determination only in the sense of further confirmation and not in the sense of further information. Nominalism and realism, as Peirce originally opposed them, cannot be reconciled by claiming that all that exists is infinitely determinate and therefore both individual and general. For only the general is infinitely determinate in a sense that does not make it merely ideal and hence unreal by Peirce's conception of reality.

Before we turn to Peirce's later views about individuals, it will help if we look briefly at his 1870 position from a slightly different standpoint. He was not concerned in 1868 with the logical analysis of singular as distinct from general statements, and he saw no problem in the fact that with the realism he proposed all statements on analysis turn out to be general. But by 1870 his work in the logic of relatives had confronted him with the need to distinguish singular statements. The need can be met only if there is some sense in which individuals as well as generals are real. His primary concern in his 1870 footnote is thus not a reconciliation of nominalism and realism but rather a reinterpretation of his realism so that in a sense individuals also become real. We noted above that in the text of his 1870 paper he proposed a distinction between the singular and the individual, a distinction he did not make in 1868.[3] The singular, he decides in 1870, is not the absolutely individual for it is not absolutely indivisible. It is only 'that which is one in number from a particular point of view' (CP 3.93).

The second Philip of Macedon is a singular from the point of view of a human lifetime. A singular statement such as 'The second Philip of Macedon is mortal' is true because the predicate applies to the subject during every moment of its existence. But 'The second Philip of Macedon is drunk' is not true of this same subject. It is true only of a certain temporal part of this subject, and is therefore true of a different singular. Accurately expressed the statement has the form 'The second Philip of Macedon at time t is drunk.' But then two statements, one affirming that Philip at t_1 is drunk and the other affirming that Philip at t_2 is sober, are about different singulars. Yet we say that they are about the same subject because 'in the logical division of substances' we 'habitually disregard' 'differences of time and the differences that accompany them' (CP 3.93).

While the determination of a singular achieved only by disregarding differences is clearly not a complete determination of the real, the other side of the coin is that this partial determination is possible only as long as there are real differences to be disregarded. That there will always be such differences Peirce takes in 1870 as equivalent to the claim that 'all that exists is infinitely determinate' and hence 'the absolutely individual.' This claim in turn he takes as equivalent to his realistic thesis that being at all is being in general, i.e.

139 Peirce's Conception of an Individual

always subject to further determination. We make a statement singular, then, when we disregard determinations beyond those we have specified, and we keep the same statement general when we seek further determinations.[4]

Looked at in this way Peirce's 1870 position may have some prima facie plausibility. But the difficulty remains that with the conception of reality he had presented in 1868, a determination of the real by further inquiry remains fundamentally different from a determination of absolute individuality by logical division.

I return at long last to Peirce's comments on the individual in his Baldwin's *Dictionary* article. I remarked that Kant's law of specification poses a difficulty for the first definition of an individual Peirce considers, which is essentially, with an important qualification I will mention in a moment, the definition he accepted in 1870. Kant's law proclaims as a regulative principle that we must always proceed as if logical division were endless. Peirce observes:

Though this law is only regulative, it is supposed to be demanded by reason, and its wide acceptance as so demanded is a strong argument in favor of the conceivability of a world without individuals in the sense of the definition now considered. Besides, since it is not in the nature of concepts adequately to define individuals, it would seem that a world from which they were eliminated would only be the more intelligible. (CP 3.612)

Peirce recognizes here that absolute individuality as the infinitely determinate is antithetical to his 1868 conception of reality. In so far as there are always differences we must disregard in order to achieve a determination of the real, it is not true that 'there is nothing to prevent our knowing outward things as they really are.' A world containing individuals in the sense in question cannot be a thoroughly intelligible world – a world in which everything is ultimately knowable. Only the approximations of a determination of absolute individuality are ultimately knowable and therefore intelligible. The determination itself as the ideal limit of the approximations, being unattainable, can never be made intelligible.

But there is a more fundamental sense in which such a world is unintelligible. Peirce, as we noted, remarks that the definition does not imply the identity of indiscernibles. One and the same absolute determination may be true of more than one individual, since individuals 'may be distinguished by their hecceities (or determinations not of a generalizable nature)' (CP 3.612). Even at the ideal of absolute determination specified by the definition, individuals remain unintelligible with respect to the relation of identity. An object x absolutely determined by all generalizable determinations is not determined as identical or not identical with an object y of which all the same generalizable determinations are true. The object is both $= y$ and $\neq y$. It is an impossible and therefore nonexistent object.

An individual by this first definition of it is thus unintelligible not only in the sense that it is an unattainable limit of logical division, but also in the sense that its individual identity would not be determined even if the limit were attained. Peirce arrives at this second sense of unintelligibility by adding a qualification to the definition as he gave it in 1870, and by introducing the notion of hecceities, or determinations not of a generalizable nature.

The qualification is that an object is an individual provided that it is 'necessitated by its mode of being' to be actually determinate in the sense specified in the definition (CP 3.611). This qualification opens the way for a modification of Peirce's extreme realism of 1868. According to his footnote of 1870, being in general contrasts with being as individual, not as one mode of being contrasts with another, but as being at all contrasts with an unattainable ideal. As we have noted, his 1870 attempt to modify his realism failed. It is inconsistent with his 1868 position to maintain that since both the general and the individual are limits of a cognitive process, being at all is both general and individual. With the distinction of modes of being, it becomes possible to maintain that both generals and individuals *are* (have being), each in their respective mode of being, even though only generals *are* as limits of a cognitive process. The reference to hecceities, or determinations not of a generalizable nature, implies that individuals have a mode of being which is not that of the limit of a process. Individuals may be distinguished by their hecceities and still be identical in all their generalizable (all their conceptual) determinations. They are thus identical by the first part of the definition in question, i.e. identical as the limit of a process of logical division, but they are not necessitated by their mode of being to be identical.

The difficulties Peirce finds with the definition thus boil down to the following. (1) The process of logical division is supposed to be the process by which an individual is rendered intelligible, yet since such a process is endless it cannot provide the means of making an individual thoroughly intelligible. (2) Since the mode of being of an individual is not that of the limit of a process of logical division, the attainment of such an ideal limit would still leave an individual indeterminate with respect to its individual identity.

The reference to modes of being and the reference to hecceities prepare the way for the second definition of an individual Peirce considers in his Baldwin's *Dictionary* article. The term 'hecceity' is of course borrowed from Duns Scotus, but it is not necessary for the present discussion to consider Scotus' use of the term or the extent to which Peirce followed it. Peirce's parenthetical phrase 'determinations not of a generalizable nature' will suffice as a preliminary explanation of what he meant by 'hecceity.'

II THE SECOND DEFINITION OF AN INDIVIDUAL

'Another definition,' Peirce says, 'which avoids the above difficulties is that an individual is something which reacts. That is to say, it does react against some things, and is of such a nature that it might react, or have reacted, against my will' (CP 3.613). If one remarks that this definition makes individuals unintelligible, Peirce replies that the remark does not provide an objection to the definition but is actually 'a merit, since an individual is unintelligible' in the sense in which the definition says it is. He also observes that the definition is 'the stoical definition of a reality,' and adds that 'since the Stoics were individualistic nominalists, this rather favors the satisfactoriness of the definition than otherwise' (CP 3.613).

There is truth after all then in nominalism. The previous definition is wrong if it is taken as implying that an individual should be thoroughly intelligible. In 1868 the definition misled Peirce so that he rejected individuals altogether and held that being at all is being in general. It also was in back of his unsuccessful attempt two years later to make individuals intelligible by assimilating them to generals. An individual, according to the second definition, is unintelligible in that 'a reaction may be experienced, but it cannot be conceived in its character of a reaction; for that element evaporates from every general idea' (CP 3.613).

While an individual is thus unintelligible in its mode of being, it does not follow that it is unintelligible (indeterminate) with respect to its individual identity. For 'everything whose identity consists in a continuity of reactions will be a single logical individual. Thus, any portion of space, so far as it can be regarded as reacting, is for logic a single individual; its spatial extension is no objection' (CP 3.613). Individual identity consists, then, not in logical indivisibility, but in the continuity of reactions in any portion of space so far as it can be regarded as reacting. The important point is that the continuity in the process of reacting differs essentially from the continuity in the process of logical division. Peirce explains the difference in his Lowell Lectures of 1903.

A 'perceptual judgment,' he says in the sixth lecture, 'is the cognitive product of a reaction' (CP 5.156). As such it is simply given and is beyond logical criticism. It thus seems like the ideal first of cognition he rejects in 1868 as nonexistent because unattainable. He then supported his rejection by arguing that just as there is no absolutely first distance Achilles must travel before overtaking the tortoise, so there is no absolutely first cognition one must have before having a given cognition. The assumption of an absolutely first cognition is no more required to account for the fact of cognition than the assumption of an absolutely first distance to be traversed is required to account for the fact that Achilles overtakes the tortoise. The 'logical difficulties' in the two cases, Peirce claims, 'are identical' (CP 5.263).

In 1903, however, he says that the case (of an absolute first in cognition) 'is quite different from that of Achilles and the tortoise' because Achilles is not required 'to make an infinite series of distinct efforts' (CP 5.157). A process of logical criticism – reasoning back to absolutely first premises or dividing a concept into logical atoms – must proceed as a series of distinct efforts of logical analysis. The series is continuous in the Kantian sense that 'every part ... is composed of parts' (CP 3.612), but the parts in this case are parts of the process itself and not of the space and time in which the process occurs. There is thus no way in which the process can be completed except by an actual procession through an infinite series of distinct efforts. In the case of physical motion, on the other hand, there is Kantian continuity, not with respect to parts of the process itself, but only with respect to what are represented as parts of the space and time in which it occurs. Physical motion is represented as having parts, not as proceeding from one part of motion to another, but only as proceeding from one part of space to another in a certain part of time. As Peirce remarks in his seventh Lowell Lecture, 'Achilles does not have to make the series of distinct endeavors which he is represented as making' (CP 5.181). He is represented as making a series of endeavours only as he is represented as moving through distinct parts of space and time and not as having a continuous unbroken motion. But there is no way of representing logical criticism or division except as a series of distinct efforts of logical analysis. It cannot be represented as a continuous unbroken process.

Reaction, including that between perceiver and perceived, is a process of physical motion. It is continuous, not in what Peirce called the Kantian sense of actually having definite parts every one of which in turn has parts, but in the sense of actually having no definite parts at all. In 1903, in a marginal note to an earlier writing, he wrote that 'a continuum, where it *is* continuous and unbroken, contains no definite parts ... its parts are created in the act of defining them and the precise definition of them breaks the continuity' (CP 6.168). The identity of an individual, then, as consisting in an unbroken continuity of reactions, is not the identity of a logical atom defined as the ideal limit of a process of logical division.

Peirce goes on to claim in his Baldwin's *Dictionary* article that with his second definition of an individual 'there is no difficulty about the truth that whatever exists is individual, since existence (not reality) and individuality are essentially the same thing' (CP 3.613). Existence is then the mode of being of an individual, while reality, as distinct from existence, is the mode of being of a general. An object qua existing is simply an object qua reacting, an object we may experience as a reaction but can never cognize in its individuality. To conceive of an object thus cognized is to conceive of it cognized as absolutely determinate with respect to every general character. But an object thus determinate is the ideal limit of logical division. It is not only nonexistent; it is also unreal, for it can never be attained in the process of cognition. The real is always something that can be so attained, but it is then always general because always subject to further determination. The dictum of Peirce's early realism remains true if 'reality,' understood as marking but one mode of being, is substituted for 'being': reality at all is reality in general.

Peirce continues in his Baldwin's *Dictionary* article, immediately after the above quotation, 'whatever fulfills the present definition [i.e. the second definition of an individual] equally fulfills the former definition by virtue of the principles of contradiction and excluded middle, regarded as mere definitions of the relation expressed by "not".'

The former definition, it will be recalled, had two clauses. An individual is that which is (1) absolutely determinate with respect to every general character and (2) necessitated by its mode of being to be so determinate. According to the second definition, an individual is that which reacts. In its mode of being it is thus a reaction and hence unintelligible, since a reaction 'cannot be conceived in its character of a reaction.' But this is not to say that we have no concept of reaction, that we cannot conceive reaction in general. We conceive it as a dyadic relation between two individuals, each of which we conceive of as being determinate in itself and not just as a term of the relation. But then we conceive of each as absolutely determinate. For if we begin by saying that x is R to y but then go on to say that x is indeterminate with respect to some general character P – that x is both P and not-P and neither P nor not-P – we make x an impossible object and therefore one we cannot conceive of as being R to y. But if x is impossible it does not exist. It cannot be the individual that is R to y. It is therefore necessitated by its mode of being – its existence – to be such that we must conceive of it as being absolutely determinate even though such determination remains for us an unattainable ideal. The unattainability here is no obstacle, since to conceive *of* an individual as being absolutely determinate is not to conceive *it* – to have a concept of it – as so determinate. I conceive *of* Philip as being an individual, as being absolutely determinate, although according to the way I conceive him (according to my concept of him) he is, say, neither drunk nor sober, but indeterminate in this and infinitely many other respects.

The two definitions of an individual are thus reconciled conceptually. Both yield the same concept of individuality. The reconciliation is achieved by taking the principles of contradiction and excluded middle as 'mere definitions' of the conceptual (logical) relation expressed by 'not.' They are not taken in their Aristotelian sense as principles of being qua being – principles true equally of every mode of being. Although the principles jointly imply, as Peirce claims in presenting the first definition of an individual, that 'whatever exists consists of individuals' (CP 3.612), they do not imply that we can ever actually determine a given individual in the way that we must conceive *of* every individual as being determinate. It follows that an absolutely determinate individual is unreal according to Peirce's definition of reality, but there is no contradiction if reality and existence are distinct modes of being.

But there remains an important respect in which the two definitions are not reconciled. While both yield the same concept of individuality, they do not yield the same concept of individual identity. We noted that the first definition does not imply the identity of indiscernibles. After explaining how the definitions can be reconciled conceptually, Peirce continues,

As for the principle of indiscernibles, if two individual things are exactly alike in all other respects, they must, according to this [second] definition, differ in their spatial relations, since space is nothing but the intuitional presentation of the conditions of reaction, or some of them. But there will be no logical hindrance to two things being exactly alike in all other respects; and if they are never so, that is a physical law, not a necessity of logic. This second definition, therefore, seems to be the preferable one (CP 3.613)

With these remarks Peirce concludes his discussion of the two definitions. The remarks hark back to those, earlier in the same paragraph, in which he proclaimed that according to the second definition individual identity consists in the continuity of reactions in any portion of space so far as it can be regarded as reacting. The continuity here, we noted, is the unbroken continuity of physical motion and not what Peirce called Kantian continuity, which characterizes logical division and provides the concept of individual identity according to the first definition. But there is an important question we passed over and one that Peirce does not touch on except for his concluding remarks. How is a portion of space to be determined, and how are we to decide when it can be regarded as reacting? That 'space is nothing but the intuitional presentation of the conditions of reaction, or some of them' suggests that the answer lies in a Kantian view of space and time, but this is all Peirce has to say on the question in his article. More is certainly required to show the acceptability of the second definition.

145 Peirce's Conception of an Individual

In a writing c 1885 Peirce discusses the question in the terms we have used here and provides a rather different answer from that suggested above. He notes that an instant of time 'is, in itself, like any other instant,' and analogously for a point in space (CP 8.41). Yet 'dates and positions can be approximately distinguished. And how are they distinguished?' 'By *intuition*, Peirce says, is Kant's answer. Kant therefore 'distinguishes Space and Time from the general conceptions of the understanding and sets them off by themselves under the head of intuition.' 'But,' Peirce continues, 'I should prefer to say that it is by volitional acts that dates and positions are distinguished.'

Volition, as Peirce explains it here, 'does not involve the sense of time (i.e. not of a continuum) but it does involve the sense of action and reaction, resistance, externality, otherness, pair-edness' (CP 8.41). In a volitional act one exerts an effort against something and experiences a sense of reaction, a sense that 'something has hit me or that I am hitting something.' Volition thus 'has an outward and an inward variety, corresponding to Kant's outer and inner sense.' The continuity of space is broken and likewise that of time. A position is distinguished from other positions and an instant from other instants. But they are not distinguished by intuition, as Peirce says Kant distinguished them. They are distinguished only by a brute sense of reaction which, unlike a Kantian intuition, is strictly noncognitive.[5]

A portion of space is determined, then, by the brute experience of a reaction, and when we cognize what we have experienced we regard the portion of space so determined as reacting against us and therefore outer (external). There are thus two factors, one purely experiential and noncognitive and another intellectual and cognitive. The experiential factor 'does not involve the sense of time'; it is 'not of a continuum.' The portions of space it determines are *hecceities*, determinations not of a generalizable (i.e. not of a cognitive) nature. The intellectual factor, on the other hand, consists in the cognitive awareness of a continuity of reactions in a portion of space through a portion of time. The identity of an individual is thus intelligible when space and time are conceived as true continua, as realities with no definite parts. Distinct portions of space and hence distinct individuals are determined by experiences of reaction that break the continuity of space and of time. Since the parts of space and time are indefinite, no part in itself being distinguished from any other, each part is itself a continuum. No matter how small or how large the portions of space and time determined by an experience of reaction, there is a continuity of reactions that constitutes the identity of an individual.

A volitional act as involving the experience of reaction thus replaces Kantian intuition as the experiential factor that distinguishes positions and dates. Space and time themselves remain purely cognitive and conceptual and our concepts of them are not to be set off from other concepts in Kantian fashion under the head of intuition. There remains, however, a parallel to Kant in that spatial relations at a given time suffice to distinguish individuals otherwise the same in all their cognitive determinations. We conceive of an individual as that which can be in but one place at one time. Peirce's remark in his Baldwin's *Dictionary* article that 'space is nothing but the intuitional presentation of the conditions of reaction, or some of them' is misleading in so far as it suggests an unqualified Kantian position. Instead of 'intuitional presentation' Peirce might have used a phrase like 'conceptual apprehension.' But while this would have muted the suggestion of similarity to Kant, it would, I believe, have been misleading in another respect, and one more serious within the context of Peirce's article.

As we have noted, Peirce held that his second definition of an individual 'seems to be the preferable one' because unlike the first definition it implies the identity of indiscernibles. The implication fails with the first definition because the definition 'does not prevent two distinct individuals from being precisely similar, since they may be distinguished by their hecceities (or determinations not of a generalizable nature)' (CP 3.612). But if an individual according to the first definition is the ideal limit of logical division, and if logical division proceeds to this limit (proceeds to infinity) only if we do not 'neglect differences of time and the differences that accompany them' (CP 3.93), an individual determined at this ideal limit would be determined with respect to all its spatial relations at a given time. The identity of indiscernibles seems to be implied by the first definition after all. But the implication here is no more than the truism that we cannot distinguish conceptually what we conceive of as indiscernible. This truism is not the principle that indiscernible individuals are necessitated by their mode of being to be identical.

According to the second definition, an individual is unintelligible in its mode of being. It is not cognized in its individual existence, it is only experienced as brute reaction. But it is cognized in its continued existence as a continuity of reactions in a portion of space. Its individual identity consists in this continuity. Distinct individuals therefore cannot be the same in all their conceptual determinations, for they will differ at least in their spatial relations at any time of their existence. Yet the spatial location of an individual is not merely another of its conceptual determinations, as the phrase 'conceptual apprehension of the conditions of reaction' might imply. The cognition of individual identity as spatial continuity presupposes an experience of reaction that noncognitively (immediately) determines a portion of space as something external within which the continuity obtains. It was with this point in mind, I suspect, that Peirce in the context of his article chose the phrase 'intuitional presentation of the conditions of reaction,' even though he did not mean to

147 Peirce's Conception of an Individual

imply that the concept of space (and perforce with it the concept of time) should be set off from other concepts under the head of intuition.

III SINGULAR TERMS

As we have noted, in 1870 Peirce distinguished singular from general statements by holding that while a singular term or proper name is always capable of logical division, we none the less regard it as denoting what is singular. For 'in the logical division of substances' we 'habitually disregard' 'differences of time and the differences that accompany them' (CP 3.93). What is singular is thus merely 'that which is one in number from a particular point of view' and not that which is absolutely individual. With this position, a singular term is no more than a general term treated as an approximation to the complete determination of an individual. As a general term it must have meaning – rational purport – specifiable with respect to the ultimate outcome of inquiry, in accord with Peirce's pragmatic theory of meaning and reality. But this pragmatic meaning can have nothing to do with approximating the complete determination of an individual, since the limit of logical division and the ultimate outcome of inquiry are approximated in fundamentally different senses.

In his later writings Peirce struggled to give an account of singular terms that accords with his second definition of an individual. In his paper 'What Pragmatism Is,' which appeared in *The Monist* in 1905, he says, 'it must be admitted that pragmaticism fails to furnish any translation or meaning of a proper name, or other designation of an individual object' (CP 5.429). But such a name or designation 'has a certain denotative function peculiar, in each case, to that name and its equivalents'; and 'every assertion contains such a denotative or pointing-out function' (CP 5.429). This function is of course what Peirce usually calls elsewhere the function of an index. It is not a function performed only by proper names, but also, more frequently, by pronouns. Every assertion contains such a function, because if it does not contain a proper name as its subject it will contain the equivalent of a pronoun or quantified variable. 'All men are mortal' is analysable as 'If anything is human *it* is mortal'; and 'Some mortals are not men' is analysable as 'Something is such that *it* is mortal and nonhuman.'

With this analysis particular statements do not represent the logical division of general terms and thus approach the determination of an individual. The fact that all humans are mortal while some mortals are nonhuman does not imply that 'x is human' is a closer approximation to the determination of an individual than 'x is mortal.' A universal statement, Peirce says in his Lowell

Lectures, differs from a particular statement in that its subject is *hypothetical* rather than *indesignative*. The subject is hypothetical because it allows 'any singular to be substituted for it that fulfills certain conditions without guaranteeing that there is any singular that fulfills these conditions' (CP 5.154). The subject is indesignative when the statement 'means that a singular of the universe might replace this subject while the truth was preserved, while failing to designate what that singular is' (CP 5.154). The difference is thus one of existential import and not one of approximating the designation of an individual. Both statements fail in their respective ways to achieve such a designation, and neither comes closer than the other to achieving it.

Peirce speaks of these failures as 'imperfections' in the subject of a statement. 'A subject,' he says, 'which has neither of these two imperfections is a *singular* subject referring to an existing singular collection in its entirety' (CP 5.154). He does not explain this statement, and I will end with the suggestion that by the crucial phrase 'an existing singular collection in its entirety' he means 'a continuity of reactions in a portion of space so far as it can be regarded as reacting.'

PART TWO: THE ASCENT OF LIFE

FRED WILSON

Goudge's Contribution to Philosophy of Science

Thomas A. Goudge once described *The Ascent of Life* thus:

This book presents issues arising from the synthetic theory of biological evolution, the mid-twentieth century descendant of Darwin's theory of 1859. An investigation is made of a number of linguistic and conceptual shifts that have taken place since Darwin, and, along with this, some of the modes of explanation and some metaphysical implications of the theory are explored. It is argued that evolutionary explanations do not all conform to the deductive-nomological model as this is exemplified in the explanations of the physical sciences. There are, in addition, other types, such as 'integrating' and 'narrative' explanations, which do not involve laws or make positive predictions. It is further argued that among the metaphysical implications of the theory, the ideas of 'direction', 'novelty', 'progress', and 'purpose', suitably specified, have a place. So do the issues of human evolution and its probable or possible future. (Goudge 1976a 100–1)

The book he thus describes received the Governor-General's non-fiction award in 1962. This award was appropriate. The book was both the first detailed exploration of the synthetic theory of biological evolution by a philosopher of science and also an important early critique of the standard logical empiricist accounts of science then current, a critique along the same lines as those associated with such names as Kuhn (1970b) and Lakatos (1970), but (one should add) a critique rather more sober in what it rejected and certainly more sober in its tone. More conversative also, in its willingness to venture into metaphysics. An earlier generation had been willing to explore the metaphysical implications of science – one thinks of Alexander, Russell, Broad – but almost everywhere positivism had triumphed and such explorations were thought to be explorations of the meaningless. Kuhn and Lakatos are hardly free of this positivist bias. Goudge's willingness to proceed into these areas now normally left to those fascinated by the pompous pseudo-religiosity of such thinkers (though that word comes to the lips only with difficulty here) as Teilhard de Chardin is perhaps due to his being situated in Canada and more particularly in Toronto, where, due largely to the influence of G.S. Brett, the earlier British tradition of metaphysical realism and an interest in science had been swamped neither by positivism nor by 'linguistic analysis.' Goudge's *The Ascent of Life*, then, is a major critique of some aspects at least of the positivist account of science, but, in its caution and sobriety is also a critique of the more radical anti-positivist positions of Kuhn and Lakatos. It is this special situation of Goudge in the philosophy of science that I shall describe. Naturally, *The Ascent of Life* will be the major text. But Goudge elsewhere develops further important points in the philosophy of science and in methodology. I shall also bring out some of these.

In his description of *The Ascent of Life*, Goudge draws our attention to *narrative* and *integrating* explanations and contrasts these with the sort of explanations found in the physical sciences and which philosophers of science have called deductive-nomological and hypothetical-deductive explanations. We can, I think, best begin our attempt to situate Goudge by examining this contrast. Two models of explanation have, I believe, dominated the thinking of philosophers of science. The right of these two models to claim to be standards of excellence need not be disputed even if one disputes their right to dominance, as does Goudge. Both these models derive from classical mechanics. Both are rightly characterized as 'deductive-nomological.' The first model is that of *process*. This derives from Newton's explanation of the behaviour of the solar system in terms of the gravitational force-function. We know the system is closed. We have a complete set of relevant variables. (In the case of the solar system: the masses, positions, and velocities of the objects in the system.) And we have a law describing how these variables interact and how the system develops over time. From a knowledge of the values of all the variables at one time, we can deduce, and thereby explain and predict, the values of the variables at any other time. And we can also deduce what would happen were values of the variables to change or be changed in any way. Accordingly, we know all the past and future behaviour of the system, given that we know its present state, and we know all the things that would occur were it different from what it is. Clearly, such knowledge provides us with a model of complete understanding, so far as individual processes are concerned (cf Bergmann 1957 ch 2; Wilson 1981 sec 2). The other model of explanation deriving from Newton is that of explanation, not of individual facts and processes, but of laws. This model is that of *abstract axiomatic theories*. In

153 Goudge's Contribution to Philosophy of Science

classical mechanics some sorts of systems are described by the gravitational force-function, others by Hooke's Law. Newton's Laws apply indifferently to these systems; they are generic, rather than specific. They apply to all systems of a genus, and succeed in doing this by abstracting generic features common to the laws describing the behaviour of the specific systems. The generic laws of classical mechanics are logically so interconnected that they can be arranged into a deductive system with Newton's three laws, and the composition law of vector addition of forces, as axioms (cf Wilson 1969b). These laws place limits upon the form which process laws take in specific systems. They must, for example, conform to the Law of Inertia. This means that when a scientist examines a specific sort of system not previously investigated, then the generic laws of the theory predict that there is a law, for the specific sort of system, there to be discovered, and that it will have such and such a form; the scientist is therefore guided in his research by the theory, and made reasonably certain by it that his search will be successful. Kuhn, in particular, has directed our attention to this feature of theories (or paradigms, as Kuhn calls them (cf Kuhn 1970b ch 4, 5; Wilson 1981 sec 3, 7). Abstractive axiomatic theories are epistemically desirable by virtue of their capacity to unify our knowledge. And they are desirable from the point of view of the researcher by virtue of their capacity to guide reasearch.

When discussing explanation, philosophers of science have tended to stress either laws for which, as in process laws, explanation and prediction amount to the same thing, i.e., the first model (cf Hempel and Oppenheim 1965), or axiomatic theories, i.e., the second model (cf Carnap 1939 ch 3). But there do in fact exist intermediate and complicating cases. Consider a generic law applied to a system of some species within that genus. Unlike a process law, it would not make determinate predictions; it will place generic restictions upon what will happen in the particular system, but will not predict specifically what will happen. The knowledge that results from applying a generic law to a particular system will be determinable and gappy, imperfect relative to what process knowledge yields.[1] However desirable generic laws are from the viewpoint of explaining laws, they are less desirable than process-type laws from the viewpoint of explaining particular facts. Nonetheless, if laws yielding determinate predictions are not available, one will have to explain particular occurrences as best one can, with gappy generic laws – though of course, scientists will aim to fill those gaps by discovering more specific laws. Another complicating sort of case can be seen when one recognizes that the use of process laws presupposes knowledge of what happens on the boundary, either that the system is closed, where nothing enters the system, or, more generally, the boundary conditions are such and such, describing what crosses into the system from the outside. To explain what happens upon the boundary one may have to invoke laws describing generically quite different sorts of systems. Thus, in order to explain why heat starts to cross the boundary of a thermodynamic system at a certain time one may cite the behaviour of something other than a system merely thermodynamic, e.g., the behaviour of an experimenter. The explanation of what happens in the thermodynamic system will involve laws explained by two quite different theories. Moreover, explanations of particular facts in terms of laws from generically different areas may use only generic, i.e., gappy or imperfect, laws from those areas.

Through their focusing upon the two dominating models, philosophers have tended to ignore explanations in terms of gappy laws and explanations in terms of laws which are part of quite different theories. There has been a tendency to concentrate on ideal cases at the expense of, if you wish, the empirical facts of explanation, i.e., that many explanations fall short of these ideals and are more complicated than the ideals suggest. It was Goudge's virtue to direct our attention to these empirical facts. The narrative and integrative explanations are types Goudge has located in the various explanations deriving from the theory of evolution. Both embody the features we have mentioned that have been ignored in the concentration on the ideal cases.

Goudge draws our attention to the existence of gappy or inexact or quasi-laws (AL 123, 75), and, in connection with this, the role of statistics (AL 123). He points out that such laws can have explanatory force, even though such explanations do not conform to any simplified deductive model of explanation (AL 123), one in which explanation and prediction are symmetric (AL 126). To see the sense of all this, consider a particular example (Scriven 1959a). Let 'F' denote a certain species of animals, let 'G' denote the property of being in the vicinity of a forest fire, let 'H' denote a degree of stamina above a certain amount, and let 'I' denote the property of outrunning the fire. Then

$$(1) \qquad (x)[(Fx \,\&\, Gx) \supset (Hx \equiv Ix)]$$

represents that a member of that species will survive the fire if and only if it has degree of stamina H. Degrees of stamina are, as 'degree' indicates, ordered, so that to have stamina below a certain amount is to exclude having it above that amount. There is, therefore, *one and only one* H such that (1) holds. This permits us to form a definite description of H, and refer to it as 'the degree of stamina sufficient to permit animals of species F to outrun endangering fires,' which we may abbreviate as 'H^*.' Since, *as a matter of fact*,

$$(2) \qquad H = H^*$$

155 Goudge's Contribution to Philosophy of Science

(1) yields

(3) $$(x)[(Fx \,\&\, Gx) \supset (H^*x \equiv Ix)]$$

which represents that an animal of the species F will outrun the fire just in case it has the stamina that is sufficient for outrunning it. (3), like (2) and (1), is synthetic. Now, we may not know an identificatory hypothesis (2); it may be that all we know is

(4) $$(\exists f)(\mathcal{H}f \,\&\, f = H^*)$$

where $\mathcal{H}$ is a genus of which the various degrees of stamina are the species. (4) is the same as

(5) $$(\exists! f)[\mathcal{H}f \,\&\, (x)[(Fx \,\&\, Gx) \supset (fx \equiv Ix)]].\,^{2}$$

If we know (5), then we know (3) obtains. But in the absence of an identificatory hypothesis (2), we have no knowledge of the sort (1). Law (3) compared to (1) is gappy. Moreover, it cannot be used to predict. Without knowing specifically what H^* is, i.e., without an identificatory hypothesis of the sort (2), we cannot determine whether H^* is present in an animal independently of that animal actually surviving a fire; for, in order to determine a is H^* we must deduce

(6) $$H^*a$$

from

(7) $$Fa$$
(8) $$Ga$$

and

(9) $$Ia.$$

With (1), we can use the presence of F, G, and H to predict I, but when our knowledge is limited to (3), we cannot predict I, that an animal will survive a fire. We can, however, use (3) to explain, even though we cannot use it to predict. Given (3), then (7), (8), and (9) permit us to deduce (6). We can now use (7), (8), and (6) as initial conditions, and the law (3) to deduce and thereby explain (9). Thus, with a gappy law like (3) we can sometimes explain ex post facto where we cannot predict.[3] There is a way of generating predictions from a law of the sort (3), however, by using statistics. If we know (3), and observe in a sample of things $F \,\&\, G$ that 30% of them are I, then we can infer 30% of them are H^*. From this we can generalize to the population, that 30% of things F & G are H^*. We can then predict that of F's endangered by a forest fire, 30% of them, by virtue of their having sufficient stamina, will survive the fire. Explanation and prediction by means of gappy laws are thus not the neat sort of business they are with the process model. Nor as neat as that suggested by deductive-nomological model as it usually presented (e.g. Hempel and Oppenheim 1965). One must note, however, that even with such ex post facto explanations from gappy laws as sketched above, there is deduction from laws, viz, gappy laws. Scriven (1959a, 1959b, 1962) has used examples similar to that analysed above to conclude that since explanation is not always symmetrical with prediction, therefore such explanation does not involve deduction from laws. As the above account shows, this argument of Scriven is invalid. Like Scriven, Goudge (AL 125–6) appeals to such examples to argue that the symmetry of explanation and prediction does not everywhere hold, but, unlike Scriven, Goudge does not draw the conclusion that there is explanation without deduction from laws.

Goudge (AL 65) introduces integrating explanations as follows and cites an example:

The combined inquiries of paleontology and historical geology, supplemented by various other special disciplines such as taxonomy, permit the broad outlines of the history of life on the earth to be reconstructed. As a result, what has taken place can be summed up in a general historical statement, an excellent example of which is the following:

'Living organisms are all related to each other and have arisen from a unified and simple ancestry by a long sequence of divergence, differentiation, and complication from that ancestry.'[4]

Call the explanation Goudge quotes '(E).' Two things should be noted about (E). First, it is a generality: it is a statement about all organisms and, therefore, about all species. Second, it involves mixed quantification: it asserts that for any organism, there are ancestors, and it places certain restrictions upon the nature of these ancestors, asserting that all individuals have a common sort of ancestor, that these earliest ancestors were simple, that the intermediate stages involve ancestors increasing in complexity over time, and that this increasing complexity is a matter of divergence, differentiation, and complication. We can think, very crudely, of this law as having the form

$$(x)[Gx \supset (\exists f, y)(\mathscr{F}f \& fy \& Ryx)].$$

Clearly, such a law will make determinable rather than determinate predic-

157 Goudge's Contribution to Philosophy of Science

tions. It predicts the *sort* of thing to expect, but not specifically what will be. Thus, one can, upon the basis of (E), expect homologies to exist. For example, the wing of a bat, the flipper of a whale, the leg of a horse and the arm of a man are all structurally similar (i.e., homologous) while being functionally quite different. (E) predicts that such homologies will exist but not that any one of these determinate homologies exist (AL 66). To that extent it predicts, though the predictions are not exact (AL 68). For that reason, explanations based on (E) are not of the deductive-nomological or process type, in which determinate predictions are made. Moreover, ex post facto explanations are often possible using the law (E).[5] Goudge (AL 68) mentions such biogeographic questions as, Why are tigers found in India but not Africa? Why are armadillos found mainly in South America? Using palaeontology and historical geology, one can infer a great deal about the nature and distribution of fauna in the past; and one can also infer the existence of such things as land bridges and deserts, the facilitators and barriers to migrations. If one now makes some futher plausible hypotheses about initial conditions, such as that of dispersal from a common centre, then one can use (E), the law of phylogenetic descent, to construct explanations answering questions of the sort mentioned above. These explanations will turn upon the existence of chains of ancestors. This is the crucial fact (E) introduces. But since (E) makes only determinable statements, the specific details of the processes involved cannot be reconstructed. Here the contrast is to process explanations of the sort Newton provided for the solar system, where, if the present state is known, the process law permits us to infer the whole state of the system at any previous time. Moreover, although (E) is generic, it is not here being used to explain laws but to explain individual facts. So the axiomatic model does not apply either.

Narrative explanations (AL 71ff) are resorted to when one attempts to explain such facts as that amphibians left their aquatic environment to become land-dwellers. The explanation of the mentioned fact is in terms of the amphibians developing limbs, and this development, and the consequent ability to live on land, '*seem, paradoxically,* to have been adaptations for remaining in the water' (AL 71; italics added) during periods of drought, when it was a favourable adaptation to migrate from an evaporating pool to some other body of water. Gradually, adaptations to non-aquatic food (insects, plants, etc) would result in land-dwelling fauna (AL 71). A narrative explanation attempts to answer the question, Why?, asked in the context of an assertion about some event E to the effect that E happened because s, and in particular where the assertion that E because s is in some way paradoxical[6] or at least implausible, judged to be so (and therefore judged as not reasonably to be accepted) on the basis of plausible and normally accepted assumptions (cf

Dray 1957, 164–6). The narrative explanation provides a *sketch* of a *fuller* explanation of how it could be that E happened because of s. It thereby removes the paradoxical aspect of the explanation (cf Hempel 1965, 428–30), showing that the assertion of E because s is, after all, not unreasonable, or, if you wish, intelligible. The narrative explanation does not rely upon any simple generality connecting E and s. This is, of course, precisely where explanations of this sort seem to conflict with the normal deductive-nomological model. The generality that would come to mind as connecting E and s when one asserts that E because s is, 'Whenever an event of the same kind as s occurs then an event of the same kind as E occurs.' But this generality is, normally, depending upon how one instantiates the free variable 'same kind,' either false or tautological (cf Scriven 1959b 471–5). If the former, then it could not be explanatory, and if the latter then it is best construed as a principle of inference rather than a premise, and it again could not be explanatory (AL 77).[7] Rather than citing a generality connecting in some simple way the explaining and explained events, the narrative explanation consists of a story which links the two events by describing other intervening events and other relevant constraining or boundary conditions (AL 74). Now, we must recognize that Newton can do the same, with his process knowledge of the solar system. In the latter case, however, it is a matter of deduction from a law; in the narrative explanation no such deduction occurs. Or, rather, the known laws impose *determinable* conditions upon the sorts of events that can be supposed to intervene, and, while the events interpolated by the narrative satisfy these determinable conditions, they are in fact given more determinate characterizations than can be deduced from known laws and initial conditions. As Goudge puts the point concerning these gappy or quasi-laws, 'The proper conclusion seems to be that law-like statements in selectionist theory serve not as premises for deductions but as components in evidential systems which render *explicanda* intelligible or rationally credible' (AL 123). Thus, narrative explanations make assertions about events that are more specific than those that can be deduced using the known laws. These assertions say more, in other words, or have greater factual content, than can be justified in terms of available lawful knowledge. So narrative explanations inevitably contain a hypothetical element.

The sequence constructed is, of course, no more than a *possible* explanation at this stage; and it is worth noting how much use is made throughout of the possibility-expressing auxiliaries, 'could', 'would', and 'might'. They indicate the conjectual nature of many of the conditions postulated by the pattern.[8]

In process explanations one can deduce all the specific links of the chain of causes and effects linking two events; in narrative explanations some of the links are literally missing, filled in only by guesswork, or hypothesizing.[9] One introduces narrative explanations in those cases where our knowledge is gappy and imperfect, determinable rather than determinate (AL 74). The aim of narrative explanations is to answer a question to the effect: how possibly could that have happened? (cf Dray 1957, 158ff) Such questions can, of course, be answered *best* by citing a process law and using it to deduce and explain the whole process. But most often process laws are not available; and certainly this is so in evolutionary theory. One must answer the question, then, as best one can, relying on the imperfect knowledge available. The result might well be a 'likely story' (AL 75), the generic framework of which is constituted by the imperfect laws one knows and the specific details of which are but plausible hypotheses.

The aim is to make the sequence of events intelligible as a relatively independent whole ... the explanatory pattern ... forms a coherent or connected narrative which represents a number of possible events in an intelligible sequence. Hence the pattern is appropriately called a 'narrative explanation.' (AL 75)

The 'intelligibility' here is, one must emphasize, *scientific* intelligibility, intelligibility in terms of laws.

We shall return to this last point directly, but a couple of other points should be made first. These, too, have to do with the inapplicability of the process and axiomatic models.

Narrative explanations are about processes. But in these explanations reference must be made to relevant boundary conditions. For example, in the case we examined, of the development of land-dwelling fauna, reference must be made to climate conditions, for it is just such conditions which meant there was accessible to the amphibians an environmental niche less hazardous to them than remaining in water (AL 73–4). The description of these boundary conditions is derived, not from selectionist theory or from biology, but from other, contributory, sciences such as historical geology. This feature is one that narrative explanations share with integrating explanations.[10] Since the sciences involved are generically different, the idea of their being unified axiomatically does not make much sense. For, that would involve more generic laws of which all these sciences would be specific cases.[11] The unity of science is indeed an ideal, but as yet we are far short of this ideal, and the facts Goudge is here directing our attention to make that perfectly clear.[12] Moreover, the relevant laws will often be imperfect. From them we will be able to deduce statements of individual fact, but the characterizations will not be specific but by means of definite descriptions (cf 'H^*' above). Since the use of definite descriptions presupposes that the descriptions are successful, in Russell's sense, the explanations will presuppose laws in a way an explanation using only specific characterizations will not presuppose laws.[13] Thus, 'an explanatory pattern of the sort we are examining is not so much a segment of a separate causal line stretching back into the past, as it is a portion of an intricate network having an enormous number of cross-connections' (AL 75).

Another complicating feature is the historical nature of any evolutionary laws. A law is historical just in case that, in order to predict the future, one needs to know not just the present state of the system but also the past states of the system.[14] (Mathematically, process laws of the sort Newton discovered are represented by differential equations, ordinary or (if the objects are fields) partial, while historical processes are represented by integro-differential equations of the sort first investigated by Volterra (cf Nagel 1961, 288–90).) Goudge points out that, at the most generic level of evolutionary theory, there are laws of the sort (E) (AL 61). These laws unify more specific laws, and thereby explain them. (We must, of course, contrast using (E) to explain individual facts, and using (E) to explain laws.) It is at the more specific level that historical features enter.

> They [i.e., historical explanations] are not just incidental items which reflect the theory's 'undeveloped' nature. On the contrary, they are essential to it, and have their foundation in the fact that organisms are literally historical creatures. Their history is built into them. Hence no scientific account of organisms can be satisfactory if it abstracts them from their concrete history. (AL 62)

This feature must not be confused with the fact that, so far as we know, evolution has occurred only once (AL 122–3). The laws of evolution have only one instantiation; but others are at least logically possible (AL 175). Neither historical uniqueness nor historicity precludes lawfulness. But both, and especially the second, considerably complicate the task of explaining, and this makes both the process and axiomatic models far distant ideals rather than useful accounts of the facts of evolutionary explanations (AL 16).[15] Finally we should note one further complication that distinguishes biology from physics, where close approximations to the two ideals are in fact to be found. Physics deals with systems where the relevant objects and their relations to other objects are well understood; in other words, in physics we can often reasonably claim to have a complete set of relevant variables and a knowledge of relevant boundary conditions. In biology, we cannot. This is, of course, simply part of the fact that in this area our knowledge is imperfect. As a result in particular of not knowing boundary conditions, biological systems, either organisms or populations, must be treated as 'open systems' rather than as 'closed systems' of the sort the physicist has managed to discover (AL 33-4). Again, the process model, exemplified by Newton's explanation of the solar system, is inapplicable (AL 16).[16] Goudge summarizes these points, concerning the historicity and openness of biological systems, in his remarks that

> ... an evolutionist has to deal with organisms and populations undergoing continuous, non-repetitive changes. These organisms and populations are open systems, ceaselessly interacting with the environment and with each other. Hence he can hardly hope to produce anything akin to astronomical forecasts. (AL 125)

Again, however, none of this means that laws are irrelevant in explanation; on the contrary, they remain essential to such explanation. Goudge is not arguing an anti-scientific thesis, one holding that laws are not necessary for there to be explanations. Rather, all he is arguing is that explanations in biology, which *are* scientific, do not conform to either of the two models to which philosophers of science have fixed their attention.

Now, this last point might be challenged as follows, by reference to narrative explanations. I have said that Goudge is not denying that explanations in evolutionary theory aim at scientific intelligibility, intelligibility in terms of laws. On the other hand, I have also pointed out that, according to Goudge, narrative explanations contain specific details which are not established fact but merely plausible hypotheses. These hypotheses are essential to the explanation; if they are deleted, the 'likely story' that explains 'how possibly' disappears, and with it *that* explanation. It would therefore seem that, for Goudge, the intelligibility he claims biology aims at is not simply scientific. For if, as Aristotle said (*Poetics* ch 10), poetry deals with the possible while history deals with the actual, then a narrative explanation, containing as it does an essentially hypothetical element, seems to be more poetry than history, at best a guess at an explanation rather than an explanation.[17] Partially, this is a quibble about 'explanation.' One starts to explain E by saying that E because s. This latter, however, turns out to be paradoxical or at least not reasonably acceptable by virtue of its apparently being highly improbable. The narrative explanation, in its hypothetical aspects, does not add further details of the explanation of E in terms of s but it *can* explain why the explanation of E because of s is not straightway to be rejected because of its apparently paradoxical or improbable nature. A mere story, based only on guesswork, may suffice to do the latter job of explaining away the apparent unacceptability of the explanation that E because s. Removing apparent paradox, rendering a proposed explanation plausible, can be an important aspect of marshalling evidence in its favour. Such moves may be particularly important in writing aimed at a non-professional audience: it is that sort of audience that would be most likely to bring to bear assumptions, which, until challenged, generate just that air of paradox the narrative explanation aims to remove. But narrative explanations can have a role with respect to a professional audience also, and once we see what this is, we will not, I think, be any longer inclined to say that the existence of 'likely stories' in narrative explanations in biology in any way argues that biology aims at an 'intelligibility' of facts which is other than scientific intelligibility, intelligibility in terms of laws. Briefly put, the role of the 'mere story' is to propose a research programme, one aimed at discovering laws, or, better, at removing imperfections; the aim of the story, then, is to bring about, in the long run, scientific intelligibility. But the point should be made in more detail.

A narrative explanation, one which renders 'E because s' plausible by suggesting certain intermediate steps which could possibly account for that change, has the merit of embodying a set of *detailed* suggestions about how the gappy explanation 'E because s' could be improved. A narrative explanation contains a set of detailed hypotheses which, if subsequent research verifies them, will provide a less imperfect explanation of E. For the practising scientist, then, a narrative explanation is both an explanation, but one which is gappy or imperfect, and a sketch of a research programme for eliminating the acknowledged gaps. This last point means, of course, that a narrative explanation, even in those parts where it is guesswork, is not mere poetry, not a mere story. It is, rather, the construction of an imagination subject to the discipline of science. The story-teller works within an established theoretical framework, viz, that constituted by the known generic or imperfect laws in the area. Within these guidelines he constructs his story. This story part of the narrative explanation is the proposal of a research programme, and as such has a legitimate place in scientific discourse (cf Lakatos 1970; Wilson 1981 sec 7, 10, 11). For, after all, if our knowledge is gappy then at least idle curiosity and perhaps also our pragmatic interests will urge us to eliminate those gaps, and the means we use so to improve our knowledge is research: research is just that activity by which we reduce the imperfection of our explanations.[18] Typically, those who defend the deductive-nomological model concentrate on the product of research, viz, explanations in the sense of their model, while ignoring the process, and the role theories play in guiding that process.[19] Kuhn, in particular (1970b ch 3), has emphasized the interaction of theorizing and research practice; essentially, this is what is involved in Kuhn's notion of paradigm (Kuhn 1970b ch 5; Wilson 1981 sec 7). Goudge is bringing out much the same point, in his idea of narrative explanations: that theoretical scientific discourse involves both elements directed at deductive explanation (the generic laws at least implicitly or contextually cited),[20] and elements directed at research (the same laws which limit the possible relevant hypotheses; and the further hypotheses which constitute the story part of the explanation, hypotheses the ultimate acceptance or rejection of which will depend upon the outcome of the research).

But why choose one imaginative narrative filling-in rather than another? Is this arbitrary? Or is there some principle at work here also, that picks out one story, one set of hypotheses or guesses, as more plausible than another? Goudge argues that, sometimes at least, such hypothesis-selection in the area of evolution is not arbitrary but guided by certain principles.

Goudge argues that inferences about whether a factor or character is an adaptation make implicit use of the concept of 'purpose' or 'proper function' (AL 97–8). There is, however, no reason to suppose either '(i) that the internal structures and processes of organisms are [literally] purposive, and (ii) that the overt, macroscopic behaviour of organisms, other than human beings, is purposive' (AL 194). Rather, 'What we are entitled to affirm is that these structures, processes and behaviour show an *ostensible* design or plan' (AL 194; his italics). Goudge (AL 196ff) argues in detail that to speak thus of purpose has none of the metaphysical implications that such philosophers as Bergson have claimed to find in the fact of evolution, nor any of the objectionable features of vitalism and finalism (AL 80ff). By 'ostensible design,' Goudge has in mind simply something along the lines of Broad's definition of 'teleological.'[21] Suppose that we have a sort or kind of system. Systems of this kind have parts which are arranged, act, and interact in certain ways. Systems of such a kind are teleological, exhibit ostensible purpose, if (a) the parts and their actions are such as might have been expected *if* the system had been constructed by an intelligent being to fulfil a certain purpose he had in mind, and if (b) when systems of this sort are further investigated under guidance of this hypothesis, hitherto unnoticed parts, arrangements among parts, and processes of interaction are discovered, and these are found to accord with the hypothesis. Broad asserts that

... the most superficial knowledge of organisms does make it look as if they were very complex systems designed to preserve themselves in the face of varying and threatening external conditions and to reproduce their kind. And, on the whole, the more fully we investigate a living organism in detail the more fully does what we discover fit in with this hypothesis. (Broad 1925, 83)

The narrative explanation we have been considering concerns the adaptation of amphibians to land by the development of limbs. The narrative consists of a hypothetical account of how this adaptation proceeded. Considerations based upon the hypothesis of ostensible purpose enable the narrator to construct a plausible story, a set of hypotheses worthy of further research. But the use of 'hypothesis' in 'hypothesis of ostensible purpose' is misleading. One is not so much reasoning on the basis of an hypothesis but rather in terms of an analogy: natural selection and human intelligence are both *problem-solving capacities*[22] – the former solves problems involved with the maintenance of individuals and the perpetuation of their kinds, the latter problems of technology – and thinking of the situation faced by biological systems we can form hypotheses about how natural selection might solve the survival problems confronting those systems. Human problem-solving capacities provide a *model* for the problem-solving capacities of natural selection. As the quote from Broad correctly indicates, the model has been a fruitful source of hypotheses, though, as Goudge (AL 195) points out, maladaptations as well as adaptations occur among the products of evolution, so that it is easy to overstate the case for ostensible design. However, this does not vitiate the utility of the purposive model, since the disanalogies between the human and evolutionary cases serve to explain the existence of the maladaptations. In particular, there is nothing in the human case that corresponds to the mutations of selectionist theory: the mutations always occur randomly relative to the needs of organisms, which means maladaptations are reasonably to be expected. When using the model or analogy to suggest hypotheses, the disanalogies must be taken into account. But this is, as Sellars (1967) has pointed out, something which, while often neglected, is one of the more important features of the use of models in scientific theorizing.

One source, then, of the hypotheses appearing in narrative explanations is, according to Goudge, models and analogies. Goudge is not the first to insist upon the role of models and analogies. N.R. Campbell was perhaps the first (1957 ch 6). Some, such as Nagel (1961 ch 6 sec 1), point out that such models are not essential to explanations. From this it is concluded that such models are not essential to theories, that they are 'merely heuristic.' But if one thinks of research as a feature of science that should as equally be accounted for as explanation, then one cannot ignore those things that systematically generate hypotheses which research comes to verify. For this, the heuristics of science – theories and models – are essential. In drawing our attention to this point, Goudge is pursuing a theme that has been taken up of late by other philosophers of science such as Lakatos (1970 sec 3a, 3b). Hesse (1963), too, has emphasized the role of models in providing reasons for hypothetically extending a theory in one direction rather than another, and using these hypotheses

165 Goudge's Contribution to Philosophy of Science

to guide research (cf Sellars 1967, 344 ff). Since an extension of a theory to include an as yet untested new hypothesis may have no basis in the evidence supporting the original theory, the choice of one extension over another must, Hesse concludes (1963, 39–40), be made for reasons internal to the theory, from which it would follow that if the use of models to select such an extension is the appropriate way to additional empirical content, then the model must be an integral part of the theory extended. Here one must qualify: all that follows is that the model is an integral part of the research tool, the hypothesis-generator, that also includes the theory. It does not follow that the model is essential to the lawful and theoretical structure which is the product of the research; to this extent Nagel is surely correct. Moreover, Hesse's claim that models are essential is too strong; not all valuable extensions flow from analogue models. Other heuristics are possible (cf Lakatos 1970 sec 3c; Buchdahl 1970), which, in fact, may be essential to the theory (as the product of a research process) in the way a model is not. For example, the composition law of classical mechanics directed one to look for Neptune when the perturbations of the orbit of Uranus did not turn out as had been predicted.[23] Still, as Hesse – and Goudge – point out, models are *often* important to research. We must note also that probably just as often they are a hindrance to research. Thermodynamics could only develop within certain limits under the assumption that caloric is a fluid. And (as we saw Goudge point out) models can contain disanalogies as well as analogies, in other words, features which would yield false rather than fruitful hypotheses if they were to be used in hypothesis-generating. We need, then, a general theory of heuristics: a theory of when it is reasonable to rely upon theories and models as generators of hypotheses to be relied upon as research guides, and when it is reasonable to reject one such set of heuristics (or part of such a set) and replace it with another. Goudge does not sketch a general account of heuristics, however. Nor does he sketch a theory of when the use of a model to generate research-guiding hypotheses is more or less reasonable. But in this last, Goudge is not without company. Kuhn, alone, has sketched such an account (it is in terms of a paradigm's 'puzzle-solving' capacities),[24] and even that is a sketch. Still, as I shall argue below, Goudge does develop a view of mind, the evolutionary view, which can provide a framework for accounting for the development and change of science, and in particular for the procedures of evaluation that determine the acceptability and unacceptability of models and paradigms. We shall see that certain insights of Peirce that Goudge has discussed show how such an account might proceed.

What emerges out of Goudge's discussion of modern selectionist theory of evolution, then, is an account of science which is in many ways similar to that of Kuhn. But Goudge omits those more baroque twists which, one fears, are responsible for much of the popularity Kuhn's theories now enjoy. Thus, in Goudge there is none of that equivocal language one finds in Kuhn about such things as the incommensurability of theories – dangerously equivocal language since it makes it appear – though Kuhn himself (1970a sec 3) protests this – that Kuhn shares the irrationalism of such philosophers as Feyerabend.[25] Nor, when Goudge attacks the standard models of explanation, does he draw the anti-scientific conclusions of such philosophers as W. Dray. Thus, when Goudge discusses narrative and 'how possibly' explanations, he is at pains to insist upon the contextual importance of laws (AL 76). The 'intelligibility' such explanations aim at[26] is *scientific* intelligibility. Dray, in contrast, insists that laws are essentially absent. When one attempts to show that an event in history is intelligible by offering a narrative explanation, the intelligibility is a matter of the behaviour being in accordance with rules, not in its being subsumable under laws.[27] Dray's case has been sufficiently demolished by others (Hempel 1965 sec 10; Addis 1975 chs 3, 6, 9; Rosenberg 1976; Brodbeck 1963); we need not go into it here. Suffice it to say that Dray's anti-scientific account of man was never part of Goudge's views. Goudge would, it goes without saying, defend the acceptability of narrative explanations in history. I suppose the significant difference for Goudge between narrative explanations in biology and those in history would lie in the source of the hypotheses constituting the story the narrative relates. In biology the source of hypotheses lies in such analogies as that of the purposive model. In history it is presumably the imaginative capacity for understanding others that we have come to call *verstehen* (cf Abel 1953). One might note, however, that if Goudge is correct, then, in biology, narrative explanations propose research rather than terminate it; in history, in contrast, narrative explanations satisfy curiosity, and thereby terminate research. But we can take this as evidence that historical narratives, unlike evolutionary narratives, are written (as Horace (*Ars Poetica* 333) says of poetry) to amuse and to edify, and are accepted upon those terms. To that extent, history is not scientific. But that is no problem, for then history is not explanatory either, but poetry. In any case, however, for Goudge there is no essential break between science and history, between man and nature: the same types of explanation apply in both. Indeed, for Goudge, man is so much a part of nature that the basic framework for understanding man is an evolutionary view of mind (cf Goudge 1973; AL 205–11). This is not to say that Goudge is a metaphysical behaviourist (he rejects Watson's position on this point (1973, 139)). Rather, if we take his approval (1973, 140) of Mead's social behaviourism as a guide, then he presumably would opt for some form of psycho-physiological parallelism (cf Mead 1934 sec 1(4)–(6)). Upon this view, mind would be reckoned emergent, but such a view is, of course, not anti-scientific (cf Bergmann 1957 ch 3). And the idea that the mental is emergent is one which Goudge seems to find congenial (1973, 146).

Biological evolution is a large-scale process which is both cumulative and involves the successive addition of new qualities and processes. A part of this process is the evolution of mind: this is a sub-process exhibiting the same features of being cumulative and of generating novelty. Scientific research is a part of this latter process, a sub-sub-process or rather organized group of sub-sub-processes which also exemplifies the features of being cumulative and of generating novelty. The emphasis that Goudge, in his discussions of explanation and of scientific theorizing, places upon the on-going process of scientific research is an emphasis no doubt deriving from his views on mind. This placing of scientific research within an evolutionary context is something Goudge shares with C.S. Peirce, about whom his first book was written. Both share what is a central theme of the pragmatists' philosophy of mind, the belief that mind is a kind of behaviour, 'a species of conduct which is largely subject to self-control' (CP 5.419). Goudge adopts '... the metaphysical view that there is an objective evolutionary process about which we have reliable knowledge, and that man, the cognizing subject, together with his knowledge, in no way "transcends" this process' (AL 207). The process of scientific research can itself be explained scientifically. And this point has important consequences when one comes to develop an account of theory-evaluation. When Goudge discusses theories and their role in explanation and research in *The Ascent of Life* he says very little concerning the evaluation of theories. He has, however, made some important remarks on this topic in *The Thought of C.S. Peirce*.

Peirce wrote a good deal about the problem of induction, and worked out an answer to this problem along the lines of what have come to be known in the current literature as vindications of induction. This answer was in terms of the idea that induction is a self-correcting process of inference. According to Goudge, the crucial feature of Peirce's answer consists of his connecting the notion of a random sample with the notion of a sustained process of inquiry:

This self-corrective tendency of induction is ... due to the fact that induction is based on samples drawn at random from the subject matter under investigation, and that each sample is free to turn up with the same relative frequency. Consequently, the objective constitution of the subject-matter *must ultimately reveal itself*, if scientific inquiry is unchecked (TP 189–90; cf Madden 1960a)

The repeated use of an inductive procedure will discover the limit, or regularity, in the long run – or, at least, it will do so *if there is any ascertainable limit or regularity there to be discovered.* But if one aims at truth – as science does – and if there is no regularity in nature, then the persistent use of induction will *not* ultimately reveal the truth. The pragmatic vindication of induction can succeed only if it presupposes that nature is regular, that limits exist, a presupposition it gives no reason for accepting. This criticism vitiates all pragmatic vindications of induction, as has recently been once again pointed out (Levi 1965). Goudge had already made the point against Peirce in 1950:

In the last analysis, his [Peirce's] appeal is to the criterion of inconceivability. We cannot imagine a world, he declares, in which inductive inference would be systematically misleading, i.e. a world totally devoid of order. Although this may be true in the sense that we cannot imagine the details of such a world, nevertheless Peirce has sufficiently stressed the fact that 'inconceivability' is no proper guarantee of truth or falsity. It cannot serve as the ground for the doctrine in question ... It seems clear, then, that Peirce's doctrine that order is a necessary attribute of existence remains a methodological postulate which the inductive reasoner has to adopt; and is thus a material assumption about the constitution of nature. (TP 193)

Peirce does not succeed, therefore, in establishing that there is a process of inquiry that will guarantee our arriving at the truth.

But in this connection Peirce makes another point of considerable importance. He notes that man does seem to have a natural capacity for discovering (so far as, within the limits of induction, we can tell) regularities in nature. He suggests an explanation of this natural capacity, and uses this explanation to provide a criterion for determining whether a theory is worthy of acceptance. The terms 'abduction,' 'retroduction,' 'presumption,' and 'hypothesis' are used almost interchangeably by Peirce: they denote 'all operations by which theories and conceptions are engendered' (CP 5.590). Peirce spent considerable time investigating these forms of inference, and connecting them to the process of induction. His mature view was that abductive inference is the only way in which we acquire ampliative knowledge. 'Abduction *suggests* the theories that induction *verifies*; abduction is the *sole source* of synthetic claims which induction tests' (Madden 1968, 40; cf TP 197–8, CP 2.776). Throughout the *Collected Papers* we find Peirce formulating the criteria a good abduction must meet (cf TP 200–1). In the *first* place, it must be 'such that definite consequences can be plentifully deduced from it of a kind which can be checked by observation' (CP 2.786). *Secondly*, in making predictions based on an hypothesis we should not restrict ourselves to a set we know beforehand will be fulfilled; rather, our choice of predictable consequences should be a random one (CP 2.786). *Thirdly*, the testing procedure must be objective and unbiased; negative instances must be honestly noted (CP 2.634). *Finally*, the hypothesis should be as *simple* as possible (CP 5.60; cf TP 200–1). In his early writings, Peirce took this to mean *logical* simplicity; the simplest hypotheses were those that went least beyond the data. This view changed, however:

> It was not until long experience forced me to realize that subsequent discoveries were every time showing I had been wrong ... that the scales fell from my eyes and my mind awoke to the broad flaming daylight that it is the simpler hypothesis in the sense of the more facile and natural, the one that instinct suggests must be preferred. (CP 6.477)

'Simplicity' has become *psychological* simplicity, that which 'comes to mind' easily and naturally. This is strange, because elsewhere Peirce makes the point that, since abduction must be fair and unbiased, one should not permit subjective elements to determine the choice of an hypothesis: 'I myself would not adopt a hypothesis, and would not even take it on probation, simply because the idea was pleasing to me' (CP 6.477). Peirce arrived at his apparently strange view about psychological simplicity being a criterion of hypothesis-seclection by reflecting upon the fact that scientists had discovered in a relatively short time more true theories than the chance success of trial and error would seem to permit. This suggested that there was a systematic answer to the question, How does a scientist ever come to discover a true theory? The answer Peirce suggests is a causal explanation of scientific discovery.

> You cannot say that it happened by chance, because the possible theories, if not strictly innumerable, at any rate exceed a trillion – or the third power of a million; and therefore the chances are too overwhelmingly against the single true theory in the twenty or thirty thousand years during which man has been a thinking animal, ever having come into any man's head ... I am quite sure that you must be brought to acknowledge that man's mind has a natural adaptation to imagining correct theories of some kind ... But if that be so, it must be good reasoning to say that a given hypothesis is good, as a hypothesis, because it is a natural one, or one readily embraced by the human mind. (CP 5.591–2)

Since *homo sapiens* has adapted successfully to his environment, he must have some true theories, at first unarticulated, of course, about the nature of reality, about the world he lives in, about himself, and about how all these things interact. Natural selection requires this to be the case (Goudge 1973, 137). Hence, the hypotheses that come to man naturally and instinctively have something to be said for them for that very reason (TP 209–11). This causal account of discovery remains fragmentary and inadequate as it stands, but it does take the strangeness out of Peirce's idea that psychological simplicity is a normative criterion for the acceptability of hypotheses: normally, subjective elements are not relevant to the selection of hypotheses, but, by means of our scientific knowledge about evolution, we can reflect upon the process of discovery, and recognize, through our being able to explain it, the regularity that it is more likely than not that psychologically simple hypotheses are true – the lawful regularity that makes it reasonable to take such simplicity as a criterion of acceptability.

There is no reason why we should rest content with some naïve version of psychological simplicity, which, in any case, hardly goes beyond our ordinary concepts of reality and cannot approach the far-from-everyday theoretical concepts of science (cf Madden 1968, 42). We noted earlier the analogy between the problem-solving capacities of natural selection and those of human intelligence. We saw that the latter could be used to suggest hypotheses about natural adaptations. Let us now reverse the analogy. We may look upon the process of hypothesis-testing as a sort of natural selection.[28] It is a process by which a mind achieves a more adequate 'intellectual adaptation' (AL 208) to the universe. The cognizing subject is motivated by the desire for truth, by the desire to be intellectually adapted to the world, and 'Not knowing one's way about is a kind of absence of adaptation' (AL 209). Abduction, hypothesis-formation, provides a set of characteristics. Research selects among these. It eliminates the false, the maladaptations. It continues until it finds one which *is* an intellectual adaptation to the world, one which successfully ends the absence of such adaptation. We wish abduction to generate hypothesis-sets which contain an hypothesis which will enable us to adapt intellectually to the world. We wish, in other words, a set of values corresponding in the context of sophisticated research to the psychological simplicity which could be appropriate only to everyday situations. These values are provided, I would suggest, by those complex structures, consisting of generic theories, models, and research techniques, that Kuhn has called 'paradigms.' Paradigms determine what hypotheses 'come naturally' to the mind of the researcher (Kuhn 1970a sec 6). But *ought* these paradigm-determined criteria to be accepted? At the very least, the norms for hypothesis-formation generated by these paradigms can themselves be evaluated relative to the goal of coming to know. We value them as a means by which we solve puzzles. They are therefore valued just so long as they yield hypotheses which are the solutions to scientific puzzles (Kuhn 1970b 76), hypotheses which can bring about our intellectual adaptation in the world where such adaptation is recognizably absent. Paradigms are therefore also subject to a sort of natural selection: failure to solve puzzles leads to the elimination of paradigms (Kuhn 1970b ch 8). As to the generation of new paradigms when the old are eliminated, this is dealt with by Kuhn in his account of revolutionary science (Kuhn 1970b ch 9; Wilson 1981 sec 11); but that is something we cannot go into here. There is, of course, a crucial difference or disanalogy between biological selection and the selection of paradigms: the latter is guided by conscious purpose, that of coming to know; it is a process which *really is* purposive and teleological. But that is simply another way of making the point made before that mind is 'a species of conduct which is largely subject to self-control' (CP 5.419). The cognizing subject or, perhaps better, the community of scientists, can reflect upon and investigate its own research methods with an aim to evaluating their worth, their capacity to improve our knowledge and fill in its gaps efficiently. In the light of this knowledge it can (if this is seen to be appropriate) modify those procedures, its own structure if you wish, so as to become *better* at problem-solving, at cognizing (cf AL 209). In this sense, the methods of science are, or can become, self-correcting, as Peirce suggested, even though such capacity for self-correction cannot, as we saw Goudge argue against Peirce,[29] provide a successful vindication of those methods. Now, Kuhn has often been accused of relativism. His emphasis upon the intimate connection of paradigms and criteria for the acceptability of hypotheses perhaps suggests this, though, as the above remarks make clear, relativism does not follow: more over-arching cognitive standards may provide criteria for justifying paradigms and their associated values. More important to the charge of relativism are Kuhn's unfortunate remarks about the 'incommensurability' of different theories, and about the 'theory-ladenness' of concepts (cf Kordig 1971). The evolutionary view of mind and of the nature of the research process provides us with a framework that enables us to incorporate the insights of Kuhn while rejecting the charge of relativism.

We saw that in *The Ascent of Life* Goudge described concretely, using the example of selectionist evolutionary theory, patterns of scientific thought and inference later described by such philosophers of science as Kuhn and Lakatos. These insights concern in particular the process of research, often ignored by those who focus only upon its product. The emphasis upon process is one Goudge shares with, and perhaps derived from, Peirce. And the evolutionary view of mind, which Goudge also shares with Peirce, provides, I think, the framework for a subtle account of the process of scientific discovery and decision-making, an account which is, moreover, free of the irrationalistic overtones that have been part of similar accounts by Kuhn and Lakatos. It is to be hoped that, as Goudge continues his work on the philosophy of mind, he will develop in more detail his views on the development – evolution – of science, and on what is an essential element of that evolution, the procedure of evaluation and the production of values.[30]

DAVID L. HULL

Historical Narratives and Integrating Explanations*

In his *The Ascent of Life,* Thomas A. Goudge distinguishes two sorts of historical explanations which he argues are irreducible to the traditional covering-law model of scientific explanation – *integrating explanations* and *narrative explanations.* The form of an integrating explanation is as follows:

From a general historical statement (that all organisms are the outcome of descent with modification from common ancestors in the remote past) an inference can be made that phenomena of a certain sort (homologous structures) are to be expected. (AL 66)

Integrating explanations of the above sort do their job without the aid of any general laws. The explanations are 'non-nomological.' Because of this fact they provide no basis for making exact, positive predictions. Present phenomena are explained by showing them to be the outcome of past sequences of phenomena; but nothing is deducible here about phenomena yet to come. (AL 68)

The structure of narrative explanations is somewhat different. A narrative explanation consists:

... not in deducing the event from a law or set of laws, but in proposing an intelligible sequence of occurrences such that the event to be explained 'falls into place' as the terminal phase of it. The event ceases to be isolated and is connected in an orderly way with states of affairs which led up to it. (AL 72)

... the explanatory force of the resulting pattern of statements resides not in any general laws which it involves, but rather in the extent to which it establishes an intelligible,

---

*I wish to thank Wim J. Van der Steen for comments on are early draft of this paper. The research for this paper was supported in part by N.S.F. grant SOC 75 03535.

173 Historical Narratives and Integrating Explanations

broadly continuous series of occurrences which leads up to the event in question. (AL 77)

Michael Ruse (1971a, 1973c) has challenged Goudge's claims about the non-nomological character of integrating and narrative explanations on several fronts: first, by complaining of Goudge's sketchy explication; second, by showing that his examples frequently do not fit his own analysis; but most important, by arguing that in every case, laws are the source of whatever explanatory force historical explanations might have. I find myself in partial agreement with both Ruse and Goudge. I agree with Ruse that scientific laws are involved in historical explanations, but I agree with Goudge that the explanatory force of historical explanations does not result entirely from deriving the event being explained from a law of nature. Laws may well be 'involved' in one way or another in historical explanations, but laws can be involved in explanations in a variety of ways and in varying degrees of importance.

In a covering-law explanation, the event or regularity to be explained is derived from a set of laws and (if necessary) statements of particular circumstances. The *explanandum* is merely an instance of the regularity referred to in the laws. Why does a particular solar planet or all solar planets travel around the sun the way they do? For the same reason that all material bodies move in the paths which they do. Given Newton's laws and few additional statements of one sort or another, one can derive statements about the relative motion of a star and its planets. Given the actual positions and momenta of particular bodies (like the sun and Mars) and a few additional statements of one sort or another, one can derive the actual path of Mars around the sun. Can such a model account for Goudge's examples? It all depends.

In his original formulations, Hempel refers to two sorts of statements to be found in the *explanans* – general laws and statements which he terms variously 'initial,' 'antecedent,' 'determining,' and 'boundary' conditions (Hempel 1942; Hempel and Oppenheim 1965).[1] The implication of the terminology is obvious. All that these statements of 'antecedent' or 'initial' conditions do is to supply values for the variables in the laws when a particular event is being explained. As such, they are necessary for the derivation but provide none of the explanatory power of the explanation. The laws do that. However, in response to various criticisms, the proponents of the covering-law model have successively expanded the sorts of statements which must be included in the *explanans*. For example, in his early discussion, Hempel claims that general laws can be derived from other, more general laws without the help of any additional statements. However, in response to objections, he later considers that even the derivation of one general law from another may well require 'additonal premises which do not have the character of fundamental laws (Hempel and Oppenheim 1965, 291). Hempel now terms these additional statements 'particular facts,' not 'antecedent conditions.' In addition, defenders of the covering-law model acknowledge the relevance of all sorts of 'background knowledge' which usually is not and need not actually be included in the *explanans* of a covering-law explanation. The end result is that any objection which cannot be met by including yet another sort of statement among the statements of fact can be countered by reference to unspecified background knowledge.

The obvious response to Goudge's claims about integrating and narrative explanations is to include the factors which he mentions either in the statements of particular fact or in the general background knowledge which every scientific explanation presupposes. I have no strong objection to either of these manoeuvres just so long as the crucial roles which such factors play in explanations are acknowledged. Both laws and the hodgepodge of additional statements lumped under the rubric 'particular circumstances' are *necessary* for scientific explanation on the covering-law model. But proponents of the covering-law model argue that all the explanatory import is supplied by the laws. For example, in discussing the relative power of scientific theories, Hempel remarks, 'Some theories seem more powerful in the sense of permitting the derivation of many data from small amount of initial information; others seem less powerful, demanding comparatively more initial data, yielding fewer results' (Hempel and Oppenheim 1965, 278). Explanations relying on powerful theories are rightly called covering-law explanations, but in some explanations the laws are so weak that they are almost incidental when compared to the extensive information contained in the statements of particular circumstances. Although such explanations have identically the same logical form as covering-law explanations, they might be termed with greater justification 'particular-circumstance explanations.'[2]

In this paper I intend to argue that defenders of the covering-law model are correct in one respect: theories are involved in historical explanations but this involvement is not simply the derivation of the event being explained. Scientific theories play a variety of roles in scientific explanation. One of them is to help distinguish between accidentally true universal generalizations and genuine laws of nature, between all the various classes which fertile minds can invent and natural kinds, and between common-sense individuals and theoretically significant individuals. An excellent sign that an isolated universal generalization is a genuine law of nature is its integration into a scientific theory. Only an investigator's imagination limits the number and variety of classes he can postulate. Which of these are scientifically important is determined by which ones function in scientific laws. Finally, the world is full of entities. Not all of them are scientifically significant. One of the functions of scientific theory is to determine which complexes are to count as individuals for that theory. More specifically, certain theories require that the individuals referred to be 'historical entities,' i.e., that these entities be spatiotemporally located, cohesive, and continuous. One interpretation of Goudge's 'integrating explanations' is that they show the consequences of organizing natural phenomena into historical entities. An item is explained by showing how it fits into an historical entity, and a complex is explained by showing that it is organized in such a way as to function as an historical entity. 'Narrative explanations' are descriptions of the continuing development of historical entities and the sequences of events in which they participate.

INTEGRATING EXPLANATIONS

Goudge explicitly mentions three classes of phenomena which are to be explained by reference to descent – homologies, vestigial organs, and biogeographic distribution. Unless whales, bats, horses, and people share a common ancestry, it is difficult to understand why their anterior limbs, which perform radically different functions, should be structurally so similar. An atrophied organ which currently performs no function to speak of (a vestigial organ) is to be explained in terms of its being a remnant of an organ which once did perform an important function in some distant ancestor. 'The diverse phenomena of homologous structures and vestiges are integrated by means of the doctrine of descent with modification and inheritance from a common ancestry' (AL 67). Goudge's third example of an integrating explanation concerns geographic distribution. The peculiarities of Australian flora and fauna are to be explained in terms of the peculiar nature of the organisms which happened to colonize the continent.

Ruse (1973c) responds to Goudge's claims about the non-nomological character of integrative explanations by pointing out the direct and extensive appeal they make to scientific theories, in particular evolutionary and genetic theories. Given what we know about the mechanisms of inheritance and the way in which organisms in successive time-slices of the same species can change, the phenomena to which Goudge refers are to be expected. If inheritance were Lamarckian, quite different phenomena would be expected. Clearly, at least some features of homologies, vestigial organs, and biogeographic distribution can be explained by deducing them from the laws of genetics and evolutionary theory.[3] But scientific theories do more than provide premises for covering-law explanations. They also determine the ontological status of the entities to which they refer, and this status need not coincide with our human perceptions or everyday conceptions. Some of the distinctions which ordinary people make concerning natural phenomena may be appropriate for science; others not. It makes no difference. The fact that space *seems* Euclidean and porpoises *look* like fish to the man on the street is irrelevant to their actual status in science.

The general point I wish to emphasize is that the ontological status of theoretical entities is theory-dependent. An atom is an individual, a particular physical element is class of such atoms, and space-time is a relation because current scientific theories confer this ontological status on them. Of course, to the extent that a scientific theory assigns an entity to an inappropriate ontological category, that theory will be unsuccessful. A more specific point is that certain scientific theories require that some of their elements be 'historical entities.' They must possess sufficient spatiotemporal unity and continuity. They cannot pass casually in and out of existence. For example, in order to derive the position and momentum of Mars at one time from its position and momentum at some other time, one must use the relevant scientific laws, but one also presupposes that Mars is an individual, the sort of individual to which these laws apply, and that it remains the same individual in all relevant respects through time. Certain features of Mars might change without affecting the applicability of the laws of celestial mechanics to it, but not others. Theories are 'involved' in such decisions but not in the sense of straightforward derivation.

The distinction which I am drawing is in no way new. For example, Mario Bunge notes, 'In the natural sciences there is a clear distinction between a law and a behavior or evolution line, such as a trajectory. A given law usually embraces infinitely possible behavior lines: The latter differ by the circumstances not by the laws' (1976, 147). The novelty is found in the claim that reference to such trajectories can be explanatory. In quite another context, B.G. Glaser and A.L. Strauss observe that when a patient is admitted to a hospital, 'the staff in solo and in concert make initial definitions of the patient's trajectory – By defining trajectories, the staff establishes for themselves a *broad-range "explanation"* of what will happen to the patient, and thereby provide an *organizing perspective* on what they will do about handling the impending flow of events' (Glaser and Strauss 1968).

The peculiar nature of integrating explanations can be understood, I think, by reference to the role of spatiotemporal unity and continuity in determining historical entities. Before the advent of evolutionary theory, the homology/analogy distinction was drawn in terms of the structure/function distinction.

177 Historical Narratives and Integrating Explanations

Two structures were homologous if they were sufficiently similar in structure, position, and ontogenetic development; analogous if they performed the same function. Four permutations are possible: the legs of horses and dogs are both homologous and analogous; the forelimbs of horses and bats are homologous but not analogous; the gills of fishes and the lungs of mammals are analogous but not homologous; while the ears of bats and the hearts of dogs are neither homologous nor analogous. This classic distinction is, of course, beset with all sorts of problems. How similar is similar enough with respect to both structure and function? The level of functional, structural, and taxonomic analysis also matters. For example, the forelimbs of horses and bats are analogous as locomotory appendages but not as wings.

Evolutionary theory supplied a theoretical justification for the notion of homology and added another dimension to it. Descent was added to the preceding considerations and the possible permutations increased accordingly. Two traits can be as structurally and/or functionally similar as one might wish and not count as the 'same' trait if this similarity was acquired independently along different lines of descent. Not only must evolutionary homologies be similar, but also this similarity must have been acquired through a single, unique event. The same observations can be made for genes and species. Two genes can be structurally and/or functionally indistinguishable and yet be classed as different kinds of genes if they arose along different lines of descent. In population genetics, this distinction is commonly marked by the terms 'independent' and 'identical.' If two alleles 'originate from different sources (from similar alleles in unrelated individuals) they are said to be *independent*, whereas they are *identical* if they are replicates of one and the same allele existing in some past generation' (Mettler and Gregg 1969, 56–7).

Exactly the same distinction is made at the level of biological species. According to most taxonomists, all taxa must be monophyletic, i.e., all taxa must be descended from a single, immediately ancestral taxon.[4] It should be noted that the principle of monophyly is a principle of taxonomy, not evolutionary theory. By fiat it seems to rule out hybrid species. However, among plants hybrid species are common. As I have argued elsewhere,[5] the relevant issue is not that all taxa stem from a single immediately ancestral *species* but that they stem from a single immediately ancestral *speciation event*. Even though hybrid species arise from organisms in two ancestral species, the origin of a hybrid species is nonetheless unique.

If some version of the principle of monophyly is adhered to, both for species and higher taxa, then it follows that monophyletic taxa are uniquely defined by their insertions in history (Vendler 1976). They are chunks of the genealogical nexus.[6] Just as Hitler cannot be born more than once, the same species cannot re-evolve. If a person were to be born today identical to Adolf Hitler, that person still could not count as Adolf Hitler. Similarly, if a species of reptile were to evolve which was identical in every respect to an extinct species of reptile save origin, it would still count as a new and distinct species. As difficult as descent may be to discover, in the conceptualization of evolutionary phenomena, descent takes priority over all other considerations.[7]

Biologists generally agree that traits must be evolutionary homologies and taxa must be monophyletic. One justification for these conventions is that biologists are interested in reconstructing phylogenies, ancestor-descendant sequences of biological entities. However, as I have argued elsewhere (1974, 1975, 1976, and 1978), a second justification exists as well. If evolution occurs by mutation and the selective retention of some of these variants and not others, then both the entities whose structure is passed on in reproduction and the entities which are selected because of their interaction with the evironment must be historical entities. As a result, the entities which evolve are themselves spatiotemporally localized. Traditionally, the level at which evolution is taking place is considered specific. Thus, on this analysis, species as lineages are spatiotemporally localized. Whether or not they possess sufficient internal cohesion to be interpreted as historical entities is still a moot question, but on no account can they be interpreted as classes. Classes are not the sort of thing that can evolve.[8]

Although Goudge notes that biologists regard populations as 'enduring in time,' he dismisses the notion because the analogy between an organism and a population is 'superficial' and liable to generate 'both linguistic and conceptual confusion' (AL 27–31). The opinions of biologists, especially those of theoretically sophisticated biologists, should not be dismissed so lightly. Genes do differ from organisms and organisms from populations, but this does not mean that they cannot all be the same sort of entity, ordinary modes of conceptualizing notwithstanding. It is certainly true that fundamental conceptual changes in science generate linguistic and conceptual confusion, but as irritating as such confusion may be, I can hardly recommend a cessation of scientific chance in order to avoid it.

Even though Goudge rejects the notion that populations are historical entities, the claims which he makes about integrating explanations make sense only if populations are interpreted in this way. For example, he remarks that the observed distribution of present-day flora and fauna might at first seem surprising. On the basis of purely ecological considerations:

... one would expect geographically similar regions, such as Africa south of the Sahara, South America and Australia, to support broadly similar animal populations. Yet this is

179 Historical Narratives and Integrating Explanations

not the case. Often a region which is ecologically suitable for a certain type of population may have few, if any, representatives of that type native to the region (e.g., placental mammals in Australia). Large groups of related animals are limited to special regions of the globe, but not always because those regions are the only places where the animals could live. (AL 68)

Evolutionary development is influenced by both ecological and phylogenetic considerations. Certain similarities between species are the result of ecological factors. For example, the Sahara and Mohave Deserts are ecologically quite similar. Hence, one should expect the species occupying particular ecological niches to be ecologically quite similar, and they are. However, these deserts are widely separated geographically and have been for some time, continental drift notwithstanding. Hence, species occupying similar ecological niches are not likely to be related phylogenetically. Conversely, certain similarities between species are the result of phylogenetic considerations. The wide variety of ecological niches occupied in other areas by phylogenetically quite distant groups are in Australia occupied by variations on the same phylogenetic theme – the kangaroo. The ecological similarities are explained by the *laws* in covering-law explanations. Other similarities can be explained only be reference to the constraints placed on evolutionary processes by the historical character of the entities which are mutating, being selected, and evolving. These similarities are explained by the statements of *particular circumstances*.[9]

One of the central points of Goudge's discussion of evolutionary theory is that the theories in chemistry and physics are 'wholly systematic and non-historical', whereas evolutionary theory 'combines systematic and historical elements' (AL 17). Ruse responds:

What has led Goudge to argue as he does? I suspect he has made a mistake which is fairly commonly made about evolutionary theory, in particular, he has confused the *theory* of evolution, the thing which tells one about the mechanisms of evolution, with descriptions of the *historical paths* taken by evolving groups of organisms ('phylogenies'). Phylogenies are obviously historical; but although we may come to learn about them and understand them through the theory of evolution, they are not themselves part of the theory which rather is, in this sense, no more historical than any other theory. One can, for example, similarly distinguish between the non-historical Newtonian mechanics and the historical descriptions of the paths planets took in the past. (1973c 67)

Ruse is surely correct. All scientific theories combine both systematic and historical elements. The solar system has a history just as much as an evolving species does. Physical theories explain how physical systems behave under certain circumstances; cosmologists reconstruct the past history of such systems. Evolutionary theory explains how the evolutionary process takes place; paleontologists reconstruct the past history of particular species. Evolutionary theory looks distinctively 'historical' only if one interprets particular species as classes and statements about particular species as 'evolutionary laws.' But claims like 'All swans are white' – even if true – are no more laws of nature than is the observation that the moon always keeps the same surface directed toward the earth. Both are simple statements of fact which could change with the course of time. The proportion of black swans could increase in a particular species of swan and the relative periods of rotation and revolution of the earth and moon might change under appropriate conditions. But laws of nature are supposed to be eternal and immutable. Hence, neither statement can count as a law of nature. Ruse might concur in this judgement because in the preceding quotation he compares the path of a particular species to the path of a particular planet, implying that he is interpreting both species and planets as historical entities.

In covering-law explanations, both laws and variety of additional sorts of statements are necessary. When the phenomenon being explained is explained by showing that it is merely another instance of the sort of phenomenon mentioned in the relevant laws, I see no reason for not claiming that the explanatory import of such explanations resides in the laws – even though numerous sorts of additional statements may be *necessary* for the derivation. Not all necessary conditions are explanatorily important. Being born is necessary for death in human beings but hardly worth mentioning in an explanation of a particular person's death. However, not all explanations consist in showing that a particular phenomenon is an instance of a general regularity. 'Why aren't the mammals indigenous to Australia eutherians? All the evironmental conditions were perfect for the establishment of such highly advanced mammals.'Because no eutherians succeeded in getting to Australia in time. It is this statement of particular circumstance which appropriately answers the preceding question. Granted, some sort of evolutionary theory is taken for granted and might appear in the background information in such an explanation, but the statement of the particular circumstance is what is explanatorily important.

In sum, I am not arguing against the covering-law model of explanation as such but against the undue emphasis placed in such explanations on one particular element in the *explanans* – the relevant laws. Both laws and statements of particular circumstances are *necessary* on the covering-law model.

181 Historical Narratives and Integrating Explanations

Sometimes one element is explanatorily important; sometimes another. In the context of evolutionary development, ecological similarities are appropriately explained by reference to evolutionary processes, other necessary elements taken for granted. Phylogenetic similarities are explained by reference to statements of particular circumstances (i.e., statements concerning the sorts of organisms which were available to adapt to the environmental exigencies), other necessary elements taken for granted. The point I wish to make in this section is that certain phenomena can be explained *only* because the entities referred to are historical entities. 'Why do human beings have an appendix?' For the same reason that a person after an appendectomy has a scar. In both explanations, the particular laws and statements of particular circumstances are different, but the role played by historical entities is the same.

NARRATIVE EXPLANATIONS

The source of the apparent explanatory power of historical narratives has long been a matter of some contention among philosophers of science. In fact, Hempel's original formulation of the covering-law model (1942) was set out in response to the claims made by historians that they could explain events in human history without recourse to scientific laws. Rarely can an event mentioned in an historical narrative be deduced from any known scientific law, let alone from the antecedent events listed in the narrative. Advocates of the covering-law model conclude, therefore, that historical narratives are at best explanation sketches and are explanatory only to the extent that they approach the covering-law ideal. However, historical narratives *seem* to be more explanatory than any reference to known laws could possibly warrant. The source of the apparent explanatory force of historical narratives can be found, I believe, in the role played in them by historical entities (Hull 1975).

Goudge characterizes narrative explanations in two different ways, first in terms of necessary and sufficient conditions, then in terms of tracing continuous strands in causal networks. In a narrative explanation, one does not give an exhaustive list of all the events in the causal network prior to the event being explained. Rather the 'necessary and contributing conditions which make up the complex sufficient condition are arranged in temporal sequence' (AL 73). Historical narratives explain an event by specifying a 'temporal sequence of conditions which, taken as a whole, constitute a unique sufficient condition of *that* event' (AL 77). Goudge also characterizes narrative explanations as the establishment of an 'intelligible, broadly continuous series of occurrences which leads up to the event in question (AL 77). However, these two characterizations are independent of each other. An event could be explained by inferring it from a set of necessary and sufficient conditions which did not form a continuous series, and conversely an intelligible, broadly continuous, sequence of events need not consist of necessary and sufficient conditions for the terminal event in that sequence.

As Ruse remarks, the examples which Goudge gives of narrative explanations do not come close to fulfilling the requirements of his own analysis. A paleontologist would be hard put to present a set of necessary and contributing conditions sufficient to explain the development of amphibian limbs in the Devonian or the descent by man's ancestors from trees to land in the Miocene or Pliocene, tasks no less difficult than presenting necessary and contributory conditions which were sufficient for Hitler invading Russia in World War II. *In point of fact*, a certain sequence of events *did* lead to each of these events, but neither Goudge nor anyone else has listed sequences of events sufficient to bring them about, let alone sequences which include at least some necessary conditions. The examples which Goudge gives of necessary conditions for amphibians to develop limbs suitable for crawling on land are either necessary be definition (e.g., that amphibians must have emerged from water to become land-dwellers) or not necessary at all (e.g., that an environmental niche less hazardous to them than remaining in the water must have been accessible).

Ruse's main objection to Goudge's narrative explanations concerns the role of general laws. According to Goudge, if laws play any role at all in narrative explanations, it is as part of the 'background knowledge' which need not be mentioned explicitly – a nice turn of the screw if you ask me. Ruse (1971a 63–4; 1973c 83) for the sake of argument admits the existence of a set of conditions (s) sufficient for the event being explained (E) but responds that the statement 'if s then E' should then surely count as a general law. Goudge anticipates just such an objection:

Whenever a narrative explanation of an event in evolution is called for, the event is not an instance of a kind, but is a singular occurrence, something which has happened just once and which cannot recur. It is, therefore, not material for any generalization or law. The same is true of its proposed explanation. What we seek to formulate is a temporal sequence or conditions which, taken as a whole, constitutes a unique sufficient condition of *that* event. This sequence will likewise never recur, though various elements of it may. (AL 77)

If I understand him correctly, Goudge is arguing that both E and s are definite descriptions and that the statement 'if s then E' is singular. If, for example, the statement to be explained is the development of limbs (as evolutionary homologies) by amphibians (as chunk of the genealogical nexus)

183 Historical Narratives and Integrating Explanations

in the Devonian (as a particular period of time), then it is as much a definite description as the statement that Germany invaded Russia on 22 June 1941. Described in this way, neither event can recur but that, of course, does not preclude their being explained by deriving them from general laws. However, Goudge claims that *s* is also a definite description. I would be the first to agree that we do not have much in the way of genuine laws governing the course of human events, but Goudge seems more than a little inclined to accept at least some of the generalizations found in the biological literature as laws. Perhaps evolutionary theory and Mendelian genetics are not axiomatic, hypothetico-deductive systems, but they are still genuine scientific theories containing genuine scientific laws. Thus, it might be the case that *E* could be derived from one or more of these laws in conjunction with the relevant statements of particular fact. But Goudge insists that *s* is just a list of sequential events sufficient to produce *E*. Each element on the list could recur but not the entire sequence. I agree (Hull 1974, 97–9) with Ruse that no matter how extensively *s* might be specified, it could recur–unless of course it too is a definite description.

The issue now becomes how *s* as a definite description can possibly explain *E* as another definite description. The problem is not that the statements are singular or that the derivation is deductive rather than inductive or probabilistic. After all, singular statements can be deduced from singular statements with ease. The problem is showing in what respect such derivations are explanatory. The solution, I think, can be found in recourse to historical entities. Goudge sees a coherence and continuity of the events leading up to the event to be explained. Ruse argues for the existence of some general law from which it can be derived. Both concentrate so intently on the *events* associated with historical entities that they overlook the historical entities themselves. I think that both Goudge and Ruse are partially correct in the claims which they make about narrative explanations. Ruse is correct in maintaining that general laws play a central role in narrative explanations, but Goudge is correct in claiming that this role is not the derivation of the event to be explained. Instead scientific theories determine which complexes are to count as individuals and what sorts of individuals. According to evolutionary theory, genes, organisms, and populations are all theoretically significant individuals. In particular, they are historical entities. Goudge is also close to the mark when he emphasizes the continuity of the events leading up to the event being explained, but he fails to identify its source. The events listed in historical narratives do possess coherence and continuity, not because they are all governed by the same law or set of laws (though in certain circumstances this may be true), not because of some intrinsic coherence of their own (the point at which Goudge terminates his analysis), but because of the unity and continuity of the historical entity.[10]

Historians write histories of particular people, families, nations, social causes, and the like. They view their task as presenting a coherent story which includes reference to the more important events and excludes the irrelevant. The search for relevant data and much of the relative importance attached to various events as causing others surely is guided by various beliefs that the historian has about people and social institutions – as low-level and poorly formulated as these beliefs may be. Some of these beliefs may even count as general laws. However, another factor is also crucial. Event follows upon event in historical narratives, not because these events can be derived sequentially from any scientific law the way that the successive positions of Mars can, but because they are all events in the life-story of the same historical entity. For example, Hitler was born in 1889 and died in 1945. Every event in which he participated is a candidate for inclusion in an historical narrative about him. People can become anti-Semitic in a wide variety of ways. There may even be laws governing the formation of such prejudices, I don't know. But I do know that we do not have to possess such laws to claim that Hitler was anti-Semitic and to include this fact in the appropriate place in his biography.

Some of the unity and continuity to be found in a good biography is supplied by various low-level sociological and psychological 'laws,' but most is supplied by the unity and continuity of the historical entity itself. In the case under discussion, the historical entity is an organism. The 'laws' which determine what is to count as an organism, a single organism, and the same organism through time are biological, not psychological or sociological. An historian need not rely on either social or psychological unity and continuity to write a biography of Hitler. After all, Germany, Europe, the entire world underwent rapid, abrupt, spasmodic changes during Hitler's lifetime, and Hitler, himself, was far from psychologically coherent. But none of this matters in writing a biography of Hitler, given Hitler as an historical entity. Of course, other historical entities also play roles in Hitler's biography, some at the same level of analysis (e.g., Hitler's parents, Eva Braun, Churchill, and Roosevelt), some at other levels of analysis (e.g., the Nazi party, Germany, the Jewish sub-culture in Europe).

Comparable observations can be made about evolving species. They too form historical entities. The sequential development of species through time can by traced in the absence of laws governing this development, though an understanding of these processes does help some. For example, certain amphibians did leave the water to become land-dwellers, just as certain primates left their arboreal habitats for the open plains. To my knowledge we

185 Historical Narratives and Integrating Explanations

currently possess no evolutionary laws concerning the conditions under which individuals change ecological niches, but the absence of the relevant laws does not preclude paleontologists deciding that such an event took place and including it in their phylogenetic reconstructions. In point of fact, evolutionary theory plays a surprisingly minor role in reconstructing phylogenies. It sets some very general constraints and that is about it.[11]

According to the position being urged in this paper, one must distinguish carefully between the processes governing natural phenomena (e.g., Newton's laws of motion), the regularities resulting because of these processes (e.g., Mars traveling in an ellipse around the sun), the individuals governed by these laws (e.g., Mars and the sun), and the histories of these individuals (e.g., the actual formation and subsequent development of the solar system). The entire process may well be law-governed: given the conditions that obtained in the beginning and some cosmic law, the entire course of the universe and everything in it might be derivable. It might also be the case that these ultimate laws make no reference to individuals, let alone historical entities. But as the situation now stands, the preceding distinctions are extremely helpful in understanding the current division of labour among scientists. Even though historians cannot derive the events mentioned in their narratives from any known general laws, they can still write coherent, cohesive historical narratives by following the course of a single historical entity or a set of interrelated historical entities, because the fundamental unity and continuity of historical narratives is supplied by the unity and continuity of these historical entities.

CONCLUSION

L.O. Mink (1970) remarks that the actions and events of an historical narrative 'comprehended as a whole are *connected by a network of overlapping descriptions.*' Because these overlapping networks are so apparent, most advocates of the explanatory character of historical narratives have looked to them for the justification of their position. To some extent, these networks of overlapping descriptions do have a coherence of their own, the sort of coherence which can be explained by derivation from general laws, but the primary reason for the events mentioned in historical narratives cohering the way they do is that they are all associated with the same historical entity, either as properties of that entity or as events in which it participates. Epistemologically we may first come to recognize historical entities because we notice a coherent cluster of properties. For example, given our period of duration and perceptual abilities, it is no surprise that we as human beings first come to recognize the species *Cygnus olor* through a particular set of traits which all swans seem to share, not through noticing the relations which integrate these organisms into a single evolutionary lineage. Nevertheless, the lineage is primary, not the cluster of traits. Similar observations can be made for particular organisms. Moses was Moses and not someone else because of the unity and continuity of his body (and possibly also his mind), not because of any cluster of properties or events associated with him.[12]

Many, though certainly not all, of the theoretically significant entities postulated by scientists are highly organized 'complex particulars' (Suppe 1974). I take integrative explanations to consist in setting out the organization of such complex particulars and showing the consequences of this organization for various natural phenomena. For example, all species are integrated into historical entities by the ancestor/descendant relation. Sexual species are further integrated by the interbreeding relation. Social species are characterized by still another level of organization. Once historical entities have been identified, they can be used as a basis for historical narratives. Events can be grouped into kinds and these kinds related in general laws. But events can also be ordered because of their association with particular historical entities in at least partial independence of any lawful connections of the preceding sort. I take narrative explanations to consist in a description of the sequential development of one or more historical entities.

The question remains, however, whether or not integrating and narrative 'explanations' are actually explanatory. At times advocates of the covering-law model seem to make all and only covering-law explanations explanatory by fiat. Anything that fulfils the requirements of the covering-law model is ipso facto explanatory, and anything which fails to meet these requirements is ipso facto totally devoid of any explanatory force, overpowering intuitions to the contrary notwithstanding. But let us look for the moment at some of the necesssary requirements for covering-law explanations. The need for both general laws and statements of particular facts is explicitly acknowledged. However, the nature of such general laws has proved to be one of the most recalcitrant problems in the philosophy of science, and the statements of particular facts have ballooned into a hodgepodge of whatever statements turn out to be necessary for the derivation. Let us assume for the moment that general laws are supposed to refer to eternal immutable regularities in nature. The existence of such regularities presupposes in turn the existence of natural kinds of some sort (events, fields, substances, quantities, etc.) which are systematically interrelated. At least in some cases, these natural kinds refer to particular individuals. In such cases, the existence of these theoretically significant individuals is just as necessary for covering-law explanations as the natural kinds and the laws relating them.

Thus, for theories couched in a thing ontology, all of the following are necessary before an event or regularity can be explained on the covering-law model: natural kinds (i.e., classes which function in general laws), theoretically significant individuals (i.e., the individuals denoted by at least some of these natural kind terms), properties of and relations between natural kinds (i.e., general laws), and a wide variety of particular facts (e.g., boundary conditions and values for the variables mentioned in the general laws). The discoveries of general laws and theoretically significant classes have always been considered important scientific achievements. A good deal less credit is given for the recognition of theoretically significant individuals, I suspect because initially common-sense individuals turned out to be theoretically significant as well. But as science progressed, scientists were forced to recognize extremely non-intuitive individuals. What makes a physical element an element, one element and not two, etc. is scientifically important. But what makes an atom an atom, one atom and not two, etc. is just as important. Why is it then that reference to natural kinds and laws is explanatory and reference to individuals is not?

Defenders of the covering-law model have an easy answer to this question: reference to individuals, the structure of individuals, the continuity through time of individuals *is* important and must be included in a totally adequate covering-law explanation – either as a general law or among the statements of particular facts. On the traditional logical empiricist analysis of science, theory reduction is merely the derivation of one theory from another; micro-reduction is the derivation of theory at one level of analysis from a theory concerning entities at a lower level of analysis. In order to perform micro-reductions, 'bridge laws' are required to connect the entities and processes at the different levels. For example, in the reduction of Mendelian genetics to molecular biology, genes are associated with segments of DNA and various patterns of inheritance with different sorts of biosynthetic pathways.

As Robert Causey has noted, structure is too important in micro-reductions to ignore:

In discussions about the reducibility of biology, the kind of reduction which is usually considered is *micro-reduction*. Roughly speaking a micro-reduction is an explanation of the behavior of structured whole in terms of the laws governing the parts of this whole. (1969, 230)

Note also that the kind of microreductive explanation I have outlined above is not a mere DN derivation. It is a very special kind of derivation which provides a genuine part-whole explanation, i.e., an understanding of the whole in terms of its parts plus its structure. (1972, 415)

Thus, the covering-law model just might be able to accommodate historical explanations if statements concerning structure can be included somehow. The fact that structure involves the part/whole relation should not pose a problem because the logic of the part/whole relation has been worked out in considerable detail, though not as completely as that of class membership. However, considerable disagreement exists about the nature of the statements which must specify structure. Some philosophers maintain that so-called 'bridge laws' are analytic, some that they are natural laws, others that they are merely contingent, still others that they are a posteriori though necessary. Only if the statements specifying structure are interpreted as general laws can they truly count as being explanatory on the covering-law model; otherwise they must be shunted to statements of particular fact or left unspecified as part of the background knowledge. At the very least, advocates of the expanded covering-law model of scientific explanation must acknowledge more fully than they have in the past the role of complex particulars and their structure in scientific explanations.

ROBERT McRAE

Life, *Vis Inertiae*, and the Mechanical Philosophy

In the *Metaphysical Foundations of Natural Science* Kant states as his second law of mechanics, 'Every change of matter has an external cause.' Then in parenthesis he gives, as its equivalent, Newton's First Law, 'Every body remains in a state of rest or motion in the same direction and with the same velocity unless it is compelled by an external cause to forsake this state' (Kant 1970, 104). Matter, according to Kant, is an empirical concept, and the external senses reveal no other determinations in it than those of external relations in space. In the *Critique of Pure Reason* he says of matter,

That which inwardly belongs to it I seek in all parts of the space which it occupies, and in all effects which it exercises, though admittedly these can only be appearances of outer sense. I have therefore nothing that is absolutely, but only what is comparatively inward and is itself again composed of outer relations. The absolutely inward [nature] of matter, as it would have to be conceived by pure understanding, is nothing but a phantom ... (Kant 1933 A277-B333)

To say that every change in matter must have an external cause is, for Kant, precisely to deny that it is self-determining or living. The 'inertia' of matter refers simply to this lack of life. Kant then provides a definition of life and some consequences of this definition:

Life means the capacity of a substance to determine itself to act from an internal principle, of a finite substance to determine itself to change, and of a material substance to determine itself to motion or rest as change of its state. Now we know of no other internal principle of a substance to change its state but desire and no other internal activity whatever but thought, along with what depends upon such desire, namely, feeling of pleasure or displeasure, and appetite or will. But these determining grounds and actions do not at all belong to the representations of the external senses and hence also not to the determinations of matter as matter. Therefore, all matter as such is lifeless. The proposition of inertia [what Kant will call 'the law of inertia'] says so much and no more. If we seek the cause of any change whatever of matter in life, we shall have to seek this cause at once in another substance different from matter, but bound up with it ... The possibility of a natural science proper rests entirely upon the law of inertia (along with the law of the permanence of substance). The opposite of this, and therefore the death of all natural philosophy, would be hylozoism. From the very concept of inertia as mere lifelessness there follows of itself the fact that inertia does not signify a positive effort of something to maintain its state. Only living things are called inert in this latter sense, inasmuch as they have a representation which they abhor and strive against with all their power. (Kant 1970, 105-6)

In the *Critique of Judgment* Kant identifies 'hylozoism' as the doctrine of 'the life of matter – this life being either inherent in it or else bestowed upon it by an inner animating principle or world soul' (Kant 1928, 43).[1]

Kant mentions Kepler as the originator of the phrase '*vis inertiae*,' but it is doubtless Newton that he has in mind for it is Newton who refers to this force in relation to his First Law, or what later came to be called the principle of inertia, and it is in its alleged connection with this law that Kant is concerned with it. In Definition III of *Principia* Book 1 Newton says,

The innate force of matter is a power of resisting by which every body, as much as in it lies, continues in its present state, whether it be of rest or of moving uniformly forward in a straight line.

In the exposition of this definition he says:

This force is always proportional to the body whose force it is and differs nothing from the inertia of the mass, but in our manner of conceiving it. A body from the inertia of matter is not without difficulty put out of its state of motion or rest. Upon which account this innate force may, by a most significant name, be called the force of inertia (*vis inertiae*). But a body only exerts this force when another force, impressed upon it, endeavours to change its state; and the exercise of this force may be considered as both resistance and impulse; it is resistance so far as the body, for maintaining its present state, opposes the force impressed; it is impulse so far as the body, by not easily giving way to the impressed force of another, endeavours to change the state of that other. (Newton 1946, 2)[2]

Here I draw attention to two points in Kant's criticism of the concept of *vis*

191 Life, *Vis Inertiae*, and the Mechanical Philosophy

inertiae: (1) that an internal cause of change is the defining characteristic of life, the active resistance to change being only a corollary of this internal causation, and (2) that the concept of life is a psychological one, for the only internal cause of which we have any experience is desire, and the only internal activity thought. Hence if we look for an internal cause of change or resistance to change we are looking for a psychic cause i.e., in substance completely different from matter.

As an historical judgment, Kant's statement about *vis inertiae* is of considerable interest, for a force of this nature, i.e., one answering to Newton's definition, but not bearing the name which he gave to it, is to be found also in Descartes, and is, indeed, the foundation of his dynamics. If we may regard Descartes as the principal spokesman for the mechanical philosophy, and if Kant is right in asserting that this concept is one of life, then hylozoism is to be found at the heart of the mechanical philosophy. It is this paradoxical historical judgment that I wish to examine, and to examine it in relation to Descartes rather than to Newton, for Newton had no objection to hylozoism, whereas, we might suppose, the mechanical philosophy is the very antithesis of hylozoism.

Newton, indeed, appears to have been powerfully attracted by hylozoism. In a draft variant of Query 22 of the *Optice* of 1706 he concludes, 'We cannot say that all nature is not alive.' *Vis inertiae*, he points out, is a passive force, and if there were no other principle than the *vis inertiae* there could be no motion in the world. What that active principle is, which is necessary for motion, and how it is related to matter Newton declares to be a mystery. But,

a man must argue from phenomena. We find in ourselves a power of moving our bodies by o^r thoughts (but the law of this power we do not know) & see ye same power in other living creatures but how this is done & by what laws we do not know. And by this instance & that of gravity it appears that there are other laws of motion (unknown to us) than those wch arise from *Vis inertiae* (unknown to us) wch is enough to justify & encourage o^r search after them. We cannot say that all nature is not alive.[3]

It is evident from this that if all nature is alive for Newton, it is not by virtue of the *vis inertiae* which he says later in Query 31 of the *Opticks* is 'accompanied with such passive Laws of of Motion as naturally result from that force' (Newton 1952, 376), namely the Laws of *Principia*, but to certain active principles accounting for such phenomena as gravitation, cohesion, fermentation, etc. However we may note that he identifies the active with the living, and like Kant, the living with the psychical. For the possibility of such an active force he appeals to the experience of our power to move our bodies by our thoughts, and in another draft variant he says, 'Life & Will (thinking) are active Principles by wch we move our bodies, & thence arise other laws of motion unknown to us.'[4]

Newton's acceptance of the possibility that all nature is alive coincides with his rejection of the mechanical philosophy as being a dogmatism in so far as it makes the claim that *all* natural phenomena can be explained mechanically.[5] So let us turn to Descartes who does accept that claim, and transfer Kant's statements on *vis inertiae* from Newton to Descartes, for not only was Descartes the first to state explicitly the principle of inertia in its full generality but he also, like Newton, invokes a force to which, while not giving it any special name, he attributes the same defining properties as Newton does to his *vis inertiae*, namely 'perseverance' in the same state of motion or rest, and 'resistance' to change of state (Descartes 1954 II 43). To give added weight to Kant's remarks about hylozoism we shall take into account Descartes' letters to Henry More, written in his last months, in which he speaks of the force by which matter is moved as possibly being the power of God himself and, moreover, as being the same kind of power as that by which the soul can move the body. In this letter he goes on to confess his reluctance to discuss this moving force for fear of 'seeming inclined to favour the view of those who consider God as a world-soul united to matter' (Descartes 1970, 257). In short, Descartes acknowledges that if he had been more open people might easily have found reasons for regarding his universe as animated. Before considering the grounds of Descartes' uneasiness let us look at this picture of Descartes and the mechanical philosophy by Richard S. Westfall in *The Construction of Modern Science: Mechanisms and Mechanics*. Westfall says of Descartes that he exercised a greater influence toward a mechanical philosophy of nature than any other man, and presented it with a greater philosophical rigour.

In the famous Cartesian dualism, he provided the reaction against Renaissance Naturalism with its metaphysical justification. All of reality, he argued, is composed of two substances ... *Res cogitans* and *res extensa* – Descartes defined them in a way to distinguish and separate them absolutely ... From the point of view of natural science, the more important result of this dichotomy lay in the rigid exclusion of any and all psychic characteristics from material nature ... In Renaissance Naturalism, mind and matter, spirit and body were not considered as separate entities; the ultimate reality in every body was its active principle, which partook at least to some extent of the characteristics of mind or spirit ... The effect of Cartesian dualism, in contrast, was to excise every trace of the psychic from material nature with surgical precision, leaving it a lifeless field knowing only the brute blows of inert chunks of matter. It was a

193 Life, *Vis Inertiae*, and the Mechanical Philosophy

conception of nature startling in its bleakness – but admirably contrived for the purposes of modern science ... The physical nature of modern science had been born. (Westfall 1971, 31)

Where Descartes most directly and explicitly opposes himself to the view that nature is animated is in his constant polemic against 'substantial forms' or 'real qualities.' His paradigm case in treating of substantial forms is gravity. To explain the motion of falling bodies by a substantial form, 'gravity,' is to treat the cause of the fall as *internal* to the body and as related to that body in the same way as our mind is related to our body when we move it. On the other hand to explain the motion of the falling body by the principles of the mechanical philosophy is to treat the cause of the fall as *external* to that body. Its fall towards the earth is caused by its continuous displacement downwards by extremely rarefied matter in the vortex. That the cause of any change of state in a body is *necessarily* external is stated in Descartes' First Law – 'every particular thing in so far as it is simple and undivided, always remains in the same state so far as it can, and never changes except through external causes' (Descartes 1954 II 37). Although the substantial form as conceived by analogy with the human mind is, like the mind, a substance, and, therefore in operating causally on a body, is acting as one substance on another substance, nevertheless this spiritual substance is not *external* to the material substance in the way one material substance is *external* to another material substance. It is thought of as operating *within* the gravitating body. Thus Descartes writes,

And although I have been viewing gravity as diffused throughout the whole of the heavy body, none the less I have not been ascribing to it that very extension which constituted the nature of body; for true bodily extension is of such a nature as to prevent any interpenetration of parts ... I also saw that while it remained coextensive with the heavy body, it could exercise its force at any point of the body ... Indeed it is in no other way that I now understand mind to be coextensive with the body, the whole in the whole, and the whole in any of its parts ... (Descartes 1911, 255)

In eliminating all substantial forms from nature except one, namely the human mind in its relation to the human body, Descartes had eliminated all internal cause of change in bodies, except one, the human mind. In every other case the cause of change in bodies is external. What is the nature of this external cause?

It must be carefully observed what it is that constitutes the force by which one body acts on another body or resists its action; it is simply the tendency of everything to persist in its present state so far as it can (according to the first law) ... Thus what is at rest has a force by which it remains at rest, and consequently of resisting everything that might change its state of rest; what is moving has a force by which it persists in its motion – in a motion constant as regards velocity and direction. (Descartes 1954 II 43)

The force by which one body acts on another is then, as Descartes says, a *consequence* of a force which is internal, the same force that Newton would later call *vis inertiae*. Is it, then, *this* internal force which could give rise to the imputation of hylozoism which Descartes feared?

Kant treated his law of inertia as the law of causality applied to matter. The causal law states that 'every change has a cause.' As applied to matter the law becomes, 'Every change of matter has an external cause.' Descartes' general causal axiom takes a significantly different form from Kant's, and it makes an important difference to his understanding of the principle of inertia. His axiom is, 'Nothing exists concerning which the question may not be asked – "What is the cause of its existence?" For this question may be asked even concerning God ... ' (Descartes 1911, 55). This means not only that the coming into existence of anything requires a cause, but also the continued existence of anything, and no less so, even should that continuing thing always have existed. In the self-examination of the *Third Meditation* Descartes discovers that he has no power by which he could bring it about that he, as existing now and possibly as always having existed, could still continue to exist in the future, for if such a power did reside in him, he would certainly, as a thinking thing, be conscious of it. He concludes that his continuing existence is sustained by the divine power. Descartes generalized his own case to apply to all things (Descartes 1911 Axiom II 56). His form of the causal axiom implies therefore that some force or power is necessary for the continued existence of bodies and certainly for their continued existence in the same state. But matter qua extended thing or as that whose essence is extension has no such power. The force of motion by which a body in motion remains in a state of motion and the force of rest by which a body remains in a state of rest are the divine power. And the force by which one body acts on another or resists its action is a *consequence* of that divine power. When bodies interact the power to persevere in the same state becomes the *tendency* or *endeavour* (*conatus*) to do so, as for example the tendency of the stone in a sling to move in a tangent to the curved path in which it is constrained by the sling to move. Descartes is much concerned, however, that this tendency or endeavour should not be taken psychologically.

When I say [bodies] 'endeavour' to move away from the centres around which they revolve, I must not therefore, be thought to be fancying that they have some thought

195 Life, *Vis Inertiae*, and the Mechanical Philosophy

from which this 'endeavour' proceeds; I just mean that their positions, and the forces that impel them to motion, are such that they would in fact go in that direction, if no other cause hindered them. (Descartes 1954 III 56)

Let us now turn to the correspondence with Henry More. In the letter of 15 April 1649, Descartes writes,

I will only add that I have not yet met anything connected with the nature of material things for which I could not very easily think out a mechanistic explanation. It is no disgrace to a philosopher to believe that God can move a body, without regarding God as corporeal; it is no more of a disgrace to him to think of other incorporeal substances. Of course I do not think that any mode of action belongs univocally to both God and creatures, but I must confess that the only idea I can find in my mind to represent the way God or an angel can move matter is the one which shows me the way in which I am conscious I can move my own body by my thought. (Descartes 1970, 252)

In the letter of August 1649 Descartes writes,

The translation which I call motion, is a thing of no less entity than shape: it is a mode in a body. The moving force may be the power of God Himself conserving the same amount of translation in matter as he put in it in the first moment of creation; or it may be the power of a created substance, like our soul, or of anything else to which He gave the power to move a body. This power is a mode in a creature, but not in God; but because this is not easy for everyone to understand, I did not want to discuss it in my writings. I was afraid of seeming inclined to favour the view of those who consider God as a world-soul united to matter. (Descartes 1970, 257)

Now compare these remarks about God's relation to motion with the account that is given in the *Principles of Philosophy*. Descartes says,

After considering the nature of motion we must treat of its cause; in fact of two sorts of cause. First, the universal and primary cause – the general cause of all the motions in the world; secondly the particular cause that makes any given piece of matter assume a motion that it had not before. As regards the general cause it seems clear to me that it cannot be other than God himself. He created matter along with motion and rest in the beginning... (Descartes 1954 II 36)

In the *Principles* motion and rest are *states* of matter. That means that God, in causing motion, is not producing a *change* in matter. Rather he is creating matter in a certain *state* – of motion or rest. Any *changes* in the world consist not in the motions of bodies but of changes in their states of motion or rest. And all changes in the states of a body are produced by secondary causes, i.e., by other bodies (ignoring for the moment that minds also can produce changes in the bodies with which they are united). It is these which make 'any given piece of matter assume a motion [qua state] that it did not have before.' The state of rest has no logical or temporal priority to the state of motion. Thus if God suspended the power by which he maintains a body in a state of motion, the result would be, not that the body would come to rest, but that it would cease to be. For God to suspend his power is not the same thing as altering a body from one state to another. It is only bodies which alter the states of bodies. God's power is not some kind of propellant moving bodies around. If a body is propelled it is always by the action of another body, although this action is a *consequence* of the divine power in the latter to persevere in its state of motion. The God of Descartes' *Principles* is not the First Mover. He moves nothing. Rather he creates matter 'along with motion and rest' and by a continuous action of creation he preserves the same quantities of these states in the world.

The situation described in the letters to More is quite different. God is here the First Mover, for he moves matter in the same way as I move my arm – 'I must confess that the only idea I can find in my mind to represent the way in which God ... can move matter is the one which shows in the way in which I am conscious I can move my own body by my own thought.' When the soul, according to Descartes, initiates motions in the body it does so by moving the pineal gland. Moreover what it does to that gland does not differ in any way from what the animal spirits (very rarefied matter) do to that gland in accounting for all the other, numerous, non-voluntary motions of the body. In other words, the soul acts on the gland as an immaterial mechanical agent does. So, if God moves matter in the same way as I move my body by my thought then he does so as an immaterial mechanical agent. In this account God is not the creator of matter in a state of motion (and rest). Rather he gives matter its first mechanical push, and if he preserves the same amount of motion in the world it would have to be by continuing to push, just as Descartes' soul would have to keep moving the pineal gland in order to keep his arm moving. Moreover if the only way in which God's moving of matter can be represented is by analogy with the way in which the soul moves the body, then the concept of God as world-soul seems unavoidable. This brings us to one of our principal questions: Is God's relation to the world one which makes the world a living thing, as the expression 'world-soul' suggests? What is a living thing for Descartes?

197 Life, *Vis Inertiae*, and the Mechanical Philosophy

Descartes retained the view, going back to the Greeks, that to have a soul and to be alive are the same thing, and, moreover, that in any living thing the soul is the source of its motion. Thus in promoting the notion that animals are self-moving machines, i.e. machines having the principle of motion in themselves, Descartes did not deny that they had souls, nor, what is the same thing, that they were alive.[6] All that he was claiming was that animal souls, unlike human souls, were material substances, not thinking substances – the material substance in question being blood.[7] Man, indeed, has both kinds of souls: 'I came to realise that there are two different principles causing our motions: one is purely mechanical and corporeal and can be called the corporeal soul; the other is the incorporeal soul, the soul which I have defined as thinking substance' (Descartes 1970, 243).

It is, I think, quite clear that since the God of the *Principles* does not have the kind of relation to the motion of bodies that either the animal soul or human soul has, the world by that account is not animated and there is no soul of the world. How then does one choose between the *Principles* and the implications of the letter to More? One chooses the *Principles*, for the account in that work is based on a consideration of motion as a *state* and not a *change* and it is this revolutionary notion which made it possible for Descartes to enunciate the principle of inertia.[8] The *vis inertiae* of Descartes' *Principles* (to borrow Newton's term) is in no respect an analogue of anything psychical. The power by which I move my body is, for Descartes, something I directly experience within myself, whereas I have absolutely no experience within myself of any power by which I can continue my existence. The force by which bodies continue to exist in the same state of motion or rest has then no analogue in the life of the soul. There is not a shadow of anthropomorphism in Descartes' conception of *vis inertiae*.

Finally let us look briefly at Spinoza, who took over Descartes' 'tendency of everything to persist in its present state so far as it can,' and, rendered it as 'the effort by which each things endeavours to persevere in its own being.' He says of this *conatus*, 'if the thing be destroyed by no external cause, by the same power by which it now exists it will always exist.'[9] The power or effort by which each thing perseveres in its being constitutes, for Spinoza, the essence of each individual thing, and that power is part of the infinite power of God. In the *Cogitata Metaphysica*, in a chapter entitled 'Concerning the Life of God,' Spinoza raises the question, What is life? He rejects the doctrine that soul is the principle of life, and that its presence, or absence, distinguishes some things as living and others as non-living. Against this he puts forward two propositions, (1) that there is nothing in matter except mechanical interrelations and operations, and (2) that there is no matter devoid of life. These apparently antithetical propositions are reconciled in the definition of life which follows:

Therefore we shall understand by this term life, *the force by which things persevere in their own being*... But the force by which God perseveres in his being is nothing else than his essence; they speak well who say that God is life.[10]

Thus the fundamental *force* of the Cartesian mechanics becomes with Spinoza *life*.

In its principal features the mechanical world of Spinoza is the same as Descartes'. Where in Descartes God creates matter along with motion and rest, in Spinoza motion-and-rest are the infinite mode which follows immediately from the divine power conceived under the attribute of extension. For both philosophers the *vis inertiae* is found in each individual body and it is by this force that bodies act on one another in the form of tendency or endeavour. For both, *vis inertiae* is the power of God in the individual body. Consequently – to use the language of Spinoza – for both philosophers God is the 'immanent' cause in everything by which it perseveres in its being, while individual bodies are the 'transitive,' external, or interactive causes which account for all changes. For both, nothing psychic characterizes the *vis inertiae* in so far as it is found in bodies. For Descartes that means that the world is not animated, and God is not its soul. For Spinoza the world *is* animated, but no more than for Descartes is God the soul of the world – for Spinoza soul is not the principle of life; *vis inertiae* is that principle, whether in God or in finite individuals. And, finally, one further great difference: for Descartes we are not conscious of the power by which we persevere in our being, for that power is divine, and therefore we have no analogue for conceiving it in other things. For Spinoza, on the other hand, to be conscious of one's self is precisely to be conscious of this power.

Since the mind, through the ideas of the affections of the body, is necessarily conscious of itself, it is therefore conscious of its effort. This effort, when it is related to the mind alone, is called will, but when it is related at the same time both to the mind and the body, is called *appetite*, which is therefore nothing but the very essence of man...
Desire is generally related to men in so far as they are conscious of their appetites and it may therefore be defined as appetite of which we are conscious.[11]

Thus with Spinoza we get a perfect synthesis of hylozoism and the mechanical philosophy but it is achieved by radically depsychologizing the concept of life.

ROLAND PUCCETTI

The Ascent of Consciousness

In the preface to his pioneering study, *The Ascent of Life*, Thomas A. Goudge warned that one question in particular, the philosophical significance of the evolution of consciousness, and in man at least of self-consciousness, opened up a subject too large and complex to be taken up there. However in a recent paper (Goudge 1976b) Professor Goudge has returned to this same question. His answer to it, broadly, is that mental phenomena have indeed had an adaptive role or function in evolution, and that theories which appear to deny this are just biologically implausible.

On some details of Goudge's argument I find myself in disagreement. For example, it is simply not true that there has been in animal evolution a continuous and cumulative progression of sense organs, nerves, and brains from simple structures to more complex ones. The lower vertebrates and even some of the simpler mammals have more complex peripheral endorgans, such as the retina of the eye, than we do (Michael 1969); while an unconsciously functioning brain structure such as the cerebellum is actually larger in proportion to the rest of the encephalon in us than in lower forms of animal life (Llinás 1975).

But these observations do not contradict Goudge's more general claim. The frog's retina has to be more complex than ours so that a large, dark stimulus will provoke it to jump unthinkingly towards blue light, hence towards water and safety (Muntz 1962); whereas a small, dark moving stimulus will provoke it equally unthinkingly to seize the passing object with its sticky tongue: if the optic nerves are crossed it misses every time (Sperry 1945). Animals with larger brains are freed from these stereotyped automatisms to do more information processing centrally, where incoming stimuli can be selectively attended to and acted upon; the enlarged cerebellum then becomes necessary to orchestrate co-ordinated bodily movements without the animal having to consciously guide each step in the movement sequence. The result is a slower but more flexible and constantly adjustable response, hence a more intelligent one. Thus in the subphylum *Vertebrata* we see less of a continuous progression than a trade-off, with phylogenetically stored subroutines of behaviour giving way to highly adaptable central neural processing that improves with practice.

The trick, of course, is to show convincingly that the conscious experience which accompanies such central processing in our own cases and, by extension, in the cases of other big-brained, reflex-poor animals like the mammals and the more highly evolved birds, is itself an adaptive feature of their behaviour. Professor Goudge has given some hints as to how to do this; in what follows I shall try to develop the argument further.

I

I begin indirectly, by drawing attention to the work recently done in paleoneurology by Jerison (1973). According to him, the beginnings of animal consciousness go back 150 to 200 million years, when archaic mammals first developed audition as a fine distance sense in the nocturnal niches they had invaded to escape large reptiles who, being cold-blooded, were immobilized at night. These early mammals still had essentially reptilian vision then, modified towards the mammalian retina with rod cells for dark-adapted seeing, and as in reptiles the retina-bound visual system directly generated a spatial 'map' of the world. However, for audition to complement vision in locating objects in space at twilight or dawn, it was necessary, first, to grow millions of neurons somewhere in the nervous system, and second, to develop new neural circuits that would translate temporally encoded auditory impulses so they would 'map' on the spatial code: otherwise the animal would 'see' prey or predator in one place and hear it elsewhere at the same time. Now if one adds to this the increasing development of olfaction as a third fine distance sense, also requiring millions of neurons and circuits to link it with vision and audition in an integrated neural network, it is clear that only vastly augmented central information processing, or encephalization, could meet these requirements. This trend would then have been enhanced during the great reptilian extinctions of about sixty-five million years ago, which allowed mammalian reinvasion of diurnal niches and spurred corticalization of vision, so that all three sensory modalities became interactive at cortical level where conscious functions presently reside; an even greater encephalization process begins about then, correlated with a tremendous increase of intelligence in primates during the past fifty million years or so.

Thus for Jerison consciousness arose as a means of integrating polymodal sensory information into a perceptual construct of the environment having both spatial and temporal characteristics. But he nowhere asks himself what was particularly adaptive about this greater central processing being done *consciously*. Rather, he simply assumes that consciousness is *equivalent* to complex information processing.[1] But surely one does not entail the other. Suppose we had one computer equipped with light sensors, a second that picks up sounds, and a third that samples chemical states of the environment. If we linked them together with a common spatiotemporal code, so that they could jointly label energetic states as coming from the same object in space at the same time, would we regard this unified information processing as *necessarily* conscious? Jerison extends his analysis only to biological systems, but in failing to distinguish between how higher vertebrates evolved the neural apparatus they did and how they evolved consciousness, he provides no explanatory link between them.

Perhaps a more promising tactic would be to select a sensory modality that is quintessentially conscious and ask what adaptive value *it* has. I would suggest that nociception, or *pain* perception, is such a modality.

II

Most of us are familiar with the sad plight of people who are insensitive to pain. Not only is there constant danger of unwitting self-injury, but insensitivity to what would be in us mild discomfort from failure to make postural adjustments leads to skeletal stress, infection, and fatal bone trauma (McMurray 1950). The adaptive value of pain perception as a system of warning us of impending tissue damage seems obvious from such considerations.

However some writers, impressed by chronic pain in terminal illnesses and pathological pain syndromes, have felt that there is a genuine 'puzzle' about pain (Melzack 1973); some have even tried to make an existential virtue out of it (Buytendijk 1961). Such writers are simply confused. It is true that if we did not have pain mechanisms that provide a warning of tissue damage we would not experience chronic or pathological pain either. But Darwinian principles apply to the species as a whole and not the fate of its individual members. A severely injured or terminally ill person, and for that matter someone afflicted with pathological pain, would probably have already ceased to contribute to the survival of his species in our race's pre-medical past; his or her suffering would then have been indifferent to the spread of the germ plasm and thus would not be the sort of contingency evolution can be expected to provide against.

Interestingly, the recent discovery of enkephalin, an opiate-like substance that binds to opiate receptors in the brain and spinal cord of all vertebrates (but not invertebrates) and provides partial analgesia following injury, *can* be understood on Darwinian assumptions (Kuhar, Pert, and Snyder 1973). Animals that evolved this biochemical apparatus would be less distracted and immobilized by pain and stand a better chance of getting to safety, and possible recovery, than those which did not. But this is a far cry from the sort of neural mechanism required, when pain no longer serves a warning function, to shut down the pain pathways and produce complete analgesia. Postulating such a mechanism, since it does nothing to promote species survival, amounts to smuggling God into the nervous system.

However, one might still ask if we could not have evolved pain mechanisms that trigger appropriate avoidance behaviour *without* our feeling pain. It is such a question that puts the whole problem of accounting for the evolution of consciousness among vertebrates in proper focus.

Consider first that in a sense we already do have 'painless' pain behaviour, for although reflex contraction of a limb threatened with tissue damage is normally accompanied by pain sensations, this contraction does not *depend on* pain perception. Hardy (1953) showed that long ago, with a paraplegic patient. He found that when one of her paralysed and insensitive extremities was subjected to thermal radiation it would withdraw from the stimulus as the skin temperature reached about forty-five degrees centigrade (the temperature at which she reported pain from the forehead), although of course she felt no pain in the contracted leg. Since forty-five degrees centigrade plus or minus one degree proved to be threshold for both limb withdrawal and pain reports in five normal control subjects, and threshold for withdrawal in three guinea pigs and three rats as well, this indicates a uniform withdrawal reflex in all higher mammals exposed to noxious stimulation, probably mediated by a spinal arc pathway.

However, this does not show that pain is unnecessary to avoiding tissue damage. In the case of the paraplegic woman, each time the withdrawal was completed her leg muscles would relax and allow the limb to swing down to be exposed to the beam of light again. Had she been blindfolded or her legs screened from sight, she might have noticed nothing as the leg was repeatedly exposed to the stimulus. This suggests that pain perception in the normal does indeed have a function: namely that of spurring the organism to take more *complex* avoidance action.

We can think of this in familiar terms. Suppose you are about to wash your hands and are thinking of something else as you let the hot water run. Now as you start to put a hand in the water you find it jerking away and out of the water *before* you actually feel pain in the fingers. (This, of course, is because the fibre pathway through the spinal cord is shorter than the one going up to the brain.) But if you had not felt pain and been sufficiently distracted, say by looking through the window at an accident in the street below, your hand could have repeatedly withdrawn and dropped back into the too-hot water, causing severe burn. Pain, however, concentrates the mind wonderfully, so you withhold the flexed arm and reach over with the other hand to turn on the cold water tap.

All of this so far may make rough sense. If it does, I want now to suggest that on no major theory of the mind-brain relation but one can we really understand why pain experience always accompanies complex avoidance-of-tissue-damage behaviour in man.

III

Consider a logically possible world in which such is not the case. Smith and Jones are walking along the street together when suddenly Jones comes to a halt.

Smith: Why are you stopping?
Jones: I'm not sure.
Smith: You're taking off your jacket.
Jones: I can see that. The question is why.
Smith: Perhaps you're warm?
Jones: No. Here, hold the jacket.
Smith: Your shirt too? People will think you're crazy.
Jones: Wait a minute. Aha, look!
Smith: A loose pin! It was sticking you in the back.
Jones: Exactly.
Smith: Did it do any damage?
Jones: I guess so. There's some blood.

What is strange about this possible world is not merely that Jones is represented as both conscious and engaging in complex behaviour without at the time understanding what it is that he is doing, but also that what he is doing is clearly what for us in the actual world would be *goal-directed* activity. Had there been a pebble in his shoe, he presumably would have found himself stopping, removing the shoe, turning it upside down, shaking it and exclaiming as the pebble rolls out. But then what supplies the *goal* in each case of removing the damaging stimulus, given that Jones perceives no localized pain?

If we had a sufficiently sophisticated computer program driving a parahuman robot to simulate complex pain behaviour of the sort Jones displays, we could expect a 'pain circuit' to supply the goal of removing a damaging stimulus without the robot feeling pain. So the natural tendancy here is to suppose that in such a logically possible world some neurological mechanism drives Jones' pain behaviour, though the functioning of that mechanism is not in fact accompanied by pain.

But in that case we can meaningfully ask, Why is Jones' world not our world? Why is it that in our world whatever neurological mechanisms subserve the driving of complex pain behaviour do not function painlessly? And following hard on the heels of that question, Which mind-brain theories intended to describe our own world do, and which do not, allow Jones' world being ours?

Let us start with those that *do* allow this. On double-aspect theory, a view not without adherents even among neuroscientists (Gray 1971), the pain one feels when a pin is sticking you in the back, and the postulated neurological mechanism, are but the mental and physical aspects, respectively, of an underlying and presumably unknowable process. But then it is not the pain that causes your pain behaviour, but the unfolding of this process. Thus if, conterfactually, pain were *not* a mental aspect of the process, your behaviour would be the same. And indeed, since in terms of this theory a complete causal account of behaviour and animal evolution is possible, in principle, with reference to the physical aspect of the process alone, it is the neurological mechanism which produces the pain behaviour. If so, Jones' world could be ours minus experienced pains, verbal reports of pain, or appropriate grimaces and moaning, etc.

Similar consequences arise for parallelism. The painful feeling of a pin sticking you in the back or a pebble in your shoe merely parallels the triggering by the stimulus of a postulated neurological mechanism. Since there are absolutely no causal relations between one and the other, your complex pain behaviour results from the neurological mechanism's functioning, not from the experiencing of pain. It is true, as some modern exponents of this view have held, that on parallelism one can express the relation between the experienced pain and the functioning of the neurological mechanism in a law-like way, so that one could not, in our world, behave as Jones does (Wilson 1974). But that presupposes, rather than establishes, the truth of parallelism. And since, as such theorists agree, if no mental state is not parallel to a neurological state one *need* never appeal to the former to explain behaviour or evolution, it is clear that the theory completely fails to rule out Jones' world becoming ours if, counterfactually, pain experience *ceased* to parallel functioning of the pain mechanism: all the physical laws of the universe remain unchanged and only one psychophysical law no longer obtains.[2]

205 The Ascent of Consciousness

The mind-brain theory most easily accommodating the assimilation of our world by Jones' is of course epiphenomenalism. For on this view the only difference between, say, Jones and me is that his pain mechanism drives just complex avoidance-of-tissue-damage behaviour, whereas mine causes that and pain sensations as well (plus exclamations and verbal reports of pain), the latter having absolutely no effect on anything physical in the universe. If so, all higher animals including ourselves could have survived just as well in Jones' world as ours, and the loss of felt pains in this actual world presents or would present no evolutionary disadvantage whatever.

The theory which appears, but only appears, to resist this implication is mind-brain identity. For its defenders hold it to be a fact that by random mutation higher animals evolved brain states, like the postulated neurological mechanism, which are of the sort that they are *identical* to mental states such as pain (Martin and Rosenberg 1976). And if this were true, they urge, it makes no sense to ask if Jones' world could be realized in our world, for in our world this identity relation contingently but strictly obtains. Thus, it is argued, the adaptive value of being driven to take complex avoidance-of-tissue-damage action cannot be divorced from feeling pain; felt pain and the functioning of the neurological mechanism, being one and the same process, are equally a state of a person or higher animal 'apt for bringing about a certain sort of behaviour' (Armstrong 1968, 82), namely complex pain behaviour.

I say this theory only appears to escape the difficulty for the following reason. Clearly if mental states like experiencing pain are 'nothing over and above' the functioning of a postulated neurological mechanism that evokes complex pain behaviour, then the *non*-identity of the former with this mechanism would not in any way detract from its causal efficacy in evoking that behaviour. After all, we do not have mental states putatively identical with toenail growth or bile secretion, yet these physiological processes are part of the organic causal network. Why could not the hypothetical neurological mechanism responsible for complex avoidance-of-tissue-damage behaviour play exactly the same role in our organic causal network, with all the adaptive advantage this has, without being identical to feeling pain in the back or foot? Thus it seems Jones' world could be realized in our own world (except, again, for verbal reports of pain, etc.) with no maladaptive consequences.[3]

Oddly enough, the four theories I have reviewed here share a common implication with one at the other end of the philosophical spectrum, namely occasionalism. For according to that view, a pin prick in the back or a loose pebble in the shoe is the occasion for God giving me pain sensations, just as my desire to end the pain is the occasion for Him to drive my complex pain behaviour (note that on this theory I do not need a brain). But since God could produce the latter without the former, my pain is gratuitous to successful removal of the damaging stimulus, and it becomes a nice question for theodicists why our world is *not* Jones' world. Similarly, if pain is only an aspect of an underlying process, parallel or epiphenomenal or contingently identical to a neurological mechanism that could evoke the same adaptive pain behaviour were there no pain experienced, it becomes a problem of evil for such views why we have pain at all.

I turn now to theories that do *not* allow for Jones' world being ours. On Rylean behaviourism, a theory without evolutionary pretensions, our world could not become Jones' because *that* world is logically impossible. And it is logically impossible since statements like 'Jones is in pain' *mean* behaving or being disposed to behave as Jones does, so it would be incoherent to say he behaves this way but feels no pain. (We can overlook the facts that Jones does not cry out or complain of pain, and indeed tells Smith he doesn't know what he's doing: he could have been an aphasic Stoic.) But I detect no contradiction in the skit I gave of Jones' behaviour, so it must be false that 'being in pain' *means* behaving or being disposed to behave as Jones did.

Finally, there is one theory that clearly explains, and in a way entirely consistent with Darwinian principles, why our world is not like Jones' logically possible one. This is the view that a postulated neurological mechanism *causes* localized pain sensations, which in turn *cause* the organism to find ways to remove the damaging stimulus. The reason Jones' world, while logically possible, is not ours is that in our world higher vertebrates evolved such that if they did not actually experience pain when exposed to tissue damage they could not undertake appropriate avoidance behaviour: this much is supported by the behaviour of humans insensitive to pain (who lack even reflex limb contractions under noxious stimulation). (See McMurray 1950 n8.) Thus mind-brain *interactionism*, while not without problems of its own, is alone among the major historical alternatives capable of accounting for mental evolution within a biological framework.

IV

I conclude, then, that Goudge was on the right track in holding consciousness to have an adaptive function and in discounting theories of the mind-brain relation which, overtly or covertly, do not acknowledge this. Without consciousness, of course, there could be no self-consciousness in *Homo sapiens* or any other animals that may be shown to possess this. The philosophical significance of the evolution of consciousness is just that without it the ascent of life itself would have no philosophical significance.

ALEXANDER ROSENBERG

The Interaction of Evolutionary and Genetic Theory*

Although biological thought played an important role in the works of philosophers of the nineteenth and early twentieth centuries, it did not assume any visible importance in the interests of the post-war philosophers of science until the publication of Thomas A. Goudge's *The Ascent of Life* in 1961. Although at least two of the figures to whom Goudge had in fact devoted considerable attention, Bergson (1949) and Peirce (TP), were heavily influenced by evolutionary thought in particular, philosophers had by and large attended to physics as a source of speculative inspiration, and a subject of logical analysis. However, the recent expansion of interest in the analysis of biological sciences owes as much to Goudge's pioneering work as it does to the revolutionary developments of that science in the last three decades. The years since the publication of *The Ascent of Life* have witnessed an ever-growing number of papers and books in the philosophy of biology, and all of them have had to come to terms with Goudge's views on evolutionary theory and evolutionary explanation. It can truly be said that most of the currently active philosophers of biology have been students of Goudge's – either directly, or through the influence of his works.[1]

I

The Ascent of Life, of course, provides the earliest philosophically sophisticated, and scientifically well-informed, post-war treatment of the theory of evolution. Naturally enough, its controversial claims about the prospects for axiomatization of the theory, and about the role of general laws in the explanations it provides, have been subject to searching criticism. But, while

---

*I am indebted to Professors Mary B. Williams and Michael Ruse for reading and commenting on an earlier draft. But agreement with my conclusions can no more be attributed to the latter than responsibility for my exposition can be attributed to the former.

disagreements have emerged on these issues, both Goudge and his critics have agreed on a particular characterization of the theory of evolution, as currently employed by practising biologists: the so-called 'Synthetic Theory of Evolution.' Goudge first brought this version of the theory to the attention of philosophers in the following terms: 'A central feature of ... developments ['in evolutionary theory which have taken place during the last few decades'] has been the extension of the Darwinian doctrine of natural selection and *the synthesis of it with the basic principles of genetics*. The result has been called "the synthetic theory" of evolutionary causality ... "modern selection theory," or simple "reborn Darwinism"' (AL 87 emphasis added). In one of the original expositions of this 'synthetic theory,' Julian Huxley wrote: 'Mendelism is now seen as an essential part of the theory of evolution. Mendelism does not merely explain the distributive hereditary mechanism; it also, together with selection, explains the progressive mechanism of evolution' (Huxley 1942). In adopting as their principal subject-matter the synthetic theory, Goudge and his successors have come to treat population genetics as the heart or core of the theory of evolution. In doing so, they and the biologists whose works they follow have, I believe, in fact obscured the actual relations between the theory that Darwin and Wallace first advanced, and the principles of genetics that Mendel discovered. This assimilation has resulted in a failure to appreciate both the important inter-theoretical relations that these two bodies of laws actually exemplify, and the historical incidents surrounding their appearance and eventual acceptance. In the present paper, by way of extending the studies of biological theory that Goudge initiated, I hope to show what relation actually exists between these distinct and independent theories. If I am correct, then it will turn out that far from only having the theory of evolution to hold up as a theoretical achievement, in virtue of its connections to genetics, biology is rich in theories of different levels.

Goudge does not explicitly argue that the basic principles of population genetics constitute the axiomatic core of the theory of evolution. One reason for this is that he doubt that evolutionary theory is open to axiomatization. But he does seem to attribute all the explanatory force of the theory or natural selection to genetic mechanisms of the sort discovered by Mendel. One of Goudge's successors, Michael Ruse, however, is more explicit. Ruse has claimed that 'all the different disciplines [of systematics, morphology, embryology, paleontology] are unified in that they presuppose a background of genetics, particularly population genetics. The knowledge that the population geneticist supplies about the way that heritable variations are transmitted from one generation to the next is presupposed and drawn upon by every kind of evolutionist ...' Ruse continues: 'By virtue of the fact that the theory of evolution has population genetics at its core, it shares many of the features of the physical sciences. The most vital part of the theory is axiomatized; through this part (if through nothing else) the theory contains reference to theoretical (non-observable, etc.) entities as well as to non-theoretical (observable, etc.) entities; there are bridge principles; and so on' (Ruse 1973, 48–9).

It is this very picture of the relation between genetics and evolutionary theory that is incorrect. The theory of natural selection does not presuppose population genetics; neither in the sense that the truth of evolutionary theory entails the truth of Mendelian genetics, nor in the sense that Mendelian genetics is logically compatible only with the Darwinian account of evolution, and not with many of its competitors. In arguing for this claim, and outlining the actual relation of these theories, I do not mean to suggest that the two theories are not intimately connected, and that population genetics does not provide an important part of the mechanism natural selection utilizes in making for evolution.

II

In retrospect, the theory of evolution is so starkly simple and is based on such accessible observations that one is surprised both by the lateness of its appearance and by the resistance which it met. The observations involved seem to include little more than the recognition that organisms reproduce themselves in geometric proportions, and yet the number of individuals in a biological population more often than not remains constant over long periods. Add to these evident facts the recognition that there are very great individual variations among organisms of the very same type, and some significant inferences may result. Among them is the conclusion that, since any population can produce more young than the region they occupy can maintain, there will be a struggle for existence among the organisms produced; and, in this competition, those organisms will survive, whose variations best suit them to the immediate environment – that is, a process of natural selection takes place. Add to these conclusions the further hypothesis that these selected variations may be inherited, and the result is the heart of Darwin's theory. Although this informal characterization of the theory has the virtues of intuitiveness, it is worth while to attempt a more precise account of it – especially if we are to determine its relation to other theories. Fortunately, there is already available an axiolatic presentation of the theory, due to Mary B. Williams, and the more precise account I shall offer owes its form, and much of its interpretation, to hers (Williams 1970). In passing, it should be noted that accepting Williams' account of the theory is tantamount to rejecting Goudge's claim that the theory of natural selection should not, and indeed cannot, be axiomatized.[2]

The formalization here involves four axioms, which may be stated with greater or lesser degrees of (set-theoretical) formality. Williams offers both a formal and an informal characterization of each, and the ones offered below are even more informal; however, what is lost in rigour is gained in exegetical brevity. Readers are urged to consult her formal version before drawing possibly valid but false conclusions from my version and attributing them to hers.

Axiom 1. There is an upper limit to the number of organisms in any generation of an interbreeding population.

This is the Darwinian assumption that owes its origin to the works of Malthus, and which identifies the environmental features that induce the struggle for suvival.

Axiom 2. Each organism has a certain amount of fitness with respect to its particular environment.

This is probably the most crucial axiom of the theory, for it introduces a theoretical term which is both overwhelmingly important and widely misunderstood. 'Fitness' is a theoretical term par excellence, and failures to understand this fact invariably lead to the mistaken supposition (current even among biologists who continually employ the theory) that evolutionary theory is a vast tautology. If 'fitness' is defined in terms of a reproduction rate, then relative fitness can provide no explanation of differential rates of reproduction, and the property to which it refers can play no causal role in evolution. To see that fitness is not to be so understood, but refers to a real property of organisms – a physical property – consider two biological entities as identical as two peas in a pod: if they are twins, the two peas will have exactly the same level of fitness with respect to their environment. (If peas cannot be twins, choose any organisms that can be.) Now, plant one, and incinerate the other. The result, of course, will be that the planted pea reproduces, while the incinerated one does not. But, if fitness meant a differential reproduction rate, then we should have to say that the two structurally identical peas had different levels of fitness. But if we say this, and seriously mean it, we must be willing to specify the physical differences between the peas in virtue of which they had differential levels of fitness, and this we cannot do, since ex hypothesi they were qualitatively identical with respect to relevant detectable biological properties. Of course, biologists can, and sometimes do, treat fitness as a disposition to reproduce, but this only postpones the necessity to admit that organisms have a non-dispositional property, in virtue of a relation between their own physical structure and environment, which is causally responsible for their differential reproduction

211 The Interaction of Evolutionary and Genetic Theory

rates. Consider the parallel in physics. We may explain the behaviour of certain pieces of metal in the presence of iron filings, by attributing to them the dispositional property of being magnetic. But this dispositional property must ultimately by explained in terms of a non-dispositional one – the physical orientation of the charges within the piece of metal.[3] Similarly, the explanatory force of the theory of evolution hinges on the recognition that fitness is a real, non-dispositional property of individual organisms, which is causally necessary for their respective rates of reproduction. Like other theoretical terms, 'fitness' is a predicate-functor whose value cannot be determined independently of the theory in which it figures. And, thus, like 'charge,' it is a term which can be characterized only through its causal role. We may offer an operational characterization of the term, but it will of course appeal to the phenomena which it is employed to account for; and such a characterization must not be confused with a definition, operational or otherwise. Thus, consider the characterization Williams offers: Amounts of fitness are measured in positive real numbers; given an organism, b, with k ancestors,

As a possible operational definition of [the fitness of b], $\phi(b)$, I might suggest the following. Let $v_1(b, k)$ be the sum over all the k-ancestors of b of the number of reproducing off-spring of each. Then let $v_2(b, k)$ be the number of k-ancestors of b. Then $v_3(b, k) = v_1(b, k)/v_2(b, k)$ is an estimate of the average fitness of the k-ancestors of b. Now let the operational definition of $\Phi_o(b)$:

$$\Phi_o(b) = \sum_{k=1}^{n} (0.5)^k v_3(b, k),$$

where n is the number of generations for which data are available. Then about 0.5 of the fitness of b is estimated by the 'average fitness' of its parents; about 0.25 of the fitness of b is estimated by the 'average fitness' of its grandparents; etc. ... $(0.5)^k$ is a factor which adjusts the importance to be attached to more remote generations ... (Williams 1970, 359)

Fitness is here (operationally, but *not definitionally*) characterized by reference to just those features of behaviour which it is intended to explain: relative reproduction rates. And this is how many theoretical terms are characterized in physical science. The concept of fitness also has another important and related feature characteristic of theoretical terms. Fitness is a functional concept. Two organic systems can realize the same amount of fitness with respect to their environment (can have the same fitness number) even though they are different organisms, and/or exposed to different environments. To attribute a level of fitness to an organism is not to make any *particular* claim about its structure, or its environment, just as calling something a 'mouse trap' implies nothing about its actual construction or operation. Failure to notice this feature of the concept of 'fitness' has led some to attach it conceptually to the notion to differential reproduction rate, and to claim that the theory is an unfalsifiable and empty one.[4]

I have dwelt on the concept of fitness at some length for several reasons. For one, to rebut the suggestion that it is only through the presence of genetics at the heart of evolutionary theory that this latter theory can claim the scientific virtues of making reference to theoretical terms, and containing bridge-principles (i.e., operational characterizations). If, as Ruse seems to suggest (see above pp 208–9), it is thought that only through the incorporation of the principles of population genetics does evolutionary theory take on a truly scientific character, then this is a mistake which reflection on the notion of 'fitness' should set right. Another reason for which I have dwelt on this notion is that philosophers of biology have in general failed to recognize that it is in virtue of the theoretical character of the notion of 'fitness' that evolutionary theory can most clearly and easily be recognized to be a non-tautologous system of general statements. Philosophers writing on the theory of evolution often have a very difficult time explaining in a brief way why the theory is not vacuous.[5] But once we separate out the theoretical and operational elements of the concept of 'fitness,' the empirical character of the theory becomes clear. Finally, a third reason that I have devoted special attention to this notion is that it will turn out that the concept of 'gene' in population genetics bears important similarities to it, and reveals that the actual relation between parts of molecular genetics and population genetics is similar to the actual relation between the latter theory and Darwinian theory.

To return to the axioms of evolutionary theory. Within an interbreeding population, consider a subclass of the members of the population which are homogenous with respect to some functionally or topographically characterizable feature (it could be an organ, a size, a colour, a mode of behaviour, a resistance to disease, etc.) and which are all related within a single descent-network. Call this subclass of the population D. Axiom 3. is a conditional statement about D:

> Axiom 3. If D is superior in fitness to the rest of the population for sufficiently many generations (where sufficiently many is determined by how much superior D is and how large D is), then the proportion of D in the total population will increase.

Given a method of observationally determining the membership of D, Axiom 3 also provides a bridge principle in the classic sense of the term, connecting 'fitness' and the observationally determinable change in the proportion of D to the rest of the population as generations succeed one another. In effect, this axiom suggests that if differences in fitness are hereditary, then in the long run

213 The Interaction of Evolutionary and Genetic Theory

a line of descent, within the population, of greater fitness, will increase with respect to the rest of the population. The last axiom of the theory is in effect a claim that the antecedent of Axiom 3. obtains: it is the assertion that fitness is sufficiently hereditary to make for this secular increase in the proportion of D within the entire population.

> Axiom 4. In every generation of an interbreeding population which is not on the verge of extinction, there is a subclass, the members of D, such that D is superior to the rest of the population for long enough to ensure that D will increase relative to the population, and will retain sufficient superiority to continue to increase, unless it comes to constitute all the living members of the whole interbreeding population at some time.

Williams points out that the clause concerning the verge of extinction must be included because 'in populations on the verge of extinction there may be no fitness differences which last long enough; therefore, the axiom's assertion ... must be worded so as to leave a loophole for such populations.' More important, she writes: [the Axiom does not] 'guarantee that all fitness differences are hereditary; it merely guarantees that there are some that are hereditary (and thus it indicates that some characteristics of the environment are reasonably stable as well that some characteristics of the organisms are passed on to their descendants)' (Williams 1970, 366–7).

Williams asserts that from the axioms on which the above propositions are based, all the characteristic statements of the theory of evolution can be derived as logical consequences, and she goes on to argue for this by deducing a large number of theorems, including ones which capture such leading Darwinian ideas as differential perpetuation, and descent with adaptive modification. But for our purposes the important features of the theory are already evident – provided the axiomatization given is accepted. What is evident is that these axioms make no claims about heredity, other than to postulate its existence, and draw its consequences. In particular, they make no claims about the *mechanism* whereby hereditary traits are transmitted from one generation to the next. As stated, Darwin's theory may cry out for laws which account for the heredity it requires, but it does not embody any such laws itself, nor does it place any restrictions on either the form or content of laws which might govern the hereditary phenomena whose existence it requires, but about which it is otherwise silent.

III

That the theory of evolution requires the existence of hereditary mechanisms is clear from inspection of its axioms. This fact is also apparent from the history of science. For Darwin himself felt impelled to provide a theory of inheritance to underlie his theory of natural selection. This was the so-called theory of pangenesis, or blending inheritance, which involved the existence of 'gemmules' as the units of hereditary function. Equally, the logical independence of this theory from the theory of evolution is evident from the fact that its obviously unsuccessful character does not shake anyone's confidence in the theory of natural selection. The real independence of the theory of evolution from any particular genetic theory is even more obvious in the fact that the earliest exponents of Mendelian genetic theory (taken by Ruse and synthetic evolutionists generally to be the core of evolutionary theory) took its results to be incompatible both with Darwin's hereditary theory of pangenesis, and with the theory of natural selection itself. Indeed, Goudge has cited one of these scientists as saying: 'Mendelism came and swept the whole theory [of evolution] away' (Goudge 1967b). Unless we are to accuse these geneticists of simple logical inconsistency, we cannot suppose that the principles of Mendelian genetics represent the axiomatic foundations of the theory of natural selection. Furthermore, Mendel's laws are in fact formally consistent with other theories, like Lamarck's, which are supposed to be logically incompatible alternative competitors to the Darwinian account of evolution. And, of course, in and of themselves Mendel's laws cannot provide the heart of evolutionary theory, for they make no mention of mutation, the vehicle of genetic change, and so do not allow for evolutionary change at all.

The central principles of Mendelian or population genetics are the law of segregation and the principle of independent assortment:

Segregation: For each sexual individual, each parent contributes one and only one of the genes at every locus. These genes come from the corresponding loci in the parents, and the chance of any parental gene being transmitted is the same as the chance of the other gene at the same parental locus.

Independent Assortment: The chances of an off-spring receiving a particular gene from a particular parent are independent of the off-spring's chances of receiving any other gene (at a different locus) from that parent (Ruse 1973c 13–14)

In and of themselves these two principles can hardly represent the core of the theory of evolution, for, as Ruse has shown, these laws entail that within an interbreeding population gene frequencies *remain the same* (ceteris paribus). That is, together with the assumptions that male-female crosses and female-male crosses are identical, that mating is random, and that the population is very large, from Mendel's laws we may deduce the Hardy-Weinberg law, to the effect that, ceteris paribus, gene frequencies and genotypes remain con-

215 The Interaction of Evolutionary and Genetic Theory

stant from generation of generation in sexually reproducing populations. But if these frequencies and proportions remain the same, then the phenotypes for which they control will remain the same, as well, over generations, and no evolution will result. In fact, it is only by introducing considerations from evolutionary theory, like fitness, and by appeal to the notion of mutation, that the Hardy-Weinberg law, and the Mendelian principles from which it stems, have a bearing on evolution. Biologists have long recognized this fact:

The Hardy-Weinberg law is entirely theoretical. A set of underlying assumptions are made that can scarcely be fulfilled in any natural population. We implicitly assume the absence of recurring mutations, the absence of any degree of preferential matings, the absence of differential mortality or fertility, the absence of immigration or emigration of individuals, and the absence of fluctuations of gene frequencies due to sheer chance. But therein lies the significance of the Hardy-Weinberg law. *In revealing the conditions under which evolutionary change cannot occur, it brings to light the forces that operate to cause a change in the genetic composition of a population. The Hardy-Weinberg law thus depicts a static situation.* (Volpe 1967, 38–9 emphasis added)

Darwinian theory requires some law of heredity or other, and Mendel's laws (suitably revised in the light of phenomena like linkage) provide correct laws of the sort that evolutionary theory requires, but they do not make for evolution itself. Indeed, they cannot, if they are to provide the sorts of laws which natural selection theory requires. What it requires is an account of how traits are passed on, *unchanged*, so that they may continue to exist in a population. It is mutation and selection which account for their spread or disappearance throughout that population. Other hereditary laws could equally well provide evolutionary theory with its hereditary requirements, and our choice of Mendel's principles is not based on the fact that they provide the assumptions of Darwinian theory, but on the fact that they are the most well-confirmed candidates among the alternative statements which could meet the evolutionary theory's requirements. This should be clear both from the fact that the laws of independent assortment and segregation have required significant modification as a result of findings in molecular genetics, without thereby producing a change in the smallest detail of evolutionary theory; and from the fact that we may construct new hereditary laws incompatible with Mendel's (and doubtless false ones) which logically could serve Darwinian needs just as well.

Even more strikingly, Mendel's laws can easily be shown to be compatible with the Darwinian theory's principal competitors as alternative accounts of evolution. Indeed, the laws of heredity that population genetics provides are as much required by this competitor's appeal to heredity as they are required by Darwin's theory. As Goudge has noted, neo-Lamarckianism is 'regarded by many as the only serious alternative to Darwin's [theory].' Goudge goes on to describe neo-Lamarckianism in the following terms:

According to it, the course of evolution is due to the following factors: (i) a changing environment which acts on individual organisms; (ii) the consequent production of new needs in those organisms, needs which must be met if the organism is to survive; (iii) the active response of some organisms to meet these needs, i.e., the establishment of new habits (or the cessation of old habits) of use of various bodily parts; (iv) a resulting change in those parts and hence in the somatic structure of the individual organisms which possess them; and (v) the transmission of the structural changes and habits from one generation to the next, so that the organisms concerned are gradually transformed into new species ... Thus the theory involves the assumption that the effects of use and disuse, or environmentally induced effects on the individual, are inherited in kind and become germinally fixed. This assumption has come to be known as 'the inheritance of acquired characteristics ...' (AL 84–5)

It is clear that neo-Lamarckianism is in as much need of laws to explain the nature of the hereditary phenomena it postulates as is Darwin's theory. In particular, factor (v) above claims that traits are hereditarily transmitted. Why should they not be transmitted by genes in accordance with Mendel's laws? All we need add to enable Mendel's laws to underlie the course of neo-Lamarckian evolution is the stipulation that the changes in somatic structure described in (v) above make for changes in the germ cells, and consequently are inherited in accordance with (v) above, and, in particular, produce adaptive mutations in the genes of the animals which make active responses to their environmental needs. It is of no consequence for this claim that there is no convincing evidence whatever for such a general statement. What is important is only that the logical conceivability of such a claim shows the consistency of neo-Lamarckianism and Mendel's laws of population genetics. It shows to be far too strong Goudge's claim that 'the idea of [neo-Lamarckian inheritance] appears incompatible with the body of knowledge so far assembled by the science of genetics' (AL 86). At any rate neo-Lamarckianism is formally consistent with Mendel's laws of segregation and independent assortment.

Nothing could more clearly show the formal independence of the theory of evolution from population genetics than a demonstration that neither theory entails the other, and that both theories are each compatible with one another's competitors. In the light of such a demonstration we must rethink the relation between them, because, though it is not that of axiom and theorem, it is an intimate relation nevertheless.

IV

How, then, are these theories related to one another? The best analogy is with the relation of bricks in a wall. Bricks at the top of a wall are supported by bricks lower down in the wall, and these lower bricks may be removed and replaced without bringing down the whole wall or dislodging the bricks at the top, so long as this is carefully done. Pursuing the analogy, in so far as the theory of evolution requires some other theory to account for the heredity which it postulates, it rests on our ability to provide such a theory. Which theory we provide will be governed by the normal canons of investigation and theory construction, but like bricks of a certain size and strength, such theories are all logically capable of filling a place in the theoretical edifice of biology; any of them will provide the sort of support which evolutionary theory requires, just so long as each of them provides an account of heredity. A theory of heredity is not the only theory upon which evolutionary theory rests. Like a brick in a wall, it rests upon a number of theories below it, each of which provides it with support, but each of which may be replaced by another theory without affecting the standing of evolution. What these theories are can be inferred from an examination of the axioms of the theory offered in section II. Axioms 3. and 4. cry out for some theory of heredity or other, while Axioms 1. and 2. demand some theories or others that will account for the anatomy, physiology, and behaviour of organisms and for their environmental conditions; theories that will explain why there is a limit in the possible size of any interbreeding population, and why a certain organism and its environment are so related as to determine a quantity of fitness. Such theories play just as central a role in the foundations of evolutionary theory as does a theory of heredity. Thus, for example, although the evolutionary fitness of some animals is a function of their skill as predators, and the theory may require some account of this skill, it is quite neutral on competing theories of operant learning or species-specific behaviour that may account for it.

Accordingly, the reduction of evolutionary theory to more basic ones is quite complicated, for it involves reduction not just to a theory of heredity, but also to a wide variety of other theories from which one or another of the assumptions of the theory follow. In this respect, reduction in biology is quite unlike some reduction in physical theory. The equation of state, for example, of a gas is deducible from one theory alone, statistical mechanics, which itself is ultimately explicable by reference to quantum mechanical considerations. The ultimate reduction of evolutionary theory (to the possibility of which the present author subscribes) will, however, not involve so direct a straight line of theories one underneath the other. Instead it will involve, first, reduction to a number of theories, of which population genetics is but one.

In the reduction, thus far accomplished, of population genetics to molecular genetics, biologists and biochemists are pursuing one of the many lines required to ultimately reduce evolutionary theory to its physical basis. Here, between population genetics and molecular genetics, there is a sort of reduction which is midway in complexity between that which characterizes evolutionary reduction, and that which characterizes thermodynamic reduction. That is, we may ultimately see how to reduce population genetics to one theory, that of the behaviour of nucleic acids, but this reduction will not be effected simply by a one to one correspondence between Mendel's genes and strings of DNA. The reductions will turn out to be of all kinds: one Mendelian gene realized by any of many molecular genes, a single molecular gene realizing a number of Mendelian genes, and so on.[6] The reason for this sort of complexity in the reduction of Mendelian to molecular genetics is that, like the evolutionary concept of 'fitness,' the Mendelian concept of 'gene' (and its successor concepts in transmission genetics: recon, muton, cistron) is a functionally characterized one. Mendelian genetics and transmission genetics offer an account of the mechanism of heredity by appeal to a theoretical item, the gene, but they tell us nothing about its nature, or about how it accomplishes (by duplication and replication) the hereditary activity the theory explains by appeal to it. More often than not the physical realization of a functionally characterized term may be one of many physically different kinds of items. Thus, just as two organisms may have the same amount of fitness in virtue of quite different relations to their environments (they may in fact be two different organisms), a single Mendelian or transmission genetical gene may reflect a diversity of different disjunctive and/or conjunctive molecular mechanisms. Although population genetics rests on only one theory, each of its general statements itself is explained by reference to disjunctions and/or conjunctions of a number of the general statements of molecular genetics. Here too, the relation is like that of one brick in a wall being sustained in position by the support of a number of others, which, again, can be gingerly removed without dislodging the bricks above it. But the bricks of our analogy are not now whole theories but various parts of them. This is why the reduction of population genetics to molecular genetics is more complicated than the reductions often available in physics, like that between the gas laws and mechanics. It is, nevertheless, simpler than the reduction which can be affected for evolutionary theory, involving as it does only one theory, and not several.

V

What may we conclude from this examination of evolutionary theory and its relation to population genetics? One thing we may conclude is that Darwinian theory has no need of an alliance with population genetics in order to lay claim to theoretical terms, bridge-principles, and the other features which logical empiricist philosophy of science *rightly* demands of a scientific theory. For it manifests such features already. Thus, an argument like Ruse's, which seeks to accord evolutionary theory respectability in virtue of an alliance with a theory that has all these features (population genetics) is really superfluous. More important, we are in a position to see what is wrong in Huxley's claim that Mendelism is an essential part of the theory of evolution, and that it explains the mechanism of evolution. Mendelism, it turns out, is not essential to the theory of evolution, except in so far as this latter theory requires some account of heredity or other, and Mendel's account (suitably revised) is the one we have adopted (on other grounds than evolutionary ones, by and large). And, similarly, Mendelism helps explain the mechanism of evolution, but only in the sense that Darwinian explanations of evolution appeal to the existence of heredity. Failure to recognize these facts leaves us at a loss to explain either why Darwin adopted a hereditary theory plainly incompatible with Mendel's, or why early Mendelians supposed that their theory was incompatible with Darwin's. It also leads to mistaken claims to the effect that embracing a particular theory of heredity logically excludes embracing any but one theory of evolution.

Perhaps most important for the philosophy of biology, we may conclude that the uncritical acceptance of the conjunction by biologists of two separate theories under one name has blinded philosophers of biology to the precise relation between the two, and to the characteristic relation between theories in biology that they actually exemplify. If the present essay has done something to encourage a more thorough analysis of these relations in the life sciences, then it will represent one more step in the path to a more sophisticated appreciation of biology among the sciences; a path opened up to this generation of philosophers by the work of Thomas Goudge.

MICHAEL RUSE

Philosophical Aspects of the Darwinian Revolution

In his writings, Thomas A. Goudge has shown a sensitive and informed approach both to historical and to philosophical aspects of the biological sciences.[1] In this essay I hope to follow in his footsteps, to this extent at least: I want to look at philosophical elements operative in a period in the history of biology, and I shall hope thereby to extract historical juice applicable to philosophical claims Goudge has made about biology. In particular, I shall argue that history suggests that Goudge's central claim about evolutionary biology, that it is in significant respects different from a physical science, is well taken, but that the major reason this is so may not be quite that argued by Goudge.

I REGULATIVE PRINCIPLES

The period of science with which I am concerned can for convenience be called the 'Darwinian Revolution.' It involves the debate over organic origins from about 1830 to about 1880, and it covers the change of world-picture from organisms as specially created to organisms as evolved, climaxed by the publication of Charles Darwin's *Origin of Species* in 1859. In dealing with this revolution I intend to isolate and consider exclusively the 'metascientific' or philosophical level of debate, and in order to do this I shall employ the metascientific notion of a *regulative principle*. This notion, championed today by neo-Kantians, draws attention to the fact that scientists frequently defend their theories on the grounds that only they satisfy the correct and adequate canons of proper scientific explanation – in short, scientists appeal to things 'governing more or less strictly the search for physical formalisms and the choice between them ... [Defining] like other normative propositions, standards of excellence, [and] standards of the intelligibility or explanatory power of physical formalisms' (Körner 1960; see also Margenau 1950 and Rescher 1973). A classic example of a debate appealing to such principles was that between Schrödinger and Einstein on the one hand and (including others) Pauli on the other hand, about whether or not scientific theories *must* be deterministically spatiotemporal in a traditional sort of way. In the Darwinian revolution, I isolate two such regulative principles which were of crucial importance. These I label the *Newtonian* principle and the *Teleological* principle.

By the *Newtonian* principle I understand the feeling that all science (including biology) must in some sense be judged by the standards of the best kind of science, that is to say physics, and more particularly, the methodology and results of Newtonian astronomy (1830s' version) (see Wilson 1974; also Ruse 1975a, 1975b, and 1979). The principle therefore, championed incessantly by the philosopher-scientists John F.W. Herschel and William Whewell, is that to do good science (including good biology), one must produce work which has significant similarities to Newtonian astronomy, and that the best methodological approach is that of the Newtonian astronomer. But, what precisely were these standards being set by Newtonian astronomy? For a start, it was taken to be a body of laws, that is universal statements whose truth was thought to be more than a matter of mere chance (Herschel 1831, 78; Whewell 1837, 2, 152–87; 1840, 1, 53–8). These laws were bound together in a hypothetico-deductive system, where certain laws were taken as basic (i.e. as axioms) and all others deduced from them (Herschel 1831, 104). The inferred laws, like Kepler's laws, made direct reference to the observable world and thus were 'phenomenal' or 'formal' laws. The axioms, like Newton's laws, however, made reference to causes and thus were not merely logically prior to phenomenal laws but in some way were epistemologically more important too. The precise notion of a cause was not always made that clear, but Herschel and Whewell were united in the belief that, as Newton himself had said the best kinds of causes are *verae causae* (Herschel 1831, 144; Whewell 1840, 2, 446) – true causes. There were taken to be two prime marks of what constitutes a *vera causa* – either when one can argue analogically from a cause of which one has direct acquaintance, or when one can locate a cause in an axiom from which many diverse areas of a science can be explained (in Whewell's words, when one has a 'consilience'). Herschel, an empiricist, tended to see the essential essence of *vera causa* in the satisfaction of the first condition, whereas Whewell, a rationalist, favoured the second condition, but both recognized that the strength of Newtonian gravitational attraction was that it satisfied either condition and was thus the *vera causa* par excellence (Herschel 1831, 148, 170; Whewell 1837, 2, 206–53; Todhunter 1876, 2, 389).

Finally, both recognized that a crucial feature of Newtonian astronomy was its quantitative nature – it yielded predictions of precise accuracy (Herschel 1831, 123; Whewell 1831, 388).

The other regulative principle is the *Teleological* principle. It was felt that in our understanding of organisms we must pay attention to (and reflect) the fact that they show organization, that they seem to have been designed for the ends of survival and reproduction. Thus, as we talk of human-made artifacts, we can and must speak of organisms having *functions* or of things happening *in order that* other things might happen. Another way of describing this sense of organization is to speak of organisms showing *final causes*, and Whewell in particular was quite explicit about the regulative import that he put on this notion:

Thus we necessarily include, in our Idea of Organization, the notion of an end, a purpose, a design; or, to use another phrase which has been peculiarly appropriated in this case, a *Final Cause*. This idea of a Final Cause is an essential condition in order to the pursuing our researches respecting organized bodies (Whewell 1840, 2, 78; Herschel 1841, 229).

Because, obviously, most people then drew a connexion between the Teleological principle and God's Design, organisms seem designed because they are in fact designed, one might be tempted to think that the principle was not really metascientific at all – that it was really natural-theological. However, although the principle was undoubtedly used to support natural theology and although, as we shall see, the principle led some thinkers to argue that certain problems lie beyond science, inasmuch as the principle demanded that one recognize in organisms a sense of organization, it seems indeed to have been fully metascientific in the sense of a criterion of good science. It was used as a heuristic principle to understand the workings of organisms, and, as a means to criticize certain proposals as bad science because they failed to satisfy it.

Two final remarks should be made about these two principles, the Newtonian and the Teleological. First, in a sense they clash. There is no place for final cause in physics. Hence, in being teleological one is being non-Newtonian (in the sense understood – Newton himself was a teleologist.) However, this was felt to cause no real tension, for the point is that inasmuch as they are teleological, organisms show that they are not just inert matter. Second, I do not claim that these two principles were necessarily the only principles functioning in biology, nor do I deny that they might be subdivided. However, I do think that as presented they were the principles most commonly invoked. Hence, by working with the two principles as presented, we enable the major pertinent metascientific or philosophical factors in the organic origins' debate to emerge.

II THE ANTI-EVOLUTIONISTS

In the 1830s and early 1840s one can distinguish two positions on the organic origins question – positions which I have elsewhere referred to as those of the 'liberals' and the 'conservatives' (Ruse 1975b). In the liberal camp I put, amongst others, Herschel and the geologist Charles Lyell. In the conservative camp I put, amongst others, Whewell and the geologist Adam Sedgwick. In this section I shall sketch the reasons behind the two positions, giving particular emphasis to the role played by my two regulative principles.

Charles Lyell (1830–3) did at least two things in his *Principles*. First, he employed the method of *actualism*, arguing that past events in world history ought to be explained in terms of causes of a kind known to us at present. Secondly, he advocated a *uniformitarian* view of world history, arguing that there were no past events which were of greater intensity than those occurring at present and that history overall shows no significant direction – that is, that it is in a kind of steady state (Ruse 1976a).[2] In the course of his argument he considered the history of the organic world, and from his treatment two claims emerge. First, employing his actualistic method he argued that we have no reason to believe at present that organisms change or can be changed from one species to another. Hence, he felt able to argue against theories of organic evolution like that of the French biologist Lamarck (Lyell 1830–3, 2, 1–65). Second, however, although he did not say so explicitly, he left the reader in little doubt that he felt that new species must come about through natural (i.e. law-bound) mechanisms (Lyell 1830–3, 2, 176–84). And about five years later this position at which Lyell was hinting was made fully explicit by Herschel, who wrote in an 1836 letter to Lyell (shortly made public) that the most reasonable belief is to suppose that even for new species God 'operates through a series of intermediate causes ...' (Babbage 1838, 226).

The liberal position therefore was that organic origins must somehow be subject to natural law. On the other hand, liberals rejected all current evolutionary theories. In arriving at these conclusions can we see either of our two regulative principles at work? That the Newtonian principle was at work was hinted at strongly by Herschel as he argued to his position that the origination of new species must by naturally law-bound, for he stated that we are led to such a position 'by all analogy' (Babbage 1838, 226) – and all analogy obviously includes that science to which he accorded highest place, Newtonian astronomy. And Lyell was more explicit. In arguing for his position (against Whewell) he invited Whewell to consider an analogy from the tides (an area in which Whewell had done extensive work and which comes under the scope of the Newtonian astronomical corpus). Suppose, asked Lyell, we were faced with some extraordinary phenomenon from the past, like a monstrous tide. Would it be reasonable to explain it in terms of miraculous creations and annihilations of the heavenly bodies (Lyell 1881, 2, 6)? Obviously not, argued Lyell. It would be more 'philosophical' to assume that such a phenomenon could eventually be explained by normal law-bound causes, particulary since phenomena like this have been so explained in the past. And, concluded Lyell, so ought one to argue in the case of geology, including the organic past. Astronomy points away from the invocation of miracles.

The Newtonian principle also arises in the other side to the liberal position, namely the refutation of existing evolutionary theories. I have explained how this was done through the actualistic methodology, and fairly obviously, Lyellian actualism is the geological counterpart to (Herschel's understanding of) the astronomical striving for *verae causae*. In both cases one is demanding that one explain from known causes, and it is clear that the great enthusiasm that Herschel felt for Lyell's methodology stemmed from this fact.[3] Moreover, Lyell himself took pride in the way he was striving for *verae causae* (Lyell 1881, 2, 3). Thus it follows that in the way evolutionary theories were attacked, the Newtonian stamp of approval was being applied. To be an actualist was to be a geological Newtonian, and it was as such a geological Newtonian that one rejected the work of men like Lamarck.

The liberals therefore thought of themselves as good Newtonians in their approach to the organic origins question. So far no mention has been made of the Teleological principle, and indeed it would seem that the Newtonian principle was all that was needed for the liberals to argue to their position. I suspect however that the liberals' position was in fact influenced by their attitude to the Teleological principle, although nothing very explicit was said. Since the conservatives were much more open about the influence of the Teleological principle I shall turn now to them, and then return briefly to the liberals at the end of this section.

Conservatives like Whewell and Sedgwick agreed entirely with the liberals' attack on evolutionary theories like Lamarck's – Whewell indeed quoted Lyell almost verbatim on this matter (Whewell 1837, 3, 575–6). But not for them were unknown causes for organic origins – causes which although unknown were definitely law-bound. They applied the Teleological principle to the problem of organic origins and they found that no natural speciation mechanism, known or unknown, could satisfy it. Organisms show organization or final causes. This we cannot and must not deny. But all our experience of the origination of things which show organization or final causes (i.e. human artifacts) is that they do not occur by blind unguided law: they require an intentional act of design and creation. Hence, organisms too must require such an intentional act. Therefore, God must have intervened directly to create new species. We must invoke miracles – that is, things outside natural law (Whewell 1837, 3, 574, 588 and Sedgwick 1833, 26–7).

It is clear that, whereas the liberals were bringing the Newtonian principle to the fore, the conservatives were bringing the Teleological principle to the fore. Does this therefore mean that the conservatives were violating the Newtonian principle in their blatant appeal to miracles? In their own eyes, this was not the case at all. They agreed fully with the liberals that to provide good scientific arguments one must explain through and only through natural law. Therefore, they argued that the question of organic origins, which must involve miracles, *must lie outside or beyond science*. The Teleological principle drove them to miracles; the Newtonian principle drove them to conclude that organic origins are extra-scientific. 'The mystery of creation is not within the range of [geology's] legitimate territory; she says nothing, but she points upward' (Whewell 1837, 3, 588).

Here then we have the positions of the liberals and the conservatives, although one should take care not to suppose the positions that far apart. Both sides were against current evolutionary theories, and even though the conservatives tried to take the whole question of organic origins right out of science because of the supposed incompatibility between the Teleological principle and a natural speciation mechanism, there is good reason to believe that the liberals worried about this conflict too. Lyell and Herschel both recognized organic adaptation, both saw a close link between this adaptation and God's Design (Lyell 1830–3, 2, 23, 41, 44, 159; Herschel 1841, 229) and, as the organic origins debate heated up, both worried about how a natural mechanism could lead to adaptation.[4] Hence, one suspects that although their Newtonianism led them to suppose that new species must occur naturally, they were relieved that their Newtonianism also led them away from all proposed mechanisms of speciation. For them, no less than for the conservatives, it would be difficult to harmonize teleology and natural species' processes.

In short, I suggest that the two regulative principles played crucial roles in the formation of the two anti-evolutionary positions, and it is clear also why some might be dissatisfied with the positions. The liberals were forced to allow that they had no *vera causa* to explain new species; the conservatives had to take the whole question of new species right out of science. Let us now turn, keeping the regulative principles as our guides, to the ways in which the evolutionists tried to escape from the problems of the liberals and the conservatives, and to the critical reactions which their proposals evoked.

III ROBERT CHAMBERS (1802–71)

Before Darwin published, the best-known British evolutionary hypothesis was that of Robert Chambers, who expounded and defended his ideas in two (anonymously published) works, *The Vestiges of the Natural History of Creation* (1844) and *Explanations, a Sequel to 'Vestiges'* (1845). Chambers began his argument by accepting the so-called 'nebular hypothesis' for the universe's origin – an hypothesis which supposes that the universe originated from clouds of gas in a law-bound, evolutionary manner – and Chambers made it crystal clear why he appealed to this hypothesis. He wanted to suggest that physical evolution makes probable biological evolution – 'We have seen ... that the construction of this globe ... was the result ... of natural laws ... What is to hinder our supposing that the organic creation is also a result of natural laws ...' (Chambers 1844, 153–4). However, Chambers made it clear also that he thought the very existence of astronomy itself was more important to his position than any mere hypothesis from astronomy. Even if the nebular hypothesis be untrue, Chambers thought physics as such, particularly astronomy, makes a biological evolutionary theory plausible – 'the basis of the entire system of nature developed in my book ... lies in the material laws found to prevail throughout the universe, which explain why the masses of space are globular; why planets revolve round suns in elliptical orbits; how their rates of speed are high in proportion to their nearness to the centre of attraction; and so forth' (Chambers 1845, 5). And, having made this claim, to round off the argument as it were, Chambers referred then to the findings of the Belgian investigator Quetelet, who had just recently shown that certain human phenomena, such as birth and death rates, sizes, and tendencies to crime, can profitably be subjected to (statistical) law-like analyses. This done, with his outer limits to the range of law set (planets and men!), Chambers felt he had provided a powerful argument showing that everything *must* be subject to the rule of law, including organic origins.

We see already, therefore, that what I have called the Newtonian regulative principle was playing a role in Chambers' position on organic origins. He thought that Newtonian astronomy (that is the astronomy of the 1830s) points not only towards a law-bound universe, but even towards a gradually evolved universe. In Chambers' opinion, the biologist cannot, or at least should not, remain indifferent to these facts. Of course, Chambers could not leave matters just like this – he had to give some reasons to back up his metascientific belief that organic origins must be due to (evolutionary) law. To this end Chambers turned next to more empirical questions – he suggested that the fossil record gives evidence of organic evolution, that many things imply that the inorganic develops into the organic, and so on. Then he came to his mechansim of evolutionary change – such as it was. Normally, he suggested, (organic) like produced (organic) like. However, he argued, sometimes laws, including those of generation, have exceptions. In support of this claim, he referred to findings by the mathematician Charles Babbage. Babbage had invented a crude kind of computer, and he showed how he could set the machine to provide a series which would follow the regular course of natural numbers from 1 up to 100,000,001, at which point the machine would produce as successor, not 100,000,002 but 100,010,002. Thus concluded Babbage (1838, 30–49) (and Chambers with him), although the laws that we hold seem to be confirmed by many instances, we should not therefore rule out the possibilities of exceptions which would show that nature is governed by laws rather different from those which we presently hold.

Chambers therefore argued that although like normally produces like it need not necessarily always do so. On rare occasions, like might produce unlike, and thus new species would be formed. But how precisely this might occur was not made too clear by Chambers. He seemed to think that in some way, sometimes, something would happen to the development of an embryo, and from the parent of one species we would get the offspring of another. Into further detail than this he did not go, for, hurrying along, Chambers was ready to bring his argument to a triumphant climax, writing that: 'The inorganic has one final comprehensive law, GRAVITATION. The organic, the other great department of mundane things, rests in like manner on one law, and that is – DEVELOPMENT (Chambers 1844, 359–60).

This claim could not make more clear the influence of the Newtonian regulative principle on Chambers' thought. He had argued at the beginning that astronomy made imperative the quest for a biological evolutionary theory – now at the end, he suggested that his theory was the biological analogue of science's finest theory, Newtonian astronomy. Chambers thought that to be a Newtonian one had to be an evolutionist, and conversely he thought that in his evolutionism he was being Newtonian.

Before turning to Chambers' critics, let us raise the question of our other regulative principle, the Teleological principle, as it relates to Chambers' thought. Apparently, in the first edition of the *Vestiges*, Chambers saw nothing in his theory which was incompatible with or in any way threatened the notion of final cause in the organic world. Indeed, he wrote simply that 'From the mandibles of insects to the hand of man, all is seen to be in the most harmonious relation to the things of the outward world' (Chambers 1844, 324), and matters were allowed to rest with this, although, as we shall see, matters did not rest very long.

Few works have had poured on them the scorn and invective it was the fate of Chambers' *Vestiges* to receive. Nothing was left untouched; no claim was unchallenged. Here I shall ignore the empirical criticisms (such as those made, for example, on the basis of the fossil record) and I shall concentrate solely on questions raised about the metascientific aspects of Chambers' theory, specifically questions relating to the connection between Chambers' theory and Newtonian and Teleological regulative principles. First, let us see what reaction there was to Chambers' claims that to be a Newtonian one must be an evolutionist and that his theory was the biological analogue of Newtonian astronomy.

Almost everyone rejected the nebular hypothesis (for example, Anon. 1850, 359; Anon. 1845, 493; Sedgwick 1845, 19; Brewster 1844, 477; and Bowen 1845, 441) – recent telescopic evidence had suggested that there were no genuine nebulae in the heavens at all (supposed clouds of gas resolved into distinct stars), and thus a crucial premise of the hypothesis vanished. By about the tenth edition of the *Vestiges* (1853), even Chambers modified his categorical support for the hypothesis, and so one of his key analogies – astronomical evolution, therefore biological evolution – crumbled if not vanished. But the critics thought that more should vanish than just the appeal to the nebular hypothesis. They disliked Chambers' whole reference to astronomy, and even more, they disliked his reference to Quetelet's findings in order to round out his general appeal to physics in support of his position (that is, that with planets subject to law at one end, man subject to law at the other, we must therefore find biological evolution in the middle). For instance Sedgwick wrote: 'That man, as a moral and social being, is under law we believe true; but when it is affirmed that this law ... is of the same order with the mechanical laws that govern the undeviating movements of the heavenly spheres, we believe the affirmation to be utterly untrue' (1850 cxlviii). In short, the critics argued that even if Chambers were able to place evolutionary laws between Newtonian laws and Quetelet's laws (and this in itself they disputed), this still would not mean that he would be able to claim that he had a genuinely scientific theory, because Quetelet's laws are not genuine laws like Newton's laws.

Next the critics turned to Chambers' use of Babbage and his findings. Again it was claimed that Chambers referred to non-genuine (i.e. non-Newtonian) laws – this time because Babbage's laws were the results of his own machinations, rather than, as in the case of Newtonian laws, something discovered and not made (Sedgwick 1845, 66). And in any case, it was argued, even were Chambers to show the organic world subject to law (in a manner similar to the astronomical world), this still proves no biological evolution. 'The laws which we study and admire, whether in the inorganic or organic world, explain the succession of phenomena, but throw no light upon the origin of the bodies in which the phenomena are observed. They tell us how things go on; they do not tell us how they began' (Gray 1846). In other words, to be a good biological Newtonian, it is enough merely to pick up the laws as we find them presently working in bodies – there is no necessity to go delving into origins.

Finally, we have the criticisms of Chambers' evolutionary 'mechanism.' It was at this point that Herschel made his attack on the *Vestiges*. Initially one might think that Herschel would not have been totally unsympathetic to Chambers' views – after all, Herschel like Chambers thought (by analogy) that species' origins must be natural. However, apparently Chambers hit a raw nerve when he had the audacity to liken his own speculations to those of Newton. In terms of the distinction between causes and phenomenal laws, it was clear that at best Chambers had provided the latter. There was certainly no question of giving a *vera causa* for new species, for Chambers could not in any way point to any known forces which lead to new species. Hence, in Herschel's eyes it was ridiculous for Chambers to claim that he had found the biological analogue of gravitational attraction. At best Chambers had provided that which ought to be explained, not that which explains (Herschel 1857, 675–6).[5]

We can see therefore that Chambers was faulted right down the line with respect to Newtonianism – his critics claimed that they did not have to become evolutionists to be Newtonian, and conversely they claimed that Chambers' theory did not measure up to Newtonian standards. But this was only one charge in the critics' barrel – there was also the question of the Teleological principle. Chambers may have been satisfied that he had left ample room for final causes – his critics soon shattered that happy dream.

Typical of the responses to *Vestiges* was that of Whewell. Whewell went after Chambers with respect to the question of teleology in two ways. First, he gathered together all the passages in his *History* and his *Philosophy* on the subject of final causes and, together with all his explicitly anti-evolutionary passages, he published them in a slim volume (1845) as being 'pertinent' to the question of organic origins! Then, just in case anyone was too naive to catch his drift, he wrote a stirring preface to the volume denying explicitly that the evidence of organic adaptation or final cause and the evidence of evolution – particularly Chambers' version of evolution – can be harmonized.

Against (the many) criticisms of this kind, Chambers, through the numerous editions of *Vestiges*, fought somewhat of a rearguard action. Basically, he did three things. First, he made very clear, as perhaps he should have done right at the beginning, that he believed that if there be final cause in the organic world, it is final cause brought about at remote control through the agency of laws (Chambers 1845, 134). But secondly, Chambers denied that the (biological) world gives much evidence of final cause anyway! (Chambers 1845, 151–2) And thirdly, against criticisms based on the Teleological principle, in later editions of the *Vestiges* Chambers added to his theory a subsidiary evolutionary mechanism – a kind of Lamarckian response to organic needs. This mechanism would, supposedly, yield many of the adaptations which we see in the organic world – adaptations which so many of his critics argued made imperative the acknowledgment of the Teleological principle in biology (Chambers 1853, 155–6).

No further argument is necessary in support of my claim that in the debate which swirled around Chambers' theory there were many metascientific elements resting on the two regulative principles which I identified in the last section. The above discussion centreing on questions to do with teleology shows that, just as much as in the context of the discussion to do with Newtonianism, Chambers and his critics were raising fundamental questions within the context of these principles about what is to count as a proper answer in biology. Let us now turn at once to Darwin and his critics to show that there too one has such fundamental questions.

IV CHARLES DARWIN (1808–82)

I have shown how Herschel and Whewell explicated in some detail the Newtonian ideal of a scientific theory. Darwin read and responded enthusiastically to works by both of these men – he was moreover somewhat of a personal protegé of Whewell. (See Ruse 1975a.) Although I doubt that Darwin ever read an actual word of Newton in his life, he knew the party line on these matters, and, from comments he made in private notebooks, there is no doubt that in his search for an evolutionary theory, Darwin was deliberately looking for a biological analogue of Newtonian astronomy!

Astronomers might formerly have said that God ordered each planet to move in its particular destiny. In the same manner God orders each animal created with certain form in certain country, but how much more simple and sublime power let attraction act according to certain law, such are inevitable consequences – let animal be created, then by the fixed laws of generation, such will be their successors. Let the powers of transportal be such, and so will be the forms of one country to another. – Let

231 Philosophical Aspects of the Darwinian Revolution

geological changes go at such a rate, so will be the number and distribution of the species![6]

But, do we in fact find this Newtonian influence reflected in the work of Darwin and particularly in the theory of the *Origin*? We should, and I think we do.

First, there is the question of the axiomatic nature of scientific theories – as many commentators have noted, Darwin set out the central arguments of the *Origin* almost self-consciously in this manner (Ruse 1971b). He posited certain axioms about potential food and population increases, and from these he inferred that there will be a struggle for existence. Then, from this conclusion Darwin went on to infer natural selection – that those organisms which survive will, on balance, be different from those which do not, and that this difference is in part a function of attributes which enable their possessors to survive in the struggle. Admittedly, much of the *Origin* does not follow this tight pattern of argumentation (a fact which, as we shall see, was seized upon by at least one of Darwin's earliest critics), but the crucial arguments at the core of Darwin's theory do.

Moreover, there is good evidence to suggest that in his quest for an evolutionary mechanism Darwin was strongly guided by this aspect of the Newtonian ideal, and that he did not feel satisfied until he had found it. Consequently the hypothetico-deductive structure of his central arguments was far from being a matter of chance. The crucial breakthrough for Darwin on the discovery of his mechanism appears to have been the reading of Malthus' *Principle of Population* in the early fall of 1838. Darwin used Malthus' work as a model for his own arguments mentioned above (Ruse 1973a). But, unless we recognize Darwin's deliberate desire to produce a theory with a Newtonian structure, his enthusiasm for Malthus is hard to fathom, for we know that before reading Malthus he knew all about the struggle for existence – he had read about it, even by that name, in Lyell's *Principles* – and he knew about natural selection – he had read about that in the writings of animal breeders (Ruse 1975c, 1973b). Malthus enabled Darwin to see things in an axiomatic structure, the kind of structure he was seeking. Furthermore, because Malthus argued that a struggle would follow from the clash between geometrical potential to increase in population number and an arithmetical potential to increase food supplies, Darwin, who borrowed these notions, was able to set things in at least a quasi-quantitative manner – that which Herschel identified as the best kind of manner and a prime feature of Newtonian astronomy. And also it is interesting to note that as soon as Darwin read Malthus he started to think of his mechanism in terms of pressures and forces,[7] just the kinds of causes for which Newtonian physics reaped the highest praise.

But this talk of causes brings me to the second way in which Darwin was being guided by (and trying to conform to) the Newtonian regulative principle. We have seen that there were two major marks of what makes a cause a *vera causa*, the best kind of cause: first, where one can argue analogically from a cause of which one has direct experience, and secondly where one can locate a cause at the centre of a consilience, Darwin tried to show that natural selection is a *vera causa* in both of these senses. He argued analogically for natural selection from artificial selection, a cause with which obviously we are directly acquainted, and with respect to the second sense one needs merely to note that one of the most distinctive features of Darwin's theory is the way it explains in so many different areas – geographical distribution, instinct, embryology, morphology, and so on. And Darwin was fully aware of this fact, constantly urging his theory on the grounds that no theory, explaining in so many different areas, could possibly be false.[8]

Let us turn next to the Teleological principle. In essential respects Darwin recognized the legitimacy of the Teleological principle no less than did strident anti-evolutionists like Sedgwick and Whewell. Darwin always thought and wrote of characteristics as serving certain 'functions' or having certain 'ends,' and in his *Autobiography* he made quite explicit the crucial role that the notion of adaptation played in his vision of the organic world. He wrote: 'I had always been much struck by such adaptations [as the ability of a woodpecker or tree-frog to climb trees, or a seed for dispersal by hooks or plumes], and until these could be explained it seemed to me almost useless to endeavour to prove by indirect evidence that species have been modified' (Darwin 1969).

The way in which Darwin tried to satisfy the Teleological principle was obviously through his mechanism of natural selection (with sexual selection as a sideline to explain beauty). If some variation helps an organism in the struggle to survive and reproduce then it stands a chance of being passed on – otherwise not. And thus from an innumerable supply of randomly caused minute variations the sophisticated adaptations which we see in the organic world are gradually built up. Natural selection working in a natural law-bound way leads to the design-like appearance of the organic world.

Let us turn now to Darwin's critics (as always, keeping our discussion at the metascientific level). As far as criticisms of Darwin's relationship to the Newtonian ideal are concerned, it will be sufficient to mention only that of the geologist and mathematician, William Hopkins. Hopkins (following Whewell and Herschel) distinguished the formal or phenomenal parts of theories from the causal parts, and did this in the context of the axiomatic ideal

(Hopkins 1860, 741). Needless to say, the finest of all sciences is Newtonian astronomy, and in Hopkins' eyes the peculiar merit of this science lies in the precision and exactness with which phenomenal laws and effects can be deduced from its causal (axiomatic) assumptions. But, argued Hopkins, in the case of Darwin's theory no such tight deductions exist between premises and (ultimate) conclusions. At best, Darwin showed that maybe things happen as his premises suggest – thus, for example, Darwin showed that it is not implausible to believe that animal instinct came about through natural selection, although he certainly did not show that things like the bee's hive-building instinct necessarily follow from his claims about selection. But in Hopkins' eyes, this 'mayby' philosophy is inadequate when judged against the Newtonian ideal. In other words, because Darwin's theory was not Newtonian in what Hopkins took to be the essential aspects of Newtonianism, Hopkins felt justified in rejecting Darwin's theory.

That there is some truth in Hopkins' charge is undeniable – Darwin's theory taken as a whole was certainly a long way from being a tight axiomatic structure (Ruse 1975d). At this point suffice it to say that, as we have seen, Darwin was not indifferent to the axiomatic ideal and that he made a conscious effort (with some success) to fit the core of his theory to the ideal.

Let us turn now to Darwin's attempt to satisfy the Teleological principle. Many, many people faulted Darwin on this score. Basically the criticisms fell into two kinds. Some, like Sedgwick, argued that the Teleological principle and evolutionism (Darwin's evolutionism in particular) cannot be reconciled in any way, and this for them was in itself sufficient to condemn Darwin's theory (Sedgwick 1860). Others, however, did not argue that Darwin's theory was necessarily incompatible with the Teleological principle, but they did think that the theory was fundamentally incomplete in that it did not adequately satisfy the principle. This was particularly the response of the liberals (see Herschel 1861) – a fact which led to my earlier supposition that behind the liberals' own treatment of the organic origins question lay a fear that no purely natural species' mechanism was going adequately to satisfy the Teleological principle.

As is well known, in the 1860s many (like Charles Lyell) thought they had found a reasonable compromise between Darwin's theory and Design (see Bartholomew 1973). They argued that God puts His Design into effect (thus leading to satisfaction of the Teleogical principle) via specially directed laws of variation, upon the effects of which natural selection then sets to work. Darwin himself however, although he was as sympathetic as possible to these would-be reconcilers, grumbled that one should not go too far to appease Paley and Co. (Darwin 1903, 1, 154). Showing great methodological acuteness, Darwin insisted that if one persisted in the belief that one must make some reference to God's direct action in order fully to satisfy the Teleological principle, then one stands in grave danger of violating the Newtonian principle. Thus, against Herschel he wrote that 'astronomers do not state that God directs the course of each comet and planet. The view that each variation has been providentially arranged seems to me to make natural selection entirely superfluous, and indeed takes the whole case of the appearance of new species out of the range of science' (Darwin 1903, 1, 191). Herschel had himself objected that attempts to relate the origins of species to God's direct intervention takes the whole question out of the range of science. Darwin argued that Herschel was hoist by his own petard.

V CONCLUSIONS: HISTORICAL AND PHILOSOPHICAL

My historical conclusions can be drawn quickly and easily. Without denying that there were many levels involved in the Darwinian revolution, scientific, religious, social, it is clear that one level was what we might call 'philosophical.' There was debate about whether certain proposals even qualified as scientific, and this debate was fought against the background of two key regulative principles. Moreover, at this philosophical level we can see why the revolution proceeded as it did. The thinkers of the 1830s failed to provide satisfactory answers to the organic origins question, and whilst Darwin's answer may not have satisfied all, it is hardly reading the present too much into the past to conclude that Darwin's attempted solution to the species problem came far closer to satisfying the Newtonian and Teleological principles than did that of Chambers.

But what of philosophical conclusions? Of course, I recognize that if one is determined to draw an absolute line between the context of discovery and the context of justification, one is going to be loath to allow any such conclusions; but, if in this post-Kuhnian age we are prepared to allow that we can at least wave to each other across the barrier, I think a couple of conclusions, pertinent to Goudge's philosophy of biology, might be drawn.

First, if one argues, as Goudge has done repeatedly,[9] that Newtonian astronomy is entirely the wrong model for evolutionary biology, then one is treading on very thin ice indeed. The claim of someone like Goudge is that the hypothetico-deductive model and all that that implies, something taken directly from astronomy, is inappropriate as an ideal for evolution theory – a claim by which Goudge seems to have meant not merely that evolution theory does not really fit the model but that talk of the model is misleading and conceals what evolutionists are trying to do. Let us as philosophers, Goudge would say, show biology as good biology, not as bad physics. In reply, history shows that Darwin certainly had astronomy as an ideal, hypothetico-deductive model and all. Hence, although history certainly shows that Darwin's theory did not exemplify the model fully, Goudge is faced with a painful dilemma. Either he must agree with an anti-evolutionary critic like Hopkins that Darwin failed entirely to do what he set out to do, rather an uncomfortable conclusion which is going to require a flat denial of any deductive nature to Darwin's core arguments and an acknowledgment by Goudge of rather an unwelcome bedfellow one would think; or he must argue that since the time of Darwin (astronomically inspired though *he* may have been) evolutionary aims have changed, in which case, why are Mayr, Simpson, and Co. so proud of being neo-Darwinians? Either way, Goudge cannot deny that one prominent Darwinian evolutionist, Charles Darwin, had astronomy as his ideal, however inappropriate Goudge may think it is.

But this leads straight to my second point. Generally speaking, what Goudge aims to do is to argue for the distinctiveness of biological understanding. Here history, in the form of the Teleological principle, comes to his rescue. As Goudge recognizes, Darwin's teleology is passed on into modern evolutionary thought – in short, what we still have is the attempt to see organisms as artefacts, as means to ends. This attempt is regarded as a necessary condition of good biology, and one suspects that not even the most hard-line reductionist would feel entirely happy with an analysis of the eye which failed completely to point out that it is used for seeing: that it has that 'function' or 'purpose.' But as history also shows, this principle being appealed to is not one used or appropriate in astronomy – the moon has no function. In other words, here, in the insistence that food biology be regulated by the Teleological principle, we have a distinctively biological way of looking at things. Hence, although Goudge may lose the Newtonian battle, thanks to teleology he wins the biological war. The Darwinian Revolution suggests that biological understanding is not just like that of astronomy.[10]

BARRY STROUD

Evolution and the Necessities of Thought

It was my good fortune to be a student of Thomas A. Goudge as an undergraduate at the University of Toronto. Whether he was explaining Moore on 'Existence is not a predicate' or Fichte on 'Das Ich setzt sich,' criticizing Quine's attack on analyticity, or inviting us to criticize Hegel's attack on formal logic, the clarity and order, and the refreshing sense that in each case there was something definite and specifiable there to engage the mind, served to dispel that feeling of disorientation, even incipient asphyxiation, some of us had been suffering from, however inarticulately, in our earlier ruminations on Plotinus, or T.H. Green, or whatever it had happened to be. But it was not simply a matter of ventilation. The fresh, clean air increased, or in my case actually generated, the conviction that there were things to be done in philosophy, and that gaining a thorough, sympathetic knowledge and keeping one's wits about one could just possibly lead somewhere. It was an important experience for me, and I will always be grateful to it.

Homo sapiens is an evolved species, and so the existence of its members with the general characteristics they have is itself a product of the evolutionary process. That process continues at present, and so there must be true evolutionary explanations of why human beings in general are the ways they are. Many will be inclined to accept this generalization with respect to straightforwardly physical characteristics of human beings, but the same point of view can be applied to complex human cultural and intellectual products as well. In particular, the theory of evolution itself can be seen as an elaborate evolutionary product with high 'adaptive value'; it enables us to adapt more easily to an environment that continually thwarted our efforts to understand it without such a theory.

Goudge discusses this idea in *The Ascent of Life*. Whether or not we agree with him that man is 'psychically immature,' that he is still in his 'racial adolescence,' and that his present knowledge is 'undoubtedly inchoate,' we must surely admit that, seen in evolutionary perspective, 'he has hardly made more than a start in the cognitive enterprise' and that, if human evolution continues for another twenty thousand years or so, 'the frontiers of knowledge will almost certainly be far beyond those of the present day' (AL 210). From all this Goudge concludes that it is absurd to suppose that 'man can establish now, once for all, the limits of possible knowledge' (AL 210).

In this paper I want to raise very briefly and at a regrettably high level of generality one of the many puzzling questions that arise quite naturally when we try to think of human beings and the development of human knowledge from an evolutionary point of view. I will not try to resolve the issue. It will be enough if I can indicate in one particular way how difficult it is to adopt the evolutionary point of view completely, or to restrict ourselves to it consistently. I think the attempt to do so has implications for our understanding of ourselves and our knowledge that are not always explicitly recognized and squarely faced.

It is undoubtedly true, as Goudge says, that we cannot set boundaries to the future scope or extent of human knowledge, but talk of 'the limits of possible knowledge' also carries with it the more Kantian idea of conditions that must be fulfilled by any possible knowledge. If there are any such conditions, and we know them to be fulfilled, then at least some of the things we now know express 'limits' or 'constraints' that no future knowledge can overstep.

In particular, if we now know some things to be not just true, but necessarily true, then since what is necessarily true could not possibly be false, it would seem that some of the things we now know do, in a sense, express 'limits' or 'constraints' on any possible future knowledge. I want to ask whether it is possible to think of human knowledge as a flexible, constantly changing product of the evolutionary process, one that might well be unrecognizably different in twenty thousand years, while continuing to believe that many of things we now know are known to be necessarily true.

I think most philosophers would be inclined to answer, 'Yes.' In *The Thought of C.S. Peirce* Goudge himself gives an affirmative answer, at least in the case of logic. He points out that, on Peirce's conception of rational human thought as itself a product of the evolutionary process, it would be 'theoretically possible to write a natural history of knowledge, one chapter of which would embrace logic as a type of human activity' (TP 128). We might then come to understand how certain ways of thinking proved themselves efficacious in the struggle for existence and thus became fixed habits and were 'eventually formulated by the logician as the principles of inference (IP 128).

Such a naturalistic, genetic account of our coming to accept or acknowledge certain principles of logic is not to be thought of as an explanation of the special 'validity' of the principles themselves. We do not regard inferences as valid simply because they proceed in accordance with principles that have survived in the struggle for existence. We have an independent notion of validity, and Goudge tentatively suggests that any adequate account of logical validity must appeal to the *necessity* of certain principles or truths (TP 134). But our having knowledge of such necessities, he holds, is not 'necessarily incompatible with a naturalistic conception of logic' (TP 135).

Goudge himself does not develop the suggestion, but I think many who endorse it would do so on the basis of a certain widespread conception of the nature of necessary truth – a conception that is thought to point the way, perhaps the only way, to a successful account of the compatibility between a sophisticated evolutionary naturalism and the necessity of many of the things we know. But it seems to me that what has come to be the more or less standard way of thinking about necessary truth and its source in fact makes it much more difficult to explain the alleged compatibility than it might look at first sight.

Certainly there is no straightforward logical incompatibility between there being a true naturalistic explanation of our believing that p, on the one hand, and its being true that p, or even its being necessarily true that p, on the other. Simply from the fact that there is a true explanation of the origin of a belief, either in an individual or in a group of individuals, nothing at all follows one way or another about the truth-value, or the modality, of what is believed. This holds even if the kind of explanation offered is one that we normally take as showing that the belief in question is false. For example, to say that someone believes that the bed before him is covered with leaves because he has just taken an hallucinatory drug certainly suggests that the bed is not covered with leaves, that he is simply hallucinating that it is, and that his belief is false. But however unlikely it is that a perfectly normal bed in perfectly ordinary circumstances should be covered with leaves, its not being covered with leaves does not follow logically from the admitted truth that someone is hallucinating that it is covered with leaves. It is at least possible for it to be covered with leaves and for someone simultaneously to hallucinate that it is. And, to take another example, if we know that it came to someone in a dream, and he subsequently believed, that Goldbach's conjecture is necessarily true, it does not follow from those facts alone that it is not the case that Goldbach's conjecture is necessarily true. Its being necessarily true and his dreaming and subsequently believing that it is is a possible state of affairs.

Similarly, if we admit in an evolutionary spirit that human beings believe what they do, or even that they believe some things to be necessarily true, because having such beliefs has great 'adaptive value' for the species and facilitates its adjustment to the world in which it lives, it does not follow from that alone that those beliefs are not true, or that none of them are necessarily true. It is at least possible for a proposition to be true, or to be necessarily true, even though people's believing it, or believing it to be necessary, is of considerable value in their efforts to survive.

But although the mere logical consistency between the truth, or the necessary truth, of a belief and a true naturalistic explanation of its origin must be admitted, to appeal only to that consistency seems to me a relatively superficial and unsatisfactory response to the challenge posed by an evolutionary theory of human knowledge.

For one thing, some explanations of the origin or retention of a belief are such that once we accept them we can no longer hold the belief in question or see it as defensible. Someone who hallucinates a bed covered with leaves and on that basis believes that the bed before him is covered with leaves does not thereby know that the bed is covered with leaves, even if it is. Someone who dreams, and thereby comes to believe, that Goldbach's conjecture is necessarily true does not on that basis know that it is necessarily true, even if it is. Therefore, even if the truth of such explanations does not logically imply the falsity of what is believed, a person accepting such explanations as an account of his own beliefs is no longer entitled to continue to hold those beliefs. His believing what he does has been to that extent discredited, even if what he believes should happen to be true, even necessarily true. The mere logical compatibility between a true naturalistic explanation of someone's believing something and the truth, or necessary truth, of what he believes is therefore not enough to show that one who accepts the explanation can reasonably continue to attribute truth, for example, to the proposition that the bed before him is covered with leaves, or necessary truth to Goldbach's conjecture. Accepting the explanations in those cases has radical implications for the way he sees his own beliefs.

A belief that is shown to be solely the result of an hallucination or a dream is thereby discredited because the truth or probable truth of what is believed plays no role in the explanation of the fact that it is believed.[1] Given the facts described, the man would have believed that the bed is covered with leaves even if it had not been so covered; its actual state had no effect on what he believed about it. And the necessary truth of Goldbach's conjecture was not in any way responsible for the man's dreaming and coming to believe that that conjecture is necessarily true. However, when the truth or probable truth of what is believed does play an essential role in the explanation of the origin or retention of that very belief, one can accept such an explanation of one's beliefs without seeing them as in any way brought into question. In fact, many would argue that that is precisely what it is to have a reasonable or justified belief. (See e.g. Harman 1973 esp ch 8.)

This leaves it open to the evolutionary epistemologist to argue that in explaining the origin and retention of the beliefs of human beings solely in terms of their usefulness or 'adaptive value,' he is not necessarily thereby discrediting those beliefs. His account could be accepted with impunity if the mode of explanation he envisages, unlike the appeal to hallucination or dreams, makes essential use of the truth or probable truth of the beliefs whose origin is being explained. It might well be argued, for example, that the survival in a species of a set of beliefs about its environment is some reason in itself to conclude that those beliefs are probably true, or approximately true, since they were acquired and retained as a result of interaction with the very environment they are about. If survival is thought to be more likely with true than with false beliefs then we might even think of ourselves as getting closer and closer to the whole truth as time goes on by getting a more and more accurate 'map' of the world we live in. On this view, appeal to the 'adaptive value' of our beliefs, far from discrediting those beliefs, actually provides positive support for them; it is argued in effect that the world would probably not have tolerated complete falsity for such a long time.

Whether an argument along these lines could be made at all plausible is a complicated question that I shall not pursue. My concern here is only with the general condition that must be fulfilled by an evolutionary explanation of our beliefs that does not detach us from those beliefs and represent them as no more likely to be true and therefore no more worthy of our assent than the products of comfortable hallucinations or congenial dreams. It will employ a notion of the truth of what is believed independently of its being believed, and it will somehow show that the truth of what is believed is implied or rendered probable by the evolutionary explanation of our acceptance and retention of the beliefs in question.

Whether evolutionary explanations fulfil the specified condition in the case of our accepting some things as true and rejecting others is perhaps an open question, but it does not even seem possible to fulfil that condition in explaining our ascribing *necessary* truth to some of the things we believe. There an appeal to the independent necessary truth of what is believed seems to no avail. An evolutionary explanation of our thought about necessity will explain how and why we have a notion of necessary truth as something objective and independent of its being believed, and why we ascribe it to some things and withhold it from others. But the facts appealed to by any naturalistic explanation of our thinking in those ways will be contingent facts about the world and its inhabitants that are held to be responsible for human beings' thinking and believing what they do. The explanation would not, and, it will be felt, could not, introduce the independent necessary truth of what is believed into the explanation of its being believed to be necessarily true.

The impossibility of any such appeal will be felt especially strongly if we think of necessary truth and its source along more or less standard lines. According to that conception, the nature of necessary truth is not to be understood by reflecting on its alleged relation to some independent domain, but only by understanding its relation to, or its source in, the ways we think and speak about the world. Necessary truths are not taken to describe or reflect any features holding necessarily of the world independently of our thought. They do not straightforwardly describe our contingent ways of thinking and speaking either, nor are they merely directives or injunctions to think or speak in certain ways; they are genuine truths, and they hold necessarily, but they are thought to owe their special status only to our thinking or speaking in certain ways, and not to the way things necessarily are independently of us. In short, necessary truths are in some sense our own 'creation' – they are true by 'convention' or 'decision,' and not by virtue of any 'facts.'

Any such conventionalist view can be made credible only if there is some way to distinguish it from the apparently rival view that necessary truths are true in virtue of the way things necessarily are. There must be something that shows that in coming to know necessary truths we are simply acknowledging and accepting the consequences of our own ways of thinking and speaking, and not discovering independent, objective, necessary features of the world. What is taken to show this is the existence of possible alternatives to our present ways of thinking and speaking which are such that, if we had thought or spoken in those ways, or if we came to do so in the future, we would solely thereby accept a set of necessary truths different from our present ones even though everything else about the world remained exactly the same. Hence our present ways of thinking and speaking, and no objective features of the world, are thought somehow to be responsible for the truth of what we now believe to be necessarily true. Not only does conventionalism about necessity imply the existence of alternatives to our present ways of thinking and speaking – alternative 'conventions' we could have 'chosen' instead – but the appeal to such relevant alternatives must be an essential part of any attempt to establish the truth of conventionalism against its rivals.[2]

There is in fact some question whether an argument for conventionalism could amout to anything more than that. Certainly there was no actual convening, and no actual agreement made, any more than an actual social contract was drawn up by our forefathers. The most that can be said is that it is *as if* there had been an explicit agreement; and even that must rest on an appeal to the existence of alternative ways of thinking and speaking.

An evolutionary view of the development of human knowledge appears most congenial if we think of necessity in this way. It appeals perforce only to contingent facts about us and the world in order to explain our believing some things to be necessarily true, and it stresses the contingency of our ways of thinking and speaking, and therefore, in that sense at least, the existence of alternatives to them. Necessary truth is therefore thought to pose no problem for an evolutionary view precisely because necessary truth, unlike truth itself, is thought to be in some sense a 'creation' of ours – a by-product of our cognitive interaction with the world that is not a representation of any objective states of affairs holding independently of our thinking in certain ways.

But whether this apparently attractive accommodation of necessity to evolutionary theory is possible depends on a question that seems to me never to have been squarely faced or adequately resolved by defenders of this standard conception of necessary truth. Precisely what relation is supposed to hold between our thinking or speaking in certain ways and the necessary truth of any of those things we now regard as necessarily true? If our thinking or speaking in those ways is in some sense 'responsible' for the necessary truth of the things we now accept as necessary, exactly how is that notion of 'responsibility' to be understood?

Consider something that we believe to be necessarily true, for example 'If all men are mortal and Socrates is a man then Socrates is mortal.' Does conventionalism, or indeed any view according to which necessary truth is in some sense our own 'creation,' imply that the truth of that statement is due solely to our present ways of thinking or speaking in the sense that if we had thought or spoken in certain other ways, or had adopted relevantly different 'conventions,' then it would not have been true that if all men are mortal and Socrates is a man then Socrates is mortal? If that is an implication of conventionalism, then to accept conventionalism would be to concede that under certain (vaguely specified but nevertheless possible) circumstances it would not have been true that if all men are mortal and Socrates is a man then Socrates is mortal. That claim in itself is difficult to understand, but it seems at least to carry the implication that there are possible circumstances in which it would not have been true that if all men are mortal and Socrates is a man then Socrates is mortal. But we take that familiar conditional to be necessarily true, and so we cannot allow that there are such circumstances – that there are in that sense alternatives to its truth. Acknowledging possible alternatives to the truth of *p* is incompatible with regarding it as necessarily true that *p*. Therefore we cannot accept what appears to be an implication of conventionalism while continuing to believe that it is necessarily true that if all men are mortal and Socrates is a man then Socrates is mortal.

There is another difficulty in the view that all necessary truths are true by convention. Even among logical truths alone there are an infinite number we now accept, and so we could not have conventionally endowed each of them singly with the truth we now attribute to them. Our explicit conventions would have to have been general ones, from which the truth of individual logical truths follows. But if one truth follows logically from another then the conditional with the latter as antecedent and the former as consequent is itself a logical truth, and so in attempting to render true by general convention some individual logical truth we make essential use of some other logical truth whose truth is so far unaccounted for. If that truth in turn had already been derived from some general convention of its own there would still be some other conditional whose truth had not been explicitly secured by convention. Any attempt to represent all necessary truths as true by convention makes use of at least one necessary truth whose truth is not accounted for by that very representation, so not all necessary truths could be shown to be true by convention.[3]

This shows that the attractive idea that in coming to know necessary truths we are simply acknowledging and accepting the consequences of our own conventional ways of thinking and speaking does not avoid all appeal to necessary truths holding independently of our having 'adopted' certain 'conventions.' That our 'conventions' have the consequences they do will not itself have been shown to be true simply by convention. But conventionalism is attractive only in so far as it purports to explain all necessary truth without appeal to any non-conventional necessities. I am inclined to think that the satisfaction provided by conventionalism is therefore illusory.

Conventionalism tries to do too much. It is supposed to explain both our acknowledging or accepting certain things as necessarily true and the very 'nature' or 'source' of the truth of the necessary truths we accept. In the latter task it fails, either by not showing the truths in question to be true in all possible circumstances, or by not showing all of them to be true by convention. But we can abandon the hope of explaining the very nature of necessary truth, and thereby abandon conventionalism, without abandoning the problem of explaining our believing or accepting some things as necessarily true. We do attribute necessary truth to some things and withold it from others, and, I think, we regard necessary truth as something holding objectively, independently of our contingent ways of thinking. How and why we do so ought to be discoverable and understandable in the same general way that any other complex contingent fact about human behaviour is to be understood. That is precisely what an evolutionary explanation tries to do.

There is no need to expect an evolutionary explanation also to provide what we could regard as a fully adequate analysis or definition of the puzzling notion of necessary truth. It might well make our possession and employment of that notion intelligible in illuminating ways without thereby yielding a helpful, non-circular explanation of the very nature of necessary truth. It is probably misguided to expect to find such an equivalence, anyway. Certainly there is not much genuine illumination to be gained from familiar explanations of necessity in terms of possibility and negation. If we find necessity and our thought of it puzzling we are not likely to get the kind of understanding we seek by tracing fairly obvious equivalences between it and other notions we find equally puzzling for the same reasons.

But the lack of a helpful analysis or definition of necessity does not leave the notion completely mysterious or obscure. We could understand a lot more than we do about our possession and employment of the notion of necessity without ever being able to define it in a way that could serve to introduce the idea to someone who lacks it. Taking our possession of the idea for granted, we can still ask why we ascribe it and withhold it in just the ways we do, what function the idea has for us, what we can do in virtue of possessing it that we could not do without, it, and even, perhaps, why we have an idea of necessity at all. These are undoubtedly difficult questions, and very general ones, but they do not obviously require for their answer a non-circular definition or analysis, in the strict sense, of 'necessary' or 'possible.' They are questions to be answered, if at all, by the naturalistic study of human beings.

There is a close parallel in the case of truth itself.[4] To explain the point or function of the notion of truth, and how and why we ascribe truth to some things and not to others, or why we believe what we do and not something else, is not necessarily to provide an analysis or definition of the notion of truth.[5] Explanations of many of our beliefs will take into account what we perceive, and what we believe or take ourselves to know already, and will explain the connection believed to hold between our perceptions or the prior set of beliefs and the belief in question. But a full and accurate statement of these 'epistemic' conditions under which we unhestitatingly ascribe truth to a particular proposition does not necessarily amount to a statement of what it is for that proposition to be true. Our notion of truth could still be a notion of something holding fully objectively and independently of what we know or believe, even though 'epistemic' considerations play an essential role in explaining why we have a notion of truth and how and why we employ it as we do.

Similarly, it might well be that 'epistemic' considerations help explain our possession and employment of the notion of necessity. Explanations of why we regard some things as necessary can be expected to take into account the special place or role of those things among all our beliefs, or the special connection believed to hold between them and certain specially placed beliefs.[6] The details of such an account would be extremely complex, and are not available at present. But a full and accurate statement of the 'epistemic' conditions under which we unhesitatingly ascribe necessity to a particular proposition does not necessarily amount to a statement of what it is for the proposition to be necessary. Our notion of necessity could still be a notion of something holding fully objectively and independently of what we know or believe even though 'epistemic' considerations play an essential role in explaining why we have a notion of necessity and how and why we employ it as we do. It is to these questions of explanation, and not to the search for a definition of truth or of necessity, that evolutionary theory might be expected to make a contribution.

But now it will appear that we are thrown back on the original difficulty about necessity. It will still be felt that, since the considerations appealed to by an evolutionary explanation of our believing some things to be necessarily true are themselves contingent, the independent necessary truth of what is believed could play no essential role in a correct naturalistic explanation of its being believed to be necessarily true. In that sense, the true explanation of our thinking about necessity would not require that thinking itself to be 'veridical'; in contrast with the notion of truth, in this case there would need to be nothing objectively answering to our notion of necessity in order for us fully to understand our possession and employment of it. And that can be seen as a repudiation of the notion of necessity as a will-o'-the-wisp or at best a confusion.

The problem is reminiscent of the case of Hume and the idea of necessary or causal connection. He held that there is nothing to be found in the world or in our experience that answers to the idea of necessary connection. Only after the repeated observation of resembling pairs of phenomena do we arrive at the 'fiction' of an objective causal connection between phenomena of those two kinds, but there is nothing more 'in' the phenomena in question than their spatial and temporal relations and their membership in classes some of whose members have been constantly conjoined in our experience. For Hume this is a genetic, naturalistic (although not evolutionary) explanation of our thinking about necessary connection, but it explains our thinking in those ways without ever supposing that such thinking is correct or 'veridical.' It is a 'fiction' or 'projection' we cannot help indulging in, but nothing more.

It is easy to see Hume's view as a sceptical argument for rejecting the idea of necessary or causal connection altogether, and making do with the regularities we find in our experience and the expectations they give rise to.[7] The genetic explanation is thought to expose the idea of necessity as superfluous, or as a confusion to be jettisoned in the name of clarity. But that was not Hume's own reaction, and to suppose that it was is to misconstrue the relation between his philosophical theory of causality and the behaviour of human beings in their ordinary and scientific pursuits. Despite his philosophical 'discoveries' Hume did not abandon the idea of causality or necessary connection when he thought about the world as a plain man, or indeed as a general theorist of human nature. He sought causal explanations of why human beings think, feel, and act in the ways they do, and in particular he thought he had found a causal explanation of our thought about causality.[8] According to that explanation, it is inevitable that human beings with certain kinds of experience will come to think in causal terms, and Hume himself was no exception to the 'principles of human nature' he discovered. It was inevitable, then, that he too would continue to think in causal terms despite his philosophical 'discoveries.'

It might well be that if we understood the source of the idea of necessity that we attribute to some propositions and not others, and could explain the point or function that idea has for us, and what we can do in virtue of possessing it that we could not do without it, the question of whether or not to jettison the idea would be as idle as it was for Hume with the idea of causality. Kant strengthened the case by showing the indispensability of the idea of causality for even so much as the thought of an objective occurrence. If our idea of necessity were found to play an equally central role in our thought, then although when we try to detach ourselves philosophically from our ways of thinking about necessity and seek their foundations in something objective and independent of our thought we inevitably fail, we would still not be able simply to give up those ways of thinking while retaining enough of the rest of our thought to make our experience intelligible to us. We might see from a naturalistic study of human beings that there are compelling reasons why we do, and must, ascribe necessity to some of the things we believe even though our best philosophical reflection can find no objective foundations for that necessity, or finds the very possibility of them obscure. This might make us at least as sceptical of the process of philosophical reflection that seems to have such devastating results as of the notion of necessity itself, especially if we had some understanding of its efficacy or indispensability in our thought.

Evolution and the Necessities of Thought

In any case, we find ourselves in an unstable position. Taking an evolutionary perspective, we must acknowledge the contingency and explicability of our present ways of thinking, and in particular our present ascriptions of necessity. But if we do regard some things as necessarily true we thereby deny their contingency and cannot countenance the possibility of alternatives to them. We must simultaneously appreciate the contingency of the fact that the limits of our thought lie just where they do while remaining unable to think beyond those limits.[9] It is not easy to hold consistently to both points of view simultaneously, and we inevitably find ourselves moving back and forth somewhat unsurely between them. That is perhaps inevitable when we try to stand outside the evolutionary process and see it as a whole, *sub specie aeternitatis*, while the terms we use to try to understand that process and our place in it are themselves products of the very process we are trying vainly to transcend.

PAUL R. THAGARD

The Autonomy of a Logic of Discovery*

In an unpublished paper, 'What Makes a Hypothesis Plausible?,' Thomas A. Goudge discusses the important question of whether conditions can be formulated which enable us to determine that a hypothesis is plausible prior to experimental tests. I shall briefly summarize this paper, then discuss it in the context of work by N.R. Hanson and others on whether there is a 'logic of discovery' distinct from a 'logic of justification.'

I GOUDGE ON PLAUSIBILITY

Goudge distinguishes reasoning about the plausibility of hypotheses from deductive, inductive, and statistical reasoning. He rejects the view, sometimes proposed by Hanson, that the act of entertaining a hypothesis exhibits a logical structure akin to that of hypothetico-deductive explanation in a scientific theory. He also argues that plausibility cannot be assessed by means of reasoning by enumerative induction. Third, he denies that plausibility is a matter of high antecedent probability.

Goudge then discusses C.S. Peirce on *abduction*, a form of reasoning involving the provisional adoption of a hypothesis on the grounds that the hypothesis would explain certain surprising facts. An investigator infers that a new hypothesis is plausible on the basis that the hypothesis explains the phenomena, or would do so if it were true.

Pursuing this topic, Goudge discusses a number of important historical examples of reasoning concerning the plausibility of hypotheses. The first is

---

*It is a pleasure to dedicate this paper to Thomas A. Goudge, in gratitude for what I have learned from his lucid writings and fine teaching. I also wish to thank him, B.C. van Fraassen, and F. Wilson for comments on an earlier draft of this paper, but it should not be assumed that they agree with all that follows.

249 The Autonomy of a Logic of Discovery

Liebig's introduction of the hypothesis of vital force. Liebig hypothesized a new kind of force to account for organic phenomena such as animal heat, growth, and reproduction. He believed that a special kind of force, comparable to but distinct from gravitational, magnetic, electrical, and chemical forces, was needed to give a mechanical explanation of organic processes. By assuming that inorganic and organic phenomena are alike in being mechanical motions, he established a conceptual link between the domain of biology and the domain of physics, where the concept of force had been employed with great success. Given this success and the link between the two domains, it was plausible for Liebig to suppose that a theory involving a kind of force, vital force, was needed. Thus the plausibility of Liebig's hypothesis derives in part from it being the same kind of explanatory hypothesis as those already proven successful. Hence Goudge concludes that even though the subsequent history of biology has not supported Liebig's hypothesis, the hypothesis was plausible at the time it was introduced.

Another example of the importance of kinds of hypotheses is the way in which the success of the 'uniformitarian principle' in geology rendered more plausible the evolutionary views of biologists such as Darwin, who like the geologists employed hypotheses involving the gradual operation of natural factors. Another case is Roentgen's hypothesis of X-rays, which had clear connections with the established concepts of light rays and cathode rays. Thus a hypothesis is judged to be plausible both on the basis that it explains certain phenomena, and on the basis that it is the same kind of hypothesis as those successful in other fields.

Liebig's hypothesis also satisfied another condition of plausibility which Goudge calls 'the amenability of the hypothesis to imaginative elaboration.' Faced with a whole new domain of phenomena lacking explanation, Liebig struggled to give an imaginative account of the operation of vital force. Roentgen's hypothesis of X-rays was open to easier elaboration through techniques already familiar from the study of related phenomena.

Goudge's final example is Huygens' use of the analogy between the familiar phenomena of water and sound waves and the hypothetical phenomena of light waves. Unlike Huygens, Liebig lacked a mechanical model to add to the plausibility of his hypothesis.

Goudge claims that conclusions that hypotheses are plausible are the result of practical reasoning, reasoning leading to a conclusion about what is to be done. He proposes that the word 'plausible' be taken to function like the word 'reasonable,' as in 'it is reasonable to do such and such.' Plausibility is then not a property ascribable to hypotheses, since it applies to steps in the process of inquiry.

Goudge concludes by stating that he does not consider the 'logic of plausibility' that he has discussed to be part of a 'logic of discovery,' since plausible hypotheses, unlike ones which have earned the title of *discoveries*, may eventually be falsified. The point about the honorific use of the term 'discovery' is correct, but since the aim, if not the necessary result, of seeking plausible hypotheses is discovery, I see no immediate reason not to consider the logic of plausibility as a possible part of a logic of discovery.

This summary shows that Goudge's paper raises many important issues concerning the plausibility of hypotheses. Some of these issues have also been discussed by philosophers such as C.S. Peirce and N.R. Hanson. Peirce, Hanson, and Goudge have all been concerned with reasoning which occurs prior to the justification of hypotheses. According to Hanson, this reasoning comprises a 'logic of discovery'; such a logic would have to include what Goudge calls the logic of plausibility. What I wish to do is evaluate Goudge's work on plausibility by considering it in this broader context, and, in particular, in the context of Hanson's question of whether there is a logic of discovery of scientific hypotheses *distinct from* a logic of justification. Hanson has approached this question by distinguishing between three sorts of reasons: (1) reasons for supposing that a sought-after hypothesis will be of a certain *kind*, (2) reasons for suggesting a hypothesis as *plausible*, and (3) reasons for *accepting* a hypothesis. I discuss these in sections II, III, and IV respectively. The different sorts of reasons are apparently used at different stages in the process of inquiry. As Hanson has it, when faced with an anomaly which demands explanation, a scientist attempts to determine what kind of hypothesis would explain it, then considers the plausibility of various hypotheses of the favoured kind, and finally comes to accept one of the hypotheses. I shall argue that Hanson's distinction between the two sorts of reasons collapses, in that reasons of the first sort are often also reasons of the second sort, and reasons of both the first and second sorts are often reasons of the third sort. It then follows that Hanson's distinction does not suffice to show that there is an autonomous logic of discovery. This critical evaluation of Hanson will provide the context for a similar critical evaluation of Goudge.

II REASONS FOR SUPPOSING THAT A HYPOTHESIS WILL BE OF A CERTAIN KIND

Scientific inquiry often begins with the recognition of anomalies, of phenomena which are not accounted for within existing theories. For example, quantum theory began with Planck's attempt to deal with the problem of black-body radiation. And Darwin's theory of evolution by means of natural

251 The Autonomy of a Logic of Discovery

selection had its origins in his observations of certain fossils and of the distribution of animals in South America and the Galapagos archipelago. As Hanson and Peirce have suggested, the invention of hypotheses in order to explain anomalies is not a random process.

Hanson claims that scientists can give reasons for expecting that a hypothesis which would explain a certain anomaly will be of a particular *kind* (Hanson 1961, 23; see also 1960, 1963, 1965a, 1965b). He attempts to establish that there is a logic of discovery by distinguishing these kinds of reasons from the kinds of reasons given for the acceptance of a scientific hypothesis. This distinction is a departure from an earlier one made for the same purpose: in early work (Hanson 1958a, 1958b), Hanson tried to distinguish between reasons for suggesting a particular hypothesis in the first place and reasons for accepting it. But he decided that these reasons differ only in degree, so that the logic of discovery cannnot be isolated by considering reasons for suggesting particular hypotheses.

What sort of reasoning is used in suggesting that a hypothesis will be of a certain kind? Historical examples suggest that the most important reason for arguing that the explanation of an anomaly A will be by a hypothesis of kind K is that hypotheses of kind K have been successful in explaining phenomena similar to A. For instance, when the orbit of Uranus was found to exhibit perturbations not explicable by existing theory, investigators supposed, on the basis of past experience with the motion of planets, that the kind of hypothesis needed to explain the perturbation involved the existence of another planet. The astronomer and philosopher of science J.F.W. Herschel wrote of the perturbation:

Of the various hypotheses formed to account for it, during the progress of its development, none seemed to have any degree of rational probability but that of the existence of an exterior, and hitherto undiscovered, planet ... Accordingly, this was the explanation which naturally, and almost of necessity, suggested itself to those conversant with the planetary perturbations who considered the subject with any degree of attention. (Herschel 1878, 539)

Leverrier and Adams subsequently formed by calculation a particular hypothesis about a planet, and this hypothesis was confirmed by Galle's observation of the planet Neptune.

Another case of reasoning involving kinds of hypotheses is that leading up to Darwin's discovery of the hypothesis that species have evolved by means of natural selection. Darwin writes of the anomalies he encountered during the voyage of the *Beagle*; 'It was evident that such facts as these, as well as many others, could only be explained on the supposition that species gradually become modified' (Darwin 1958, 42). As Goudge pointed out, Darwin's conclusion that the explanatory hypothesis needed would involve evolution was in part based on the success of 'uniformitarian' explanations in geology. And Darwin's reasoning as to the kind of hypothesis needed did not stop there. He perceived an analogy between the changes in natural species and the modification of domestic species by breeding – artificial selection. Concluding that selection was therefore in some way involved in the evolution of all species, he set out to discover how. Finally, after reading Malthus, he conceived of natural selection resulting from competition for food and land in the face of increasing population. Thus Darwin's discovery of the hypothesis of evolution by natural selection resulted from at least two occasions of reasoning to a kind of hypothesis.

The structure of this sort of reasoning is:

(S_1) Phenomena $P_1 \ldots P_n$ have been explained by hypotheses of kind K.
Anomaly A is similar in certain respects to the P_i.
∴ It is likely that the hypothesis we are looking for to explain A will be of kind K.

The above examples show that such reasoning is used in science previous to the evaluation of particular hypotheses. However, I shall show that reasoning concerning kinds of hypotheses can also be used to help show that a hypothesis is plausible or even acceptable.

III REASONS FOR JUDGING A HYPOTHESIS TO BE PLAUSIBLE

After determining that a desired explanatory hypothesis is likely to be of a particular kind, we naturally attempt to construct an individual hypothesis. Reasoning of the sort described in II narrows the problem of inventing a hypothesis considerably, by selecting certain concepts or mathematical techniques. We have seen that there is a kind of logic to this selection. Is there a kind of logic involved in the discovery of an individual hypothesis? An affirmative answer is not incompatible with the fact that discovery is affected by any number of sociological and psychological factors, ranging from what research funding an investigator is able to get, to what he or she eats for breakfast. One might claim that the actual occurrence of a particular hypothesis to an investigator is not governed by logical considerations, and that the 'act of insight' which Peirce describes (CP 5.181). is merely a matter for psychology. However, this neglects the fact that inference has both psychological and logical components. The process in which I infer q from $(p \, \& \, (p \supset q))$ is a psychological one, but nevertheless the legitimacy of the inference is a

253 The Autonomy of a Logic of Discovery

question of logic. Similarly, the existence of psychological determinants of the process of forming an explanatory hypothesis does not show that it is not a logical operation.

An investigator wants to construct hypotheses which are sufficiently plausible to warrant further investigation. As we shall see, determination that a hypothesis is plausible is a logical matter. However, it still can be claimed that the formation of a plausible hypothesis is not a matter of logic, since a hypothesis must be formed before it can be judged to be plausible; just as in order to infer that x is G from the propositions that all F are G and that x is F, we must first have constructed the proposition that x is G. This is basically correct, but the psychological act of forming a proposition and the logical act of deriving it from others are intimately connected nevertheless, since we might not have come to form the proposition in this situation if it were not derivable. Similarly, it is often true that if a proposition were not capable of being evaluated as plausible, we would not have formed it. In this extended sense, we may consider the formation of a plausible hypothesis as a logical operation.

Plausibility is defined informatively, although awkwardly, by Peirce as follows:

By Plausible, I mean that a theory that has not yet been subjected to any test, although more or less surprising phenomena have occurred which it would explain if it were true, is in itself of such a character as to recommend it for further examination, or, if it be highly plausible, justify us in seriously inclining toward belief in it, as long as the phenomena be inexplicable otherwise. (CP 2.662)

Explanation is here central to judgments of plausibility. Shortened and simplified, the form of reasoning is:
(S_2) We are presented with phenomena $P_1 \ldots P_n$.
    Hypothesis H explains many or all of $P_1 \ldots P_n$.
∴ H is plausible.

Similar schemata are to be found in Peirce, Hanson, and Goudge (CP 5.189; Hanson 1958b 86; Goudge unpub 9).

According to Goudge, plausibility is not a property which belongs or fails to belong to hypotheses (Goudge unpub 15; see also Goudge 1966, 623).[1] He proposes that, rather than speaking of hypotheses as plausible, we should prefer the locution 'it is plausible to entertain H.' The import of this is practical, signifying that further investigation of H is in order. Now, although the plausibility of H implies that it should be investigated, I see no reason to assume that this practical consideration exhausts the meaning of 'plausible.'

Goudge points out similarities in the use of 'plausible' and 'reasonable,' as in 'it is reasonable (plausible) to entertain H.' But a more common use of 'plausible' is akin to that of 'true' or 'probable,' as in 'it is true (probable) (plausible) that ... ' Truth is assumed to be a property of propositions, as is probability on some interpretations, so the same may hold of plausibility. Inferences conforming to schema (S_2) lead to conclusions that a hypothesis possesses plausibility, and it is in virtue of this possession that further investigation of H is warranted.

There is thus more to plausibility than practical considerations concerning the next step in the process of investigation; it is also true that not all practical considerations are based on plausibility. Peirce emphasizes the importance of matters of *economy* in deciding what hypothesis to subject to testing (CP 7.220). He says that abduction, a kind of inference which leads to the provisional adoption of a hypothesis for the sake of inductive testing,[2] should take into account the *cost* of a hypothesis, that is, the expense in money, time, energy, and thought involved in testing it. The second economic feature mentioned by Peirce is the *inherent value* of a hypothesis, which is partly a matter of instinct and partly a matter of 'likelihood,' the extent to which a hypothesis accords or discords with our previous ideas. The third category of factors of economy concerns the relation of what is proposed to other projects. Here the desirable qualities are 'caution,' 'breadth,' and 'incomplexity.' A cautious hypothesis is one which is broken up into its smallest logical components, so that if the hypothesis fails, the defective component can be isolated. A broad hypothesis is one whose elementary parts involve explanations which can be generalized, resulting in gains of economy from the avoidance of repetitious work. Finally, incomplexity is desirable because a simple hypothesis, although proven to be inadequate, may be instructive concerning new hypotheses.

All but two of these economic considerations are irrelevant to plausibility, in Peirce's sense. Of those that are not, likelihood is too subjective to be of much importance; but breadth, which is similar to what Hanson calls 'explanatory fertility,' represents an important element in judgments of plausibility. It is a central feature of scientific theories that they *grow*, that is, that they be extended to apply to new phenomena. For example, the Newtonian theory of gravitation was shown to explain such phenomena as the motion of comets and the tides, in addition to the motion of planets; and the quantum theory became much more attractive when it was shown to account for phenomena other than black-body radiation, such as the specific heat of solids and the photoelectric effect (Thomson 1969, 57–8). Hence the potential of a theory for expansion and generalization contributes towards its plausibility.

255 The Autonomy of a Logic of Discovery

This factor is similar to Goudge's condition of the amenability of a hypothesis to imaginative elaboration (Goudge unpub 22).

Gary Gutting claims that hypotheses are evaluated as plausible according to 'regulative principles' (Gutting 1973, 386). These include heuristic principles, involving for example simplicity and analogy; principles of definition, requiring for instance that hypotheses be verifiable or quantitative; and cosmological principles, such as determinism, making assertions about the physical world. This classification has some features in common with one proposed by Wesley Salmon, who describes three kinds of criteria for plausibility (1966, 125). Formal criteria concern deductive relations that a hypothesis has to accepted ones: if H' entails H, then H is at least as plausible as H'. Pragmatic criteria include the source of a hypothesis; to appeal to the source of a hypothesis is not to commit the genetic fallacy, since we know for example that a successful hypothesis is more likely to come out of M.I.T. than out of Maharishi Mahesh Yogi U. Finally, Salmon discusses material criteria. These include analogy, which we use in judging as plausible hypotheses which are similar in relevant respects to ones which have been successful. Other material criteria are simplicity, kinds of causal processes introduced, and kinds of assumptions made about the nature of space.

Kenneth Schaffner, in a study of the history of regulatory genetics, describes four types of logical considerations that play a role in the development of theory. These are experimental adequacy – the degree to which a new hypothesis accounts for relevant experimental results; sufficiency with respect to background theories; simplicity; and a principle of the unity of fundamenal biological processes (Schaffner 1974, 375–6). Mary Hesse describes James Clerk Maxwell' recommendations for using 'physical analogies' to 'deduce' hypotheses from phenomena, where such an analogy is a 'partial similarity between the laws of one science and those of another which makes each of these illustrate the other' (Hesse 1973, 89). Maxwell's physical analogies are much like what Hanson calls 'formal symmetries' (Hanson 1961, 26).

It is evident from the above that explanatory considerations are not the only ones relevant to the plausibility of hypotheses. However, explanation is a factor interrelated with many of those just mentioned. A scientist may adhere to a regulative principle requiring that explanations be complete, deductive, or causal, rather than statistical or teleological, and accordingly assess the plausibility of a hypothesis on the basis of what kind of explanation it affords. For example, contemporaries of Darwin such as William Hopkins criticized him for not meeting the deductive standards of explanation found in such mathematical sciences as astronomy. Or a hypothesis may be judged to be plausible on the grounds that it affords the same kind of explanation which has been successful in an analogous field. But the explanatory considerations most directly in favour of the plausibility of hypothesis are those fitting schema (S_2): a hypothesis is plausible if it explains the facts. One might also aruge that a hypothesis is *implausible* because it fails to explain a certain phenomenon, or because another hypothesis offers a better explanation.

Lavoisier argued in accord with schema (S_2) when he put forward his theory of combustion in opposition to the phlogiston theory. He writes in an important paper of 1777:

I venture to propose today to the Academy a new theory of combustion, or rather, to speak with the reserve which I impose on myself, a hypothesis, by whose aid are explained in a very satisfactory manner all the phenomena of combustion, of calcination, and even, in part, those which accompany animal respiration. (Lavoisier 1862, 2, 225, my translation)

He stresses in this paper that he does not claim to show that his theory should be substituted for the accepted phlogiston theory of Stahl, but only that it is a more probable (= plausible) hypothesis (Lavoisier 1862, 2, 233). However, in his 1783 paper, 'Réflexions sur le Phlogistique,' he uses much the same kind of reasoning to argue that phlogiston does not exist and that his own theory of combustion, based on the 'principe oxygine,' should be accepted (Lavoisier 1862, 2, 623). Because of the similarity between the way in which Lavoisier argues for the plausibility of his hypothesis and the way in which he argues for its acceptance, this example brings into question whether there is in fact any logical difference between arguments for plausibility and those for acceptance.

But first recall the issue of whether the reasons for holding that a hypothesis will be of a certain kind are different from the reasons for holding that a particular hypothesis is plausible. The distinction breaks down, because the fact that there is reason to believe that a hypothesis will be of kind K imparts plausibility to particular hypotheses of kind K. Schematically:

(S_3) It is likely that the hypothesis we are looking for is of kind K.
   (conclusion of (S_1)
   H is of kind K.
∴ H is plausible. (conclusion of (S_2)).

For example, we saw above that Darwin reasoned from the analogy between domestic and natural species that a hypothesis of a kind involving selection was needed. But Darwin also used the analogy to contribute towards the plausibility of his finished theory, by devoting the first chapter of *On the Origin of Species* to variation under domestication. Similarly, Galle and other astronomers presumably found the hypothesis of Leverrier and Adams concerning an unknown planet to be plausible because it was of an appropriate kind.

257 The Autonomy of a Logic of Discovery

Thus reasons for supposing that a hypothesis will be of a certain kind are also reasons for the plausibility of a particular hypothesis of that kind. But does this mean that there are no logical differences between reasoning to a kind of hypothesis and reasoning to a particular one? Not exactly: the obvious logical difference is in the form of the conclusion. However, the difference is not sufficiently great to conclude that there are two distinct logics, one a logic of discovery involving kinds of hypotheses, and the other a logic of plausibility. (S_1) and (S_2) characterize distinct kinds of inference, but because of the link shown by (S_3), there is no reason to speak of different logics.

However, there may still be logic of discovery which *is* distinct from the logic represented by these schemata. This would be a 'logic of the economy of research,' concerning the way in which various practical considerations affect the choice of hypotheses for testing. It would deal with such matters as cost and utility; these are not reasons of any of the three sorts distinguished by Hanson, so further discussion of them here is unnecessary. (For more on this topic see Rescher 1976.)

IV REASONS FOR ACCEPTING A HYPOTHESIS

I referred above in II to Hanson's conclusion that the difference between reasons for suggesting a particular hypothesis as plausible and the reasons for accepting it is only a matter of degree. Wesley Salmon has suggested an even more direct way in which the distinction between plausibility and acceptance criteria breaks down. He claims that Bayes' theorem represents the reasoning used in favour of scientific hypotheses, and plausibility considerations play a major role by establishing the prior probabilities which are essential to the calculation of the probabilities of hypotheses (Salmon 1966, 124). Now, although it is a moot question whether scientists ever actually use Bayes' theorem, Salmon's discussion is important in suggesting that the evaluation of a hypothesis as acceptable or not is often based on combination of (1) factors establishing its initial plausibility and (2) results of experimental tests. Moreover, empirical testing is also relevant to plausibility. An investigator may do a trial test, one involving only a limited sample or set of observations, and on this basis conclude that a hypothesis is plausible and worthy of further investigation. This provides additional reason to believe that there is no major logical difference between claims for the plausibility and claims for the acceptability of hypotheses.

Because of this and the link between plausibility and kinds of hypotheses described in the last section, it is likely that the distinction between reasons for suggesting that a hypothesis will be of a particular kind and reasons for accepting a hypothesis also collapses. Hanson argues against this collapse by saying that analogy and symmetry, which constitute reasons for supposing that a hypothesis will be of a particular kind, are not sufficient to establish particular hypotheses; inductive arguments *are* sufficient, so they must differ from arguments of the analogical or symmetrical sorts (Hanson 1961, 27). This argument has two main flaws. First, the paradoxes of confirmation and the lack of a satisfactory confirmation theory make it unclear whether inductive arguments of the traditional sort are in fact sufficient to establish hypotheses. Second, even if they are sufficient in some cases, it is possible to point to myriad other cases where a hypothesis is argued for in a more complicated manner. An argument for the acceptance of a scientific hypothesis often uses a combination of such considerations as experimental tests, explanatory power, simplicity, analogy, and compliance with regulative principles. Darwin, as well as using the analogy between domestic and natural variation as a reason for suggesting a kind of hypothesis and for the plausibility of his hypothesis, also cites it as one of the general considerations which are the ground of his belief in natural selection (Darwin 1887, 3, 25). According to Schaffner, the logical considerations mentioned above which were used in regulatory genetics preceding justification were also used to assess and defend hypotheses (Schaffner 1974, 384). Therefore, the consideration that a hypothesis H is of an appropriate kind K can be used as partial support for the acceptance of H. This is especially clear if one follows Salmon and supposes that K locates H in a reference class and thereby makes possible a judgment of the prior probability of H.

V CONCLUSION

I have argued that there is no sharp distinction between reasons for supposing that a hypothesis will be of a certain kind and reasons for judging a hypothesis to be plausible. Moreover, there is no sharp distinction between either of these sorts of reasons and reasons for accepting a hypothesis. Therefore, Hanson's attempted distinction between the three sorts of reasons cannot be used to establish the autonomy of logic of discovery from the logic of justification.

Nevertheless, as Peirce, Hanson and Goudge have urged, much reasoning *does* occur that is not immediately directed towards the justification of hypotheses, but rather towards either selecting a likely kind of hypothesis or establishing a hypothesis as plausible. My point is not that there is no such reasoning, but only that it is not substantially different from reasoning used to help justify hypotheses. There is thus no *autonomous* logic of discovery based upon the sorts of reasons in Hanson's classification. An autonomous logic of discovery would have to consist of something like the 'logic of the economy of research' mentioned above.

259 The Autonomy of a Logic of Discovery

Recognition that Hanson's logic of discovery and Goudge's logic or plausibility fall withing the logic of justification is hindered by the narrowness of the logical empiricist model of justification. The logical empiricists drew a sharp distinction between the context of discovery and the context of justification. They contended that the former context is of purely psychological interest, while the latter context is the province of the hypothetico-deductive method, in which an axiomatized theory is confirmed by deducing observation statements from it and checking that they are true. This excluded from the context of justification such reasoning as that involving kinds of hypotheses. The logical empiricists also claimed that explanation has the same deductive structure as prediction; Goudge has strongly challenged the deductive account of explanation in biology (AL 68–9, 74), and perhaps we should see Goudge's defence of a distinct 'logic of plausibility' as part of a more general critique of the narrow logical empiricist model of justification.

Recently, alternatives to the hypothetico-deductive account of theory justification have been proposed. One of these is based on what Gilbert Harman calls 'inference to the best explanation' (Harman 1965). On this account, acceptance of a theory is justified if it provides a better explanation of the evidence than is provided by competing theories. Darwin's use of this kind of inference is illustrated by the following passage from the sixth edition of *The Origin of Species*:

It can hardly be supposed that a false theory would explain, in so satisfactory a manner as does the theory of natural selection, the several large classes of facts above specified. It has recently been objected that this is an unsafe method of arguing; but it is a method used in judging of the common events of life, and has often been used by the greatest natural philosophers. (Darwin 1962, 476)

I shall now try to show how reasoning concerning the plausibility of hypotheses falls within the logic of inference to the best explanation. This will serve to locate Goudge's work on plausibility within the context of these other recent critiques of the logical empiricist model of the logic of justification.

I argue elsewhere that in inference to the best explanation the best explanatory theory is selected by means of the criteria of consilience, simplicity, and analogy (1978, 76–92). The notion of consilience is derived from the writings of William Whewell (Whewell 1967, 65 ff). Consilience is intended to serve as a measure of *how much* a theory explains, so that we can use it to tell when one theory explains *more* of the evidence than another theory. Roughly, a theory is said to be consilient if it explains at least two classes of facts. Then one theory is *more* consilient than another if it explains more classes of facts than the other one does. We show one theory to be more consilient than another by pointing to a class or classes of facts which it explains but which the other theory does not. Plausibility, as determined in schema (S_2), is like a weakened version of consilience: a theory is plausible if it explains *some* classes of facts.

Plausibility is also related to a notion I call 'dynamic consilience,' which is similar to Imre Lakatos' notion of theoretical progressiveness (Lakatos 1970, 116ff). A theory is more dynamically consilient than another if over a period of time it has been more successful in adding to the set of classes of facts which it explains. Then explanatory fertility and amenability to elaboration can be treated as indicators of potential dynamic consilience. Such treatment again places the logic of plausibility within the logic of inference to the best explanation.

Simplicity puts a constraint on consilience in the following way. We want a theory to explain all the relevant facts, but we do not want it to do so only by introducing in an ad hoc manner a host of auxiliary hypotheses in its explanations (Lakatos 1970, 175). One theory is simpler than another if it requires fewer auxiliary hypotheses in its explanations. Simplicity appears unconnected with plausibility.

The last principal criterion of the best explanation is analogy. A theory is supported by this criterion if the hypotheses in it and the explanations it gives are similar to ones which have proven successful in related fields. Other things being equal, the explanations afforded by a theory are better explanations if the theory uses hypotheses of the same *kind* as accepted ones. A hypothesis is of the same kind as another if it introduces the same sorts of mechanisms or entities; employs the same sorts of concepts, mathematical techniques, or models; and figures in the same patterns of explanation. Gerd Buchdahl has described how such 'regulative ideas' and 'preferred explanation types' often figure in the evaluation of hypotheses (Buchdahl 1970). The criterion of analogy supports theories using accepted kinds of hypotheses, and thereby incorporates the reasoning used in schemata (S_1) and (S_3). The latter schema licenses the conclusion that a hypothesis is plausible on the grounds that it is of the appropriate kind, which is equivalent to saying that the hypothesis is supported by the criterion of analogy.

Thus the main considerations cited by T.A. Goudge and others as affecting the initial plausibility of a hypothesis fall within the criteria used in accepting a theory as the best explanation. This result gives further support to the thesis that there is no autonomous logic of discovery.

R.E. TULLY

Emergence Revisited

The impact of Darwinism on philosophy rivals that of Relativity. Numerous philosophers, seeking to develop a metaphysics of evolutionary theory, have fashioned concepts like vital force, entelechy, mind as an instrument, process, and emergence, but the indispensable basis for appraising such ideas is a precise review of the scientific claims which support evolutionary theory, and far fewer philosophers have undertaken that important task. Thomas A. Goudge, because of his careful work on the philosophy of biology, belongs to this smaller group. One position to which he has given special attention is that called 'emergent evolutionism' (AL 166–7; 1967a) and it is against the background of Goudge's discussion that I want to examine an epistemological aspect of the concept of emergence. I shall try to show how two early emergentists deployed evolutionary theory in order to argue some basic issues in the theory of knowledge.

It is a prevailing opinion in contemporary philosophy of science that the terms 'explainable,' 'reducible,' and 'emergent' form an inconsistent triad. So far, many observable phenomena have not been brought under satisfactory covering laws, while many of the ones which have still lack a law-like characterization in terms of particle theory, but to those philosophers who confidently await the unification of the sciences this state of affairs is only temporary. In the fullness of whatever time it takes for Unified Science to be realized, they hold, there *are* no emergents. Classical emergentists like Broad and Alexander have thus been condemned either for failing to comprehend the nature of scientific explanation, for having insufficient faith in the potentiality of science, or for promoting a metaphysical outlook which at best is simply irrelevant to the proper study of scientific method. The list of accusers, a long one, includes names like Hempel, Putnam, Nagel, Feigl, Smart, and Armstrong, for some of whom Unified Science has represented the apotheosis of materialism (Armstrong 1968, Feigl 1958, Hempel and Oppenheim 1953, Nagel 1965, Oppenheim and Putnam 1958, Pap 1962, Smart 1963). Not surprisingly, the combined assault has tended to suppress enthusiasm for the concept of emergence.

This is a fate which the early emergentists helped bring upon themselves. Broad and his contemporaries[1] were as severely critical of vitalism and related notions as later philosophers have been, and much like their successors they were convinced that philosophy should aim for harmony with science; yet their principal work was completed well before the concepts of explanation, law, and reduction, not to mention the details of part-whole logic, had received careful analysis. Although they used these notions in a pre-systematic way both to explain emergentism and to defend it, their doctrine has been judged by later standards – in effect, by what they *ought* to have meant. This has been the source of much misunderstanding, and not a little unfairness. My first task, then, will be to expose some of the confusions which have deprived the concept of emergence of a better reputation, and my second will be to develop a philosophical theme about emergent qualities which helps to clarify the aims of Broad and others. Despite the accepted view that they intended to set limits to the possibility of reductive explanation which are artificial and unnecessary, I believe they understood emergentism to be essentially a doctrine about the limits of *theoretical* descriptions of phenomena. Whether such an idea is relevant to philosophy of science is a matter to which I shall return briefly at the end.

I

What *kinds* of things are emergents? To this question there never was an official answer. Unlike the positivists in the thirties, emergentists did not form themselves into a movement to advance their ideas, and whatever the bonds of sympathy among them there was never anything like collaboration. They did show a mutual respect for science, however, and saw themselves as interpreting its results in a way which stressed an evolutionary development of biological and psychological phenomena from less complex 'levels' of matter. Emergents were variously spoken of as entities, compounds, properties or characteristics, processes, and qualities; even laws were sometimes classified as emergent. The most comprehensive list was given by Alexander who regarded time as emerging from space, secondary qualities from matter, life from inorganic compounds, mind or consciousness from brains, and deity from the remainder of the universe. Nevertheless, just one item from this list will be sufficient to show where emergentism has been judged unfairly, as well as what its real weaknesses and strengths are. The traditional secondary qualities were certified by every emergentist as prime examples of their kind and will therefore be the focus here, despite the fact that their exact ontological status was a disputed question.[2]

Critics of emergentism have challenged three well-known theses about secondary qualities which they attribute to its defenders:

(1) secondary qualities are such a novel type of phenomena that their first occurrence in space and time could not have been predicted, even by super-scientists;

(2) secondary qualities are absolutely unexplainable by means of purely 'mechanical' or physicalistic theories; and

(3) the successful explanation of such qualities requires a peculiar variety of law essentially distinct from the laws of physical theory.

Broad has been most often singled out as the chief exponent of these three claims, and indeed it is quite clear that he thought them to be closely related. If the occurrence of secondary qualities could not have been predicted, no matter how sophisticated our knowledge of the behaviour of elementary particles, then any theory which tries to explain what makes them occur, and consequently the laws belonging to that theory, would have to be enriched by a class of emergent terms. While 'Pure Mechanism' (as Broad called it) might well give a complete account of the realm of unobservable matter, it would fail to give 'the whole truth about the external world' (Broad 1925, 47).[3] His critics have typically opened their attack on the second thesis. They assume, to start with, that the distinction between predicting and explaining phenomena is basically a pragmatic one which depends on whether a deduced fact has become already known or has not yet been recorded by an observation. Then the argument begins: Phenomena related to perception are either capable or incapable of being explained by means of the physical theory at our disposal. If the former, then the theory is adequate also to *predict* the occurrence of perceptual phenomena, so that it will hardly need to be supplemented by the 'unique and ultimate' laws of which Broad spoke. If the latter, then the incapacity is only provisional, and it would be both foolhardy and wrong for anyone to argue on a priori grounds that physical theory would be forever unable to take the full measure of perceptual phenomena. Either way, the argument concludes, the emergentist who insists on talking about novel or logically unrelated qualities is really promoting a version of epiphenomenalism, which no philosopher of science need take seriously.

This line of argument seems to have originated with an article by Pepper in 1926; it was refined by Henle just over fifteen years later, and has been repeated with modifications many times since, receiving perhaps its clearest formulation in the Hempel-Oppenheim studies on the concept of explanation (cf Pepper 1926; Henle 1942; Hempel and Oppenheim 1965, 260–4). The central flaw in all these versions, however, is the assumption that Broad and his fellow emergentists were using the term 'predict' in a way which is roughly convertible with 'explain.' Allowing for imprecision in the emergentists' understanding of what constitutes a law-like explanation, the net weight of evidence is clearly against this assumption.

(i) Both Alexander and Broad made use of the unpredictability argument (thesis 1), though with very different aims. Alexander seems to have been more interested in it as a basis for arguing about human freedom in a world causally determined by physical events (cf Alexander 1966 II 327), while Broad's purpose seems to have been wholly epistemological, despite his frequent recourse to examples from science. Unfortunately it is the examples, rather than Broad's assessment of them, which have remained in the minds of his critics. 'Most of the chemical and physical properties of water,' he wrote, 'have no known connexion, either quantitative or qualitative, with those of Oxygen or Hydrogen. Here we have a clear case where, so far as we can tell, the properties of a whole composed of two constituents could not have been predicted from a knowledge of the properties of these constituents taken separately, or from this combined with a knowledge of the properties of other wholes which contain these constituents' (Broad 1925, 63). His most famous example concerns the limitations which a mathematical archangel would face when presented with a sample of ammonia. With the theory of Pure Mechanism at his disposal, the archangel 'would know exactly what the microscopic structure must be; but he would be totally unable to predict that a substance with this structure must smell as ammonia does when it gets into the human nose. The utmost he could predict on this subject would be that certain changes would take place in the mucous membrane, the olfactory nerve and so on. But he could not possibly know that these changes would be accompanied by the appearance of a smell in general or of the peculiar smell of ammonia in particular, unless someone told him or he had smelled it for himself' (Broad 1925, 71).

Broad clearly overreached himself in the first of these examples. Scientists should be the ones who analyse the properties of water and its molecular structure, and if some details of the analysis happened to be uncertain fifty years ago it was not for Broad to declare this to be a permanent state of affairs. Instead of showing the concept of emergence to have lasting interest, he appears to have been using it merely to temporize. More than that, many of his examples fail to bring out the specific differences between emergent and non-emergent properties of chemical compounds in a way that makes scientific sense. One of hydrogen's properties is that of combining with a demi-portion of oxygen to form water, but Broad's habit of limiting the notion of property so as to exclude any reference to the manner in which one molecular constituent can potentially combine with another was certain to appear arbitrary to later philosophers of science. They have felt that such a narrow definition guarantees Broad's claim about the separate status of emergents at the cost of robbing it of any real interest; they have also felt that his treatment neither reflects the actual procedures of science nor makes it intelligible that we should ever speak of deducing the properties of a whole from those of its parts. Herein lies one of the main reasons why emergentism has been derided as no more than an a priori proposal regarding the use of expressions like 'part-whole,' 'property,' and 'isolation.'

However, the example of the archangel indicates the kind of general point Broad was really trying to make. The properties which he wanted to call unpredictable are *observational*, by which he meant *perceptible*. The (visual) transparency of water and its (tactual) fluidity, along with observably related characteristics, all belong to a compound but not to its components, and all of them he considered to be emergents, not because they are superadded to the compound in some mysterious way but because they make it directly known. The secondary qualities we encounter is perception enable us to identify and describe material objects; it is they, rather than molecules themselves, which reveal the physical presence of a compound to an observer. Naturally, many instances of chemical compounds would lack them. A microscopic sample of ammonia, for instance, would fail to have an emergent quality: complexity of structure is not what endows a compound with such a quality but rather the compound's being encountered in sufficient amount to allow an observer to make sensory contact with it. The dividing line between part and whole is reached, accordingly, when the parts themselves cease to be perceptible, for it is at the level of non-perceptual reference that one can avoid altogether any mention of emergent qualities. While such qualities are the 'macroscopic properties of matter' which we come to know through experience, the perceptually observable represents only a small portion of what we can say about the external world.

It thus seemed obvious to Broad that emergent and theoretical descriptions do not *mean* the same, since they characterize matter in terms of separate but complementary levels of organization. And this is basically what he sought to convey by using the unpredictability argument. The archangel could provide a complete description of the microscopic properties of ammonia without making any 'essential reference' to secondary qualities. If the meaning of emergent terms is given in experience, as Broad believed, then prior to becoming acquainted with the referents of such terms any predictions about them would have been vacuous. By hypothesis, the only referential expressions which would be meaningful to the archangel are the quantitative terms of Pure Mechanism, so that his predictions would have come no closer to secondary qualities than the microscopic events on which their occurrence causally depends. Broad's argument has an unassailable logical core, as long as one is prepared to grant his semantical assumption about the way in which emergent terms fundamentally gain their meaning.

Regrettably, the unpredictability argument itself does more to prejudice than defend that core, for the emphasis appears to be on the cognitional capacities of observers. Indeed, if we regard predicting as an activity of skill leading to the acquisition of new knowledge, then Broad's argument implies the unwanted result that any increase in the amount of perceptual knowledge obtained by observers would automatically diminish the number or even kinds of qualities which can be considered emergent. Nowadays there are fairly accurate standards for determining how much of a compound must be present in order to be minimally visible, and what qualities would be seen to pervade it under normal conditions of perception. Worse still, just which qualities should be treated as emergent at any one time is likely to vary from observer to observer. The cumulative effect of the emergentists' talk of unpredictability has been to make them appear preoccupied with the conditions of ignorance. Broad's own preference for illustrations from science, along with his characterization of emergentism as a 'theory,' was also bound to mislead later philosophers into thinking that he intended to offer an analysis of scientific method, and it may well be that Broad himself never clearly distinguished the logical core of his position from the encumbering detail. But, whatever his practice, emergentism is not a scientific theory at all. The hidden strength is linguistic. If secondary quality terms are indefinable in terms of microscopic particles, then this is a logical fact about the language we use to describe the world, and one which does not depend on considerations of predictability, absolute or otherwise.

Pap has been one of the few recent philosophers of science to recognize the basis of Broad's position to be the semantic truism that 'some descriptive terms must be given meaning by ostensive definition if any descriptive terms are to be given meaning by verbal definition' (cf Pap 1962, 369–71).[4] Still, he objected, which terms are to be designated as emergent owing to the way they are defined is a matter of choice rather than compulsion. As in a calculus, the selection of a set of primitive descriptive expressions would never be uniquely determined, and if there happen to be any constraints on our choice they are psychological only, 'specifically concerning possibilities of imagination and identification of qualities that have not been previously perceived.' What this shows, however, is that Pap still regarded unpredictability as an *inseparable* part of Broad's position. Pap himself realized that even if some quality terms could be introduced on the basis of simpler ones it would hardly follow that either sort of term could be *defined* by reference to microscopic processes alone – and yet just this was Broad's central point. Whether secondary qualities are interdefinable within the limits of a certain level of language is strictly irrelevant to the question of whether definition is possible between levels, and the fact that the qualities for which they stand resist being defined as microscopic elements was enough for Broad to count them as emergents.[5]

(ii) Emergentist ideas concerning scientific explanation (thesis 2) have occasioned severe criticism, not all of it deserving. Pap's judgment of Alexander, for instance, was more than just wide of the mark; it was unjust. He has accused Alexander of metaphysical obscurantism, suggesting that he suffered from 'a subconscious hostility against the faith in the omnipotence of science.' The reason for this condemnation is that Alexander had once declared that the existence of emergent qualities 'admits no explanation' but must be accepted with 'natural piety' as a brute empirical fact (cf Pap 1962, 365–6 and Alexander 1966 II 45).[6] Those sentiments were indeed metaphysical but were never intended to be anti-scientific. They were meant only to express a sense of philosophical wonder at why the universe should contain emergent qualities like consciousness at all. Alexander's notion of the unexplainable was certainly cloudy, and the elevated Edwardian tone of his remarks is apt to strike the more pragmatically minded as condescending and unprofessional – but reactionary he was not. The reduction of emergents and their complete empirical explanation was a prominent feature of the elaborate system he had constructed: 'Each new type of existence when it emerges is expressible completely or without residue in terms of the lower stage, and therefore indirectly in terms of all lower stages; mind in terms of living process, sense quality like colour in terms of matter with its movements ... '(Alexander 1966 II 67). The very least that Alexander meant by 'expressible's here is that emergent qualities can be causally explained by means of simpler spatiotemporal elements. Since he still insisted on their status as emergents, he plainly considered scientific analysis to pose no threat.

The charge against Alexander of bad faith towards science is easily cancelled, and while it is impossible to absolve Broad completely of this charge as far as physicalistic explanation is concerned it seems fairer to reduce it to one of negligence. Broad seriously doubted, for example, whether one could ever establish a unique correlation between patterns of neural events and individual experiences, though his reasons were paltry. He apparently supposed that any detailed knowledge of neural processes could only be achieved if the living brain were literally opened to inspection, and that in any case these would always be too minute to be perceived. He also speculated that a complete causal explanation of mental events might have to mention psychical as well as material conditions (cf Broad 1925, 121-5, 169-70). Curious as it sounds, Broad was actually trying to avoid being doctrinaire. He wished to discuss the question, 'What empirical reasons are there for believing in a unique correlation between mental and physical events?,' and he surmised that philosophers who do posit such law-like correlations are more apt to be influenced by some 'general metaphysical theory of the nature of matter and mind' than by purely scientific considerations. His point applies as much to the parallelists of his own day as to the central state materialists of ours. Nevertheless, he was willing to concede at least the *logical* possibility of detailed correlations, and he fully granted that the emergent properties of a chemical compound such as silver chloride were all causally determined by its constituents (Broad 1925, 64). In Broad's time the complete physical explanation of all observable phenomena lay much further in the future than it does today, but even so he was willing to anticipate its attainment. As far as explanatory power is concerned, the contemporary notion of Unified Science is little more than Broad's mathematical archangel secularized.

Hempel and Oppenheim have accused emergentists like Broad of holding that the properties of chemical compounds are 'absolutely unexplainable' prior to the observation of samples. In rebuttal, they cite Mendeleev's prediction of the existence of germanium, together with some of its properties, about fifteen years prior to its actual discovery (cf Hempel and Oppenheim 1965, 262 inc fn). Yet Broad himself was doubtless aware of this accomplishment and might have used it to give an illustration of emergence (admittedly, not a very compelling one). He could have pointed out that the Russian's predictions pertained only to the atomic properties of germanium and to observable properties of just those sorts with which one was already familiar, but that if there were a question of germanium having a unique or novel smell then no prediction would have been able to depict its character. Like the archangel, the scientist might have been able to foretell that the atomic element would have a pleasing or irritating effect on the olfactory sense of an observer, but any expression intended to serve as a name for the smell itself would have been meaningless on its own until an instance of the quality had been supplied in experience. The fact that Mendeleev's achievement concerning undiscovered properties had nothing whatever to do with unique qualities in Broad's sense only shows what a tenuous connection there is between the concepts of explanation and emergence. But nowhere, in fact, did Broad hold that emergent qualities are absolutely unexplainable. The real point of his illustration would have been that our ability to frame predictions about qualities is partly a function of the resources of our language, which is not an empirical point at all.

Broad can be faulted, however, for having borrowed some of the technical vocabulary related to scientific explanation without precisely indicating the different context he had in mind. It is worth repeating that the grounds of his distinction between emergents and non-emergents concern what can be immediately known and said on the basis of perceptual experience, and although these grounds are much too narrow to permit an adequate account of the critical notion of scientific observation this affects only the latitude of Broad's distinction, not its integrity. His view seems to have been mainly guided by logical considerations of the use of language, and this may help explain why he thought it so obvious to hold that emergents are *irreducible* to the properties of microscopic particles. To Broad, 'x reduces to y,' 'x is identical to y,' and 'x is definable in terms of y' were all interchangeable expressions; consequently, his claim that secondary qualities are irreducible amounts to the narrow and innocuous thesis which even strict materialists have been willing to admit: as used in our language, 'red' (for example) does not *mean* the same as 'the longest wavelength in the visible spectrum.' Apart from his careless borrowing of scientific vocabulary, which must be acknowledged as no slight flaw, the real weakness of Broad's position lies not so much with what it says as with the fact that it ventures little. Later and more technically improved senses of reduction, whether in terms of biconditionals (Woodger) or according to a pragmatic criterion of successful replacement (Kemeny-Oppenheim), specifically concern only the theoretical explanation of phenomena and hence are neutral with regard to Broad's restricted notion of reduction. Broad was not talking at cross purposes with later philosophers of science. But neither does he have very much to say to them.

(iii) On the question of emergent laws (thesis 3), the situation is somewhat different. Broad's position has been articulated more carefully by later philosophers of science and indeed has been restated without explicit mention of emergence, but it has not been substantially altered. Emergent or 'transphysical' laws (as he sometimes called them) are really just correspondence rules or laws of composition, statements which correlate events or properties of disparate types. Broad has not been alone in contending that such statements should be interpreted not as definitions but as law-like indications of the necessary and sufficient conditions on which the occurrence of certain phenomena depend, conditions which must be known if their occurrence is to be reliably predicted.

A few, like Pap, have explicitly supported Broad's position (cf Pap 1962, 363–4), while others, curiously, have supposed that the ability of scientists to frame composition laws somehow strikes a blow at the heart of emergentism and its pretence of ultimate laws. This is the view taken by Hempel and Oppenheim. If we want to be able to predict the optical activity of a solution of sarco-lactic acid, they say, then our theory will have to contain a law that provides for the effect the atomic structure of this solution has on polarized light; but if our theory fails to include 'this micro-macro law, then the phenomenon is emergent with respect to that theory' (Hempel and Oppenheim 1965, 262). Broad would have applauded this example as a cogent instance of a trans-physical law – with one significant reservation. He considered a phenomenon to be emergent *even after* a correlation with its set of necessary and sufficient conditions had been established: it does not yield its status because it has been successfully explained or because a theory has become powerful enough to predict its occurrence unerringly. All that has happened to the emergent is that it has become nomologically bound. And so it is with all of Broad's emergent qualities that the laws of their occurrence are not threatened by the possibility of reductive explanation. So clear was Broad about this point that it is hard to understand how he could have been misunderstood as using emergentism to subvert the aims of orthodox explanation. 'If the existence of the so-called "secondary qualities," or the fact of their appearance, depends on the microscopic movements and arrangements of material particles which do not have these qualities themselves, then the laws of this dependence are certainly of the emergent type' (Broad 1925, 71–2).

All the same perhaps Broad should never have spoken of *emergent* laws in the first place, for, however blameless his notion, he was inviting the interpretation that emergents are discovered to exist in the course of research and are of so unique a nature that they must be specially accommodated in any theory designed to explain them. Ironically, the more Broad tried to argue for emergentism in the context of science, the more suspicious and sceptical the response. He would have stood on less shifting ground had he expressly defended emergentism as a doctrine about the nature of descriptive language, and particularly of that sector of language which is used to express elementary perceptual knowledge claims. The secondary qualities of experience to which he was so partial do not await isolation in a laboratory and neither does their fate depend on acceptance by the scientific community. The terms we use to refer to them face no such test at all, because they are already among the most entrenched words in our language, and this, though Broad seems not to have fully grasped it, is the underlying reason why his 'laws of dependence' will remain of the emergent type.

II

Like many other views which have been called doctrines, emergentism lacks a creed. Its chief proponents offer us fragmentary ideas about what it is to be an emergent, and sometimes one of their arguments against a rival view such as behaviourism will yield an important clue, but on the whole what has passed under the name 'doctrine' is really a cluster of themes which they manipulated in argument but rarely paused to analyse, and not all of these are peculiar to emergentism even if typical of it. Two such themes are particularly worth considering because they bring together certain important ideas about language and experience which the emergentists clearly favoured but did little to refine. The first is the claim that secondary quality terms are immune from replacement by the theoretical terms of science, since they are incapable of being *defined* by those expressions. The other theme is more general, and despite the fact that it is of greater importance for the defence of emergentism it was never critically developed beyond the embryonic state. It is the idea that, among instances of secondary qualities, those which occur in *perceptual* experiences form a privileged class. Both Alexander and Broad recognized that the demarcation between emergent and non-emergent could not rest on subjective claims about privacy or the primacy of experience, for they looked upon perceptual qualities as comprising an objective part of the world and as therefore within the province of public knowledge and scientific study. Theyalso believed that any account of matter which failed to mention them would be incomplete.

These two emergentist themes are complementary, but, as they stand, of unequal merit, thanks to the manner in which the indefinability claim has been presented. Its leading advocate was Broad; yet, when Alexander's remark is recalled that emergent qualities are 'expressible completely or without residue' in terms of lower level elements, there hardly seems to have been a united front. However, there was no clash of principles. Alexander also held that while an emergent is 'determined infallibly by its structure' it is nonetheless 'not mechanical in the sense of being purely material' (Alexander 1966 II 65).[7] Generally, he was more concerned to rebuke parallelists and epiphenomenalists for failing to appreciate the causal dependency of emergent qualities than he was to explain their distinctness from material compounds; the fact that they were qualities and *therefore* indefinable in quantitative terms was assumed to be a self-evident feature of his metaphysical system. In any case, Broad's own use of the indefinability claim was both narrow and ininspired. Only nonsense results, he thought, when one tries to substitute a description of some microscopic process for the name of a secondary quality (cf Broad

1925, 612–25), and there have been many anti-emergentists ready to concede the point. But perhaps this is mainly because it has been thought to present no obstacle whatever to materialism. They have urged the sufficiency of a weaker claim, that a relation of synthetic rather than logical identity connects secondary quality terms with descriptions of microscopic events or processes. Thus outflanked, Broad is left defending a position which no one has found it necessary to attack.

Even so, his position is not invulnerable. What Broad assumed was that secondary quality terms, *no matter how used*, are indefinable in theoretical terms, and that emergence is thus a feature of entire classes of terms in our language. Not only is this a simplistic assumption, it is very probably wrong. Secondary quality terms can be used in a great variety of ways: to describe what we perceive, indicate how objects merely seem to us, narrate the sensory impressions we have dreamed or imagined, and express what we are undergoing during an hallucination or illusion. And these are only a few of their uses, for besides making secondary qualities the objects of descriptions, reports, and avowals, we can also use them in the course of explaining and predicting, giving an illustration, or teaching a linguistic rule. Is it so implausible that a theoretical description could be made to approximate any of these uses? Let us consider just one example which concerns Broad's challenge to behaviourism (in the strict Watsonian sense). Assume that a sensory psychologist is investigating the physical conditions of perception with the ultimate aim of being able to establish a comprehensive and unique theoretical correlate for any instance of perceiving a secondary quality. It is natural to assume that at the beginning of the programme the psychologist would depend heavily on the avowals of his subjects, since he would want to discover exactly what qualities they regularly experience, given the conditions and stimuli which are both under his control. But with the accumulation of reliable evidence of law-like correlations, it is also likely that the psychologist's dependency would begin to decline and perhaps cease altogether when he has reached the stage of being able to *tell* his subjects precisely what sort of quality they are experiencing at a particular time. He will have become able to substitute his theoretical reports for their sensory ones. Being able to provide matched pairs of physical descriptions and sensory avowals hardly amounts to having established full synonymity, of course, but the importance of such correlations lies in the use to which they might be put: they are the raw material of a theoretical reduction of avowals, for they anticipate the elimination of secondary quality terms from perceptual theory.

The unique theoretical correlate for a given quality as experienced stands to be an enormously complex description of events beginning at or near the surface of the object perceived, and ending – as far as the experience goes, if not the verbal response to it – with neuronal events in the cortex of the subject. The correlated set of theoretical events would thus span a far greater region of physical space than that comparatively small region in which the quality itself is perceived to lie, a fact which makes it unlikely that the reductive explanation of perception could ever be successfully based on a criterion of extensional identity. But there is another criterion available. A theory of perception that uses secondary quality terms might be supplanted by a purely physical theory for which those terms have been constructionally defined, in Goodman's sense (Goodman 1966, 3–29),[8] according to a criterion of extensional isomorphism. Such a theory of perception would therefore take the form of a constructional system in which the laws themselves relating to objects of perception would be purely theoretical in character: secondary quality terms, at least in their avowal use, would have become a superfluous commodity. The practical effect of adopting a set of constructional definitions would be to enable the sensory psychologist to dispense with any mention of secondary qualities when describing perceptual experience in physical terms, and within limits his freedom of description might be extended even further. By resorting to the notions of a standard observer and standard conditions of observation, the scientist could probably avoid explicit reference to the secondary qualities of things; he need mention only those particular sorts of events, such as electromagnetic radiation, which in the presence of an observer would initiate a causal sequence that culminates in the seeing of a quality. This in turn would partially vindicate the materialists view that secondary quality terms would 'have no place' in the language of an ideally finished science. With such a language at his disposal, even Broad's archangel might nearly overcome the deficiency of never having been blessed with a nose.

The possibility of advancing constructional definitions for secondary quality terms obviously forces a breach in the indefinability claim which Broad had thought so secure; more than that, it strips away his confident belief that a class of emergent terms would have to be added via unique laws to the model of Pure Mechanism. But it would not force the total abrogation of his indefinability claim. Broad could still argue that there are other, more crucial uses of secondary quality terms which have nothing at all to do with the context of scientific explanation and which therefore could not be accommodated by a constructional system. Avowals, he might say, scarcely typify our ordinary use of secondary quality terms: they are merely sensory responses, and much like our expressions of anger and bodily feelings their role is primarily to report different kinds of events happening to ourselves. They do not aspire to be perceptual claims. And yet the prevalent occurrences of secondary qualities in our experence are perceptual in character. We most often encounter such qualities as external, factual elements of the world, typically judging them to be properties of material objects, so that most often we use secondary quality terms in an objective and public way to describe not simply how an object appears to us, which primarily interests the sensory psychologist, but how it *is*. When we say that a sample of ammonia has a pungent smell, we are not emitting a subjective response but instead are committing ourselves to the objective claim that any other observer in the same situation and with a normal nose would confront the very *same* smell. Perceptual claims such as these may imply avowals but are not founded on them – nor should they be confused with them. Constructional definitions are therefore likely to impress the emergentist as a peripheral achievement in spite of their importance for reductive explanation.

So Broad's indefinability claim, duly qualified, can be left standing. But this idea hardly sums up emergentism; it is no more than a detail in a picture of knowledge which the emergentists left unfinished. Other details can be discerned in – of all things – the unpredicability argument which has created so much antipathy. Since Broad recognized the possibility of causally accounting for emergent phenomena, his purpose in dwelling on unpredictability may become less puzzling if we suppose that he was conducting a kind of thought experiment, imagining the structure of the world at a time when it contained no sentient life of any kind. For such a world he wanted to ask what sort of descriptive language would be adequate, and the answer which appears to have satisfied him was a 'physical$_2$,' language, one that would (in the words of Sellars and Meehl) 'describe completely the actual states though not necessarily the potentialities of the universe before the appearance of life' (Sellars and Meehl 1956, 252). The issue is not whether the pre-vital world contained secondary qualities, though Broad apparently believed it did not, but whether the world at this stage of development could be intelligibly portrayed in purely mechanistic terms, and Broad raised no objections against supposing that it could. There are no interactions between observer and object to be accounted for yet, and he was careful to avoid retrojecting any common-sense assumptions into this description of the world as an assemblage of 'material particles.' But with the advent of observers there comes a radical change. Mechanistic descriptions will have ceased to express everything that can be said about the world, for they will be incapable of expressing the perceptual claims which observers are entitled to make on the basis of what they experience. One of the potentialities of the universe will have been realized: the qualities met with in perception can be *shown* in language, not simply identified mechanistically. And between these two different functions of describing and explaining the qualities we perceive Broad had no trouble deciding which takes philosophical precedence. The world we inhabit, as he conceived it, is a material world perceptually understood.[9]

Neither Alexander nor Broad regarded perceptual qualities as something added on to the world, the mere trappings of matter; rather these *are* matter, or at least as much of it as we are able to discern through perception. Like many philosophers before and since, they were deeply influenced by physical science as a model of objective knowledge and they sought to devise a comparably objective role for emergent qualities within that model. The basis of their solution was to characterize perceived secondary qualities as the way in which matter is immediately known to us. Just how such qualities depend on material compounds and what their causal relationship is to the perceiver were differently interpreted by the emergentists, but the differences are far less important than their common insight that any account of objective knowledge requires that *some* instances of secondary qualities, though obviously not all, must by treated as common points of reference in an external world. Otherwise there would be no sense in speaking of an observer as making perceptual claims or confirming those made by others; there would be no point in speaking of there being an *observer* at all. Being able to adduce our perceptions and to accept or reject the perceptual judgments of other observers is part of what constitutes objective knowledge.

Wisely, the emergentists never tried to legislate what credentials must be given to a secondary quality in order for it to be regarded as perceptual. They simply relied on common-sense knowledge, on the fact that we assume ouselves to belong to a community of observers having regular access to identical instances of secondary qualities which enable us to identify material objects. But despite this assumption, which they considered essential to emergentism, they stopped far short of incorporating many other common-sense beliefs about material objects. This is particularly the case with Broad, though even Alexander preferred to consider any material object of the usual garden variety as a 'synthesis of sensa, percepta, images, memories, and thoughts or plans of configuration' (cf Alexander 1966 II 183–5). They shied away from calling perceptual qualities parts of material objects or enduring properties of such objects, as favoured in the common-sense view of the world, and it is for just this reason that the expression 'emergent' was pressed into service as a technical term. Because they subordinated the common-sense conception of material objects to a scientific one, the only *parts* which the emergentists were inclined to speak of were the constituents of material compounds, ultimately the microscopic elements of matter whose processes determine the various patterns of causal interactions, both in perception and apart from it. Emergent qualities were not considered to be a new kind of constituent of matter, since nothing like them is discovered in the scientific analysis of compounds. It is we, the observers, who locate them in compounds by means of perception. To the emergentists, it is a brute empirical fact that matter affects us perceptually and that we have no more direct way of confronting the external world.

Despite the judgment of posterity, Alexander and Broad have a certain right to be ranked among materialists.[10] What chiefly distinguishes their basic viewpoint from those of recent materialists like Armstrong and Feyerabend is not that perceptual qualities can be reductively explained, for that is not a point of disagreement at all, but the emergentists' conviction that these should be retained in any epistemic account of the world. Perceptual quality terms coincide in their ascriptive role with that of theoretical descriptions, but do not overlap. We can say of the same region of space both that it is red and that it radiates an electromagnetic wavelength of 450 mµ, but neither kind of description is exhaustive. Nevertheless, the hypothesis of referential identity between perceptual and theoretical descriptions was used by the emergentists as a device for representing a kind of unity to knowledge. The clearest expression of this was given by Alexander who looked upon the emergent quality of consciousness as being identical with a complex of neural processes. It is deceptive, he argued, to speak of two different but correlated processes, for in reality there is only one: 'That which as experienced from the inside or enjoyed is a conscious process, is as experienced from the outside or contemplated a neural one' (Alexander 1966 II 5). The sort of view he was advocating, often called a 'double aspect theory,' has sometimes been misleadingly interpreted to set a polarity between subjective and objective varieties of knowledge, the assumption apparently being that experiential knowledge should be regarded as a form of introspection. What this interpretation ignores is that probably the majority of our experiences are taken to be perceptual in character, and that while the act of perceiving coincides with a series of events in the brain this hardly converts the object perceived into an introspected one. Whatever form it might take, the 'contemplation' of a set of neural events would be a *perceptual* activity. The orientation of secondary quality terms is similarly perceptual. Contemporary materialists who have declared that such terms are gaps waiting to be filled by behavioural criteria, or else expressions capable of undergoing fundamental shifts in meaning, have cast themselves unwittingly into the role of the mathematical archangel who symbolized for Broad a naive materialist. Having preferred to classify experiences of secondary qualities as so many sensory effects, they have seriously underestimated the role these qualities have in the formation of knowledge.

277 Emergence Revisited

Sellars and Meehl (1956, 250) have conjectured that the introduction of emergents 'might be "forced" upon us by theoretical necessities (insofar as we are ever forced to make theoretical sense by the postulation of hypothetical entities).' Yet the possibility of constructionally defining secondary qualities makes it unlikely that they would ever be needed to fill a scientific role. To suggest that emergents be treated as hypothetical constructs is in any case a bizarre twist and only perpetuates the erroneous idea that in order to merit any respect the concept of emergence must first be legitimized by science. Emergent qualities are nothing like the hypothetical raw feels mentioned by Sellars and Meehl; there is nothing remote about them, for they are the kinds of things which our experience, especially perceptual experience, makes us most familiar. And in their role as ingredients of objective knowledge emergent qualities have their strongest defence against the old charge of epiphenomenalism. Although Alexander and Broad rightly stressed the importance of perceptual qualities among emergents, they have left most of the interesting work yet to be done. They never attempted an analysis of the more complex types of observational term, including those (like 'hydrodynamic force') whose meaning shows the enrichment of physical or chemical theory, nor did they ever examine the perceptual framework of our language within which more abstruse sorts of theoretical term (like 'unit of positive charge') are meaningfully applied. Still another lack was a logically precise battery of concepts for describing the relationship between emergent qualities and the microscopic processes on which they depend. But the early emergentists have offered us a basic insight regarding the distinction between observational and theoretical terms, and it is the need to develop their insight which can provide emergentism with an opportunity to renew its career.

MARY B. WILLIAMS

Is Biology a Different Type of Science?*

Is the structure of biology intrinsically different from the structure of physics? Many philosophers have contended that it is. They have pointed out the existence of forms of explanation (AL; Beckner 1969), definition (Beckner 1959), prediction (Scriven 1959a), and laws (Smart 1963) which appear to be important in biology but which apparently do not have the same logical structure as the explanations, definitions, predictions, and laws of physics. These philosophers all concluded that the theories of biology do not have the deductive, axiomatic structure which characterize the theories of physics. In recent years, however, analyses of biological explanation (Ruse 1973c; Williams 1976; Wimsatt 1972; Wright 1973), definition (Ghiselin 1974; Hull 1976; Ruse 1973c), prediction (Kochanski 1973; Williams 1973), laws (Ruse 1973c) have concluded that they can be shown to have the usual logical form. These analyses, together with an axiomatization of Darwinian theory (William 1970), support the conclusion that the structure of biology is not intrinsically different from the structure of physics. My purpose in this paper is to show that these analyses can be unified, and the appearance of peculiarity explained, by an understanding of the nature of the referents of biological laws.

The referents of many (perhaps all) biological laws are individuals whose size and duration is vastly greater than the size and duration of the phenomena that our perceptual apparatus is designed for. Because of this the relationship between biological theories and our perceived reality is significantly different from the relationship between physical theories and our perceived reality; the differences which Goudge, Beckner, etc. attributed to intrinsic differences in the theories (that is, to differences in the structure of the internal principles)

*Work on this paper was partially supported by NSF grant GS 41878.

Is Biology a Different Type of Science?

are due to significant differences in the relationships between human beings and the theories (that is, to significant differences in the bridge principles required to connect the internal principles of the theories to *our* perceptions). Both in order to avoid detailed presentations of arguments already available in the literature and in order to give an intuitive feeling for the way in which differences in relationship to the phenomena affect our view, I will present views of the phenomena under discussion as they would be seen by an atom-sized scientist (whose relationship to the phenomena studied by Galileo is similar in significant respects to our relationship to the phenomena studied by Darwin) or a planet-sized scientist (whose relationship to the phenomena studied by Darwin is similar in significant respects to our relationship to the phenomena studied by Galileo). So I must first introduce you to Micro Angstrom and Macro Lightyear.

Micro Angstrom is a small (about the size of an atom) scientist with a life-span of about a microsecond. For Micro phenomena at the quantum level are immediately perceivable; he does not consider quarks strange particles – their charms are immediately apparent to him. Quantum theory was the first solid scientific theory developed by the scientists of his world. Micro is living in a particularly exciting period of science, for the scientists of his world have just discovered the existence of solids, liquids, and gases; the discovery of the existence of these different types of statistical regularities among collections of neighboring atoms has stimulated new types of scientific questions, as a result of which Micro and his colleagues are on the verge of discovering Galileo's laws of terrestrial mechanics.

Macro Lightyear is a large (about the size of a planet) scientist with a life-span of eons. For Macro phenomena at the species level are immediately perceivable; she[1] perceives a species by its biochemical signature, and the direct perception of the evolutionary changes in species, genera, etc. is as much a part of her ordinary experience as the direct perception of the growth and development of individual organisms is of our ordinary experience. (Her major perceptual mode is similar to the sense of smell; hence she recognizes a species when it is large enough for the concentration of its characteristic biochemicals to cross her perceptual threshold.) Individual organisms are, of course, too small for her to perceive with her unaided senses, but the recent invention of a microsnope has opened up exciting new vistas for the scientists of her world, allowing them to perceive the subdivision of species into subclans which have slightly different biochemical signatures, and to recognize that evolutionary changes are caused by the expansion of certain of these subclans at the expense of the others. When the causal factors underlying these expansions are simple biochemical phenomena (e.g., when the expanding subclan produces a biochemical which neutralizes a poison) the causal relationships are easily recognized by Macro and her colleagues; so they recognize that the changes in species are adaptive, and they are on the verge of discovering the Darwinian laws.

CONCEPTS

The first peculiarity to be discussed is the apparent peculiarity of the concepts of biology, for it is a mistake about the status of the concepts which has caused the appearance of peculiarity in laws, predictions, and explanations. The mistake was first verbalized by Beckner (1959), who examined the way in which taxonomists characterize particular biological species and pointed out that these sets of organisms are not defined by sets of severally necessary and jointly sufficient characteristics; the species description typically is a list of characteristics such that each organism in the species has most of the characteristics and each characteristic is possessed by most of the organisms in the species. That taxonomists don't give well-formed definitions is neither accidental nor due to systematic stupidity; biological theory tells us that taxonomic descriptions cannot be well-formed definitions. For example, genetic theory tells us that mutations occur regularly; and the species concept is such that, except in special (and rare) circumstances, no single mutation could determine that an organism was not a member of the same species as its parents; therefore there cannot be any necessary characteristics. Beckner proposed that we should resolve this problem not by rejecting these species definitions on the grounds that they fail to meet normal logical standards, but rather by accepting them as definitive of a new, logically peculiar, type of definition.

In recent years philosophers of biology (Ghiselin 1974; Hull 1976) have begun to assert that the reason particular species cannot be defined is that the species is an individual with respect to the processes described in evolutionary theory and that, therefore, in biological discussions species must be treated as individuals. (Hull argued cogently for this conclusion by examining the philosophical tradition concerning individuals and showing how the concept of species as individual fits into this tradition.) But if a species functions as an individual in any biological discussion, then the reason that definitions of species are logically peculiar is very clear: an individual cannot be defined (though it may be described); names are attached to individuals by christening, not by definition. Naturally any attempt to define something which cannot be defined will result in logically peculiar 'definitions.'

The idea that species are individuals is still controversial; I wish to give

281 Is Biology a Different Type of Science?

further reason for accepting it not by arguing directly for it but rather by showing how it resolves the appearance of logical peculiarity in biological laws, predictions, and explanations. But first I will try to give an intuitive feeling for what it means to think of a species as an individual.

We (humans) perceive organisms as individuals, while Micro Angstrom perceives organisms as collections of molecules; similarly Macro Lightyear perceives species as individuals, while we perceive them as collections of organisms. This perception of individuality is not a function of whether one can see spaces between constituent parts but of whether the supposed individual functions as a unit in the processes of practical significance to the perceiver. Thus Macro perceives a species as an individual not because her sense organs are such that she does not perceive the spaces between organisms (although she does not), but because her life-span is such that the Darwinian processes of adaptation, species splitting, adaptive radiation, etc. cause significant changes during her life in the world with which she interacts. In these Darwinian processes species are cohesive, coherent units which are changed significantly by the processes, so as a part of dealing with their world Macro's people had to be able to recognize these units. When Macro watches a species reacting to a threat to its survival she sees it reacting as a whole – acquiring new traits (whose biochemical traces change the biochemical signature of the species), or dwindling in size until it is extinct; when Macro watches an isolated population of a species reacting to an opportunity and becoming a new species, she sees the original species maintaining its identity while the incipient species establishes its own distinctive biochemical signature. Thus Macro sees a species as a spatiotemporal whole extending over millennia. She also sees other spatiotemporal wholes (populations, genera, phyla, etc.) that are units with respect to some Darwinian processes; because sexually reproducing species are the most cohesive of all these evolutionary units, the children in Macro's world begin their study of biology by concentrating on this simplest example of an evolutionary unit; they then learn that many of the generalizations learned in their study of species are also true of these other evolutionary units, and they learn that the unit which seemed simple and 'solid' is itself made up of smaller units. As we learn that to the scientist a table is not solid but is a collection of atoms interacting somewhat independently, so Macro had to learn that to the scientist a species is not 'solid' but is a collection of organisms interacting somewhat independently.

We, of course, do not perceive species (or the other evolutionary units) as individuals, and we do not need to in our everyday life; but when we are studying Darwinian processes we find that we cannot make sense of the data when we treat them as sets. And when we begin to treat species and other

Darwinian subclans as individuals the appearance of chaos – of logical peculiarity – begins to disappear.

LAWS

How does the idea that species are individuals change our expectations about the form of the laws derived from evolutionary theory? We have expected biological laws to be instantiated by individual organisms; now we should expect them to be instantiated by individual species (or other Darwinian subclans). Thus, instead of 'All swans are white' the evolutionary law would be 'All species fulfilling certain conditions [e.g., living in a situation in which whiteness provides significantly useful protective colouration] are white.' From this type of law, and knowledge that *Cygnus olor* is a species fulfilling those conditions, we could derive the statement '*Cygnus olor* is white.' Instead of 'Albinotic mice always breed true' (J.J.C. Smart's candidate for the status of biological law), the evolutionary law would be 'Populations in which a recessive gene is fixed breed true for that gene.' (Of course 'breed true' must have been defined to allow for a normal amount of mutation from the recessive gene.) From this law, and knowledge that albinism is a recessive trait and has been fixed in a particular population, we could derive the statement 'This population always breeds true for albinism.' Another law, which can clearly be derived from the laws of natural selection (see Fisher 1958, 158–60 for a sketch of the derivation and the conditions), is 'All sexually reproducing species satisfying certain conditions have a 1 : 1 sex ration at puberty.' From this law, and the knowledge that *Homo sapiens* satisfies those conditions, we could derive the statement '*Homo sapiens* has a 1 : 1 sex ration at puberty.' Thus the types of statements that have traditionally been viewed as candidates for the status of biological law are in actuality candidates for the status of instantiation of a biological law.

These laws, however, contain serious ambiguities. Is whiteness a property that can be predicated of *Cygnus olor*? Is puberty a property that can be predicated of *Homo sapiens*? When biologists make statements like this they are not committing the error of thinking that the properties of members of a class are properties of the class (since a species is not a class) or the error of thinking that properties of parts of an individual are properties of the individual. They either are coining an additional technical meaning (for a character term like 'white') or are using a term (like 'puberty') as shorthand, which they are confident will be understood, for the necessarily complex explicit statement of what they mean. They are forced to do this because our language does not have words to describe characteristics of species-as-individuals. (For

Macro Lightyear the odour of white pigment characterizing *Cygnus olor* would be a single impression embedded in the odour of *Cygnus olor*, and her language would have words for such characteristics. Similarly, hormonal characteristics of puberty in organisms could identify for her the organisms undergoing puberty as the pubertal part of the species, and she would have a word for that, just as we have words for the parts of organisms.) Such new technical meanings of terms are often introduced and used without explicit definition; since such a definition would necessarily contain both words from our observation language and words from the theoretical language of evolutionary theory, it would be a bridge principle connecting our observations of organisms with the characteristics of species-as-individuals. Since at least some of these bridge principles must fail to be eliminable, at least some of these 'definitions' of terms for the characteristics of species would fail to be real definitions. This examination of the ambiguities in these examples of biological laws shows in yet another way that the attempts to state biological laws in observation terms (i.e., in terms of organisms) will inevitably give logically peculiar results.

PREDICTIONS

I have previously (Williams 1973) discussed Scriven's (1959a) claim that no real predictions can be derived from evolutionary theory. Within the context of a recognition that evolutionary laws are about species and other Darwinian subclans this discussion can be summarized briefly. Scriven claims that the best evolutionary theory can do is to make 'hypothetical probability predictions' of the form 'If X occurred then organisms with trait Y would be likely to survive.' This is a statement about the probable fate of individual organisms. The related evolutionary law would be 'If (1) the Darwinian subclan S has a sub-subclan, S_Y, with trait Y, (2) organisms with trait Y have a greater than average probability of survival/reproduction in conditions X, (3) conditions X occur regularly in the environment of S, and (4) trait Y has no equivalent disadvantages in other regularly occurring environmental conditions of S, then S_Y will expand until it is fixed in S.' (This law could, if stated slightly more precisely, be derived from the axioms of Darwinian theory given in Williams 1970.) While explanations in evolutionary biology frequently involve instantiations of this law, the predictions made by evolutionary biologists usually use other laws, because our life-span is too short for us to check on predictions which involve instantiations of this law. However, we do not need to examine examples of actual evolutionary predictions in order to discover the source of the peculiarity pointed out by Scriven: this peculiarity is the result of attempts to find evolutionary laws that can be instantiated by organisms.

Recognition of this fact resolves the problem as posed – evolutionary predictions no longer appear logically peculiar – but it raises a new problem: how can the predictions of evolutionary theory be translated into predictions about human-sized phenomena? Some examples of such translations are given in Williams (1973), so it is clear that biologists have an intuitive grasp of at least some of the necessary bridge principles. These and similar examples provide the raw material with which analysis of the bridge principles can begin.

EXPLANATIONS

Failure to find laws of the expected type (i.e., laws about organisms) has not led to a general belief that biological theory gives no explanations, but it has forced some philosophers (AL; Scriven 1959a) to conclude that some scientific explanations contain no law-like statements and to defend the thesis that 'one cannot regard explanations as unsatisfactory when they do not contain laws' (Scriven 1959a). Because there are examples of explanation in biology which are generally accepted paradigms and which seem peculiar, the problem for philosophers has been to provide a satisfactory analysis of these explanations. The recognition that the referents of evolutionary laws are spatiotemporal wholes such as species illuminates both the problems and the successes of these analyses.

Narrative Explanations

Goudge (AL) pointed out that narrative explanations, which apparently play no role in physics, play an important role in evolutionary biology. His analysis, which includes claims that these explanations are of unique events and make no reference to general laws, has been criticized by Ruse (1973c) and Hull (1974). The axiomatization of Darwinian theory permits these criticisms to be made even stronger, since it legitimizes reference to Darwinian laws. Consider the following example, which Goudge quotes (AL 71) from Romer:

'The Devonian, the period in which the amphibians originated, was a time of seasonal droughts. At times the streams would cease to flow ... If the water dried up altogether and did not soon return ... the amphibian, with his newly-developed land limbs, could crawl out of the shrunken pool, walk up or down the stream bed or overland and reach another pool where he might take up his aquatic existence again ...

'... Instead of immediately taking to the water again, the amphibian might learn to linger about the drying pools and devour stranded fish.'

Is Biology a Different Type of Science?   285

In this passage Romer alludes to two instantiations of the evolutionary law I sketched in the section on predictions. The phrase 'newly-developed land limbs' indicates that condition (1) was satisfied; the second two sentences indicate that condition (2) was satisfied; and the first two sentences indicate that condition (3) was satisfied. While the adaptive value indicated in this first instantiation is the value of reaching another pool and thus avoiding death from lack of water, the adaptive value indicated in the second instantiation (alluded to in the sentence quoted from the second paragraph) is the value of an increase in the food supply; Romer does not directly mention mutations that allow the amphibian to remain out of water for longer periods of time, but an implicit reference to such mutations would be inferred by biologists and would indicate that condition (1) was satisfied; the adaptive value of an increased food supply would indicate that condition (2) was satisfied; and the first two sentences of the first paragraph indicate that condition (3) was satisfied. Thus this narrative explanation (and, I assert, all narrative explanations) consists of a sequence of descriptions, each one of which describes, or alludes to, a character trait and the environment which made that trait advantageous; each narrative explanation consists of a sequence of instantiations of this Darwinian law.

This disposes of the claim that narrative explanations make no reference to general laws; it does not, however, dispose of the claim that explanations of unique events play a role in biology which has no counterpart in physics. A glance at some difficulties experienced by Micro Angstrom in his attempts to formulate and test the laws of terrestrial mechanics should clarify the nature of the difference between physics and biology which is revealed by our use of narrative explanation in biology.

A solid as large as, e.g., a billiard ball would take several life-spans to cross Micro's path. Hence Micro and his colleagues at first thought of solids as stationary. However, the invention of a device which allowed the study of solids at great distances led to the discovery of various traces (a line of chipped rocks ending in a rock with matching scars, a rock hotter than its neighbors at the end of a heated path, etc.) which led Micro and his colleagues to theorize that these solids were moving. Micro happened to live at a time of a minor rock slide in his vicinity, and consequently found many moving rocks. From the heat due to friction he could estimate the speed of the rock and from chips and other historical traces he could estimate its direction. Micro theorized that solids moved in a straight line unless deflected by hitting another solid, and that the speed and direction after such a deflection was related by simple laws to the solid's previous speed and direction. The historical reconstructions which Micro and his colleagues laboriously built up play an important role in their science of terrestrial mechanics.

Such historical reconstructions play an important role in sciences in which the life-span of the scientist is much shorter than the duration of the processes studied. In these circumstances the scientist cannot set up experiments with simple conditions and must rely on God-made experiments; therefore the complexity of the simplest available examples of the processes is much greater than it is when scientist-made experiments are possible. The reconstructions are descriptions of these God-made experiments; their role is to provide the data from which laws are discovered and to provide tests of those laws. Neither the greater complexity of these God-made experiments nor the fact that it is necessary to use them indicate any philosophically interesting structural differences in the internal principles of the theory. The fact that it is necessary to use them is not determined by the nature of the internal principles (that is, it is not determined by the interrelationships of the phenomena themselves); it is determined by the relationship between the duration of the phenomena and the life-span of the scientist. Consequently we do not use historical reconstructions in studying terrestrial mechanics, though Micro Angstrom must; and Macro Lightyear does not use historical reconstructions in studying evolutionary biology, though we must.

Teleological Explanations

Biologists frequently give explanations that use concepts such as goal, purpose, and function; these seem to have no analogue among the explanations given by physicists. On the surface a teleological explanation is an explanation of the past in terms of the future (goals) or of the part in terms of the whole (functions), while ordinary scientific explanations are of the present in terms of the past or of the whole in terms of the parts. The common occurrence of teleological explanations in the biological literature, together with the belief (see, e.g., Beckner 1969) that teleological statements cannot be translated into non-teleological statements, has led to the conclusion that there are fundamental differences between the structures of biology and physics. In recent years many philosophers (Wimsatt 1972; Ruse 1973c; Wright 1973; Hull 1974; Brody 1975; and Williams 1976) have claimed that these explanations should be understood as assertions that the trait explained was selected for because of the effect claimed to be its goal or function. That is, these philosophers are asserting that teleological explanations of the form 'The purpose of trait T is to cause the effect U' are explanation sketches for covering law explanations in which the laws used are the Darwinian laws. (A schema exposing the covering law structure of these explanations is given in Williams

1976.) Since the referents of these laws are populations rather than organisms, the conclusion of such an explanation must be 'Therefore trait T is a characteristic of species S' rather than 'Therefore trait T is a characteristic of organism b.' (This distinction appears even in common usage: note that we say 'The function of *the* heart is ...' rather than 'The function of John Smith's heart is ...' If John Smith had been born without arms, we would say 'John Smith's toes *function as* a prehensile organ' rather than '*The function of* John Smith's toes is to serve as a prehensile organ.') Philosophical analyses that take teleological statements to be explanations of particular instances (of behavior or of traits) in particular organisms, or that assume a simple relationship of instantiation between species traits and organism traits, confound the structure of the relationship between organisms and species with the structure of the form of explanation. When we have the bridge principles that will allow us to separate the explanatory portion of the teleological explanation from the portion that connects the theoretical conclusions with our perceptions, it will be clear that the explanatory portion of these biological explanations has the same form as the explanatory portion of physical explanations.

We do not now have those bridge principles, but we can get a clearer intuitive picture of the covering law structure of teleological explanation by considering how these traits would be explained by someone who directly perceives the trait as a species trait: For Macro Lightyear the phenomena that require explanation are the changes in species traits, just as for Newton the phenomena that required explanation were the changes in the state of motion of bodies. If a gene for a new mate-attracting pheromone begins to spread through an insect species, Macro perceives its spread as a gradual change in the biochemical signature of the species. In order to explain this change she uses the microsnope to discover that the male/female mixture in subclans with this gene is spatially more homogenous than in subclans without the gene (i.e., the average distance between a female and the nearest male is smaller), and that there is a positive correlation between amount of male/female homogeneity and subsequent increases in the size of the subclan. Further investigation with an ultramicrosnope reveals that the greater homogeniety is caused by the males' tendency to fly toward the source of the pheromone which is being exuded by the female; this increases the probability of mating for females which exude the pheromone and this increases the number of offspring produced. Further investigation with an ultra-ultramicrosnope reveals the morphological basis for this male tendency. So Macro explains the trait T (the portion of the biochemical signature associated with use of the mate-attracting pheromone) as follows: The presence of morphological structures t_1 and t_2 in, respectively, males and females of subclan D* causes females to exude the pheromone and males to fly toward the source of the pheromone, which makes the percentage of unmated females in D* smaller than the percentage of unmated females over the whole species D, wich causes the proportion of D* in D to increase in each generation, which causes the biochemical signature of the species to change.

This explanation explains the present in terms of the past and the whole in terms of the parts; it is an ordinary causal explanation. Because the necessary conditions involved (conditions (1) through (4) of the law sketched above) are conditions *about subclans* and not conditions about organisms, the problems that have been created by the search for necessary conditions *about organisms* simply do not arise; e.g., in the above example some organisms of D (the females) have the trait of pheromone production and others do not, some organisms which produce the pheromone fail to attract a mate, and clearly a female that was born without the capacity to produce the pheromone might nevertheless succeed in mating. Similarly, the problems that have been created by attempt to cash out, in terms of usefulness *to organisms*, the intuition that a functional effect must be useful effect do not arise when we recognize that it is usefulness *to subclans* which is relevant; even in the case of the peacock's tail, which is suspected of being detrimental both to the individual peacock and to the species, the more ornamental tail is useful to the subclan. Thus the fact that Macro Lightyear considers functional explanations to be ordinary causal explanations is not due to the fact that she has ignored problems that exist but rather to the fact that she has not created problems by trying to use laws about Darwinian subclans to make statements about organisms.

Much more analysis of these insights into teleological explanation is needed. But even this rough, intuitive discussion indicates a causal relationship between our attempts to analyse teleological explanations as explanations about organisms and the peculiarity which these analyses reveal. Teleological explanations *are* peculiar, but the peculiarity does not reside in the explanatory portion of the explanation; it resides in the connection between the referents of the laws used and the individuals that our perceptual organs recognize.

CONCLUSIONS

An important part of the structure of the problems discussed in this paper can be illustrated by an analysis of the following syllogism: 'All swans are white. Ralph is a swan. Therefore Ralph is white.' According to the analysis present above, the first premise of the syllogism must be changed to '*Cygnus olor* is white' and the second premise becomes 'Ralph is part of *Cygnus olor*.' Now

289 Is Biology a Different Type of Science?

'Ralph is white' clearly is not implied by these two premises any more than 'Joan's heart is black' is implied by 'Joan is black.' But in biological discussions '*Cygnus olor* is white' does imply something about Ralph. The peculiarity problems in philosophy of biology have arisen, in one way or another, from the problem of explicating what such statements do imply while incorrectly assuming (1) that the statement underlying biologists' discussions of the whiteness of swans is 'All swans are white' (and, in general, that the statements underlying such discussions are about organisms), and (2) that the relationship between Ralph and *Cygnus olor* (and, in general, between an organism and the relevant Darwinian subclan) is set membership.

Thus Smart's objection that, since mice are specifically terrestrial organisms, 'Albinotic mice always breed true' is not unrestricted in space and time no longer applies when the underlying law is properly formulated. The same fate overtakes Scriven's conclusion that, since there are no real predictions in biology, biological explanations are intrinsically different from physical explanations; when the underlying laws are properly formulated the hypothetical probability predictions become real predictions connected to perceptual reality by as yet unformulated bridge principles. Goudge's discussion of narrative explanation brings to our attention explanations which appear to be simply narratives about sets of organisms; the realization that Darwinian subclans, not organisms, are the referents of biological laws allows us to recognize the instantiations of laws which are implicit in these narratives. And the recognition of the complex and poorly understood nature of the relationship between organisms and species pinpoints the cause of much of the peculiarity of teleological explanations.

In this paper I have contended that the law-like statements underlying biological discussions are always statements about sets of Darwinian subclans (e.g., statements about all species which fulfill given conditions) and that bridge principles which capture the relevant part of the relationship between individual organisms and species have not yet been formulated. It will be relatively easy to express the underlying statements as statements about Darwinian subclans, but the problem of formulating the necessary bridge principles is non-trivial and may prove to be a seminal problem which will stimulate significant new insights into the structure of science.

NOTES

LIST OF WORKS CITED

THE PUBLISHED WORKS OF
THOMAS A. GOUDGE

NOTES ON CONTRIBUTORS

INDEX OF NAMES

INDEX OF SUBJECTS

Notes

SAVAN / THE UNITY OF PEIRCE'S THOUGHT

1 Although Goudge consulted the unpublished manuscripts (see TP *Preface* viii), his text refers almost exclusively to CP, to some of Peirce's reviews for the *Nation*, and to some of the published correspondence with William James.
2 Smith (1952) and Leonard (1951) declare themselves unconvinced. Neither, however, attempts a detailed examination.
3 Peirce may well have been influenced by Darwin (1872).
4 In his later period Peirce stresses not nearness to ultimate truth, but self-control and self-correction of methodological error as central to scientific method.
5 Professor Max Fisch has recently pointed out that Peirce, like Locke, followed the Greek spelling and wrote 'semeiotic.' I will adopt this spelling when I write about Peirce's distinctive theory of signs.
6 See my 'Peirce's Semiotic Theory of Emotion', in *Proceedings of the International Peirce Congress* (forthcoming).
7 In Peirce's vocabulary, *vagueness* is not pejorative, since all new ideas are more or less vague and are gradually clarified in the course of action, testing, and observation. He said of some of his metaphysical ideas that they were general and vague (CP 1.276–8, 1.410–12). He also wrote, concerning his cosmology, that 'it predicts many more things which new observations can alone bring to the test. May some future student go over this ground again, and have the leisure to give his results to the world' (CP 6.34; cf 6.35)'.

BURBIDGE / PEIRCE ON HISTORICAL EXPLANATION

1 See CP 7.162–255. Volume 7 of CP appeared only in 1958, the same year as Wiener (1958) which contained 'Letters to Samuel P. Langley and "Hume on Miracles and

Laws of Nature"' (279–321) [reprinted from *Proceedings of the American Philosophical Society* 91 (April 1947)]. Therefore, much of the relevant material was not readily available to earlier students of Peirce.
2 See Bibliography CP 8 G–1901–2 and Wiener 1958, 275–9. Peirce recognized that Hume was not in fact arguing 'against miracles, in general, but only against *historical* miracles ... Precisely, the same line of reasoning can be, and frequently is, applied to extra-ordinary events that are not miraculous' (Weiner 1958, 294).
3 Cf CP 8.65: 'To explain any process not understood is simply to show that it is a special case of a wider description of process which is more intelligible.'
4 While this quotation specifically refers to *scientific* explanation, Hempel argues that historical explanation is analogous.
5 Cf Hempel (1949, 459): 'By a general law, we shall here understand a statement of universal conditional form which is capable of being confirmed or disconfirmed by suitable empirical findings.'
6 Cf CP 7.166: '[The theory of balancing likelihoods] involves the notion that the different arguments have likelihoods, that they are quantities upon an algebraical scale, and that they are to be combined as independent.'
7 Therefore he introduces *explanation sketches* as an interim operation.
8 Cf Wiener 1958, 313: 'I may mention, however, among the objections to that method, that it confounds two totally different things; objective probabilities, which are statistical facts, such as form the basis of the insurance business; and subjective probabilities, or likelihoods, which are nothing more than the expression of our preconceived notions.'
9 Cf 306: 'The activity of becoming aware of this continuity is what is called thinking.'
10 This does not mean that events never are independent. When we expect a randomness and irregularity which actually happens surprise does not arise.
11 Psychics includes the study of history. Despite the suggestion of a methodological unity of empirical science (see Hempel 1949, 471) the implication of this text is that Peirce's method is quite distinct from that of Hempel.
12 See Collingwood (1946, 217): 'This at least is certain: that, so far as our scientific and historical knowledge goes, the processes of events which constitute the world of nature are altogether different in kind from the processes of thought which constitute the world of history.'
13 Collingwood (1946, 241) also calls the procedure an act of '*a priori* imagination.' Such a term suggests a contingency analogous to that in Peirce's term 'guessing.'
14 See Collingwood (1946, 242): 'The historian's picture of his subject, whether that subject be a sequence of events or a past state of things, thus appears as a web of imaginative construction stretched between certain fixed points provided by the statements of his authorities; and if these points are frequent enough and the

threads spun from each to the next are constructed with due care, always by the *a priori* imagination and never by merely arbitrary fancy, the whole picture is constantly verified by appeal to these data, and runs little risk of losing touch with the reality which it represents.' Because 'these data' refers to 'certain fixed points' it appears that it is the evidence given prior to reflection, rather than evidence discovered after, which is the basis of verification.

15 See Hempel (1949, 462): 'In view of the structural equality of explanation and prediction, it may be said that an explanation ... is not complete unless it might as well have functioned as a prediction.'

16 Wiener (1958, 300 n 45). Interesting is his explicit rejection of Collingwood's thesis: 'I do not *define* the reasonable as that which accords with men's natural way of thinking, when corrected by careful consideration; although it is a *fact* that men's natural ways of thinking are more or less reasonable.'

17 The precise relation between deduction and the continuity of thought is a problem which cannot be handled within the limits of this paper. However, since deduction establishes the validity of the major premiss, the latter is not simply a scientific law. Observed correlations are quite distinct from inferred continuity.

18 See CP 6.606n1: 'All history is of the nature of hypothesis; since its facts cannot be directly observed, but are only supposed to be true to account for the characters of the monuments and other documents.' Compare here a passage from Kant: 'To know a thing completely, we must know every possible [predicate] and must determine it thereby, either affirmatively or negatively. The complete determination is thus a concept, which in its totality can never be exhibited *in concreto*.' *Critique of Pure Reason* A573/B601. The purpose of the proposed middle term is to find a predicate from which we can determine other possible predicates either affirmatively or negatively. These possible predicates must include the predicates already known, and more. The 'more' makes possible the inductive procedure examined below.

19 In the second procedure, a single negative instance can decisively disconfirm the theory being tested, even though there be an indefinitely large number of positive cases.

20 The need to weigh the significance of successful predictions leaves the final evaluation to the mind of the researcher, which is what Collingwood maintained all along.

FISCH / THE 'PROOF' OF PRAGMATISM

1 CP 5.415, dated September 1904. The only merit Peirce claimed for the name 'pragmaticism' was that it was 'ugly enough to be safe from kidnappers' (CP 5.414). So it proved to be, as 'pragmatism' had not. But it was also too ugly for constant use

by Peirce himself. For example, nearly three years later, in what the editors of the *Collected Papers* mistitle 'A Survey of Pragmaticism,' Peirce's only title is 'Pragmatism' (CP 5.464*), and in the more than five hundred manuscript pages (PP 317-24) the uglier name does not once occur. Except in quotations, I shall seldom use it myself, but I trust the reader will understand that the pragmatism whose proof is in question is Peirce's, not anybody else's, and not pragmatism in general. See note 9 below.

2 Draft letter to J.S. Engle, 14 February 1905, PP L 133.
3 Next year (1904) Peirce was invited by Hugo Münsterberg to present a paper or two to the Congress of Arts and Sciences at the St Louis Exposition. Peirce responded that he could not afford to go to St Louis but would gladly send two papers: for the logic section, 'Classification and nomenclature for triadic relations'; and for the methodology section, 'Pragmatism as the Methodeutic of Metaphysics (Outline of the proof of it. Definition. General character of its tendencies. Precautions to be observed in its employment.)'. Münsterberg replied that only papers to be presented by their authors in person could be accepted. PP L 308; Boston Public Library Mss Acc. 2031, Peirce to Münsterberg, 19 July 1904. See note 9 below.
4 See however PP 453, 32-3; 462, 34-46; 466, 19-20.
5 Letters from Putnam of 15 and 21 December 1903, formerly in the Peirce Papers but no longer there and not listed in Robin (1967) under L 364.
6 Peirce to Schiller, draft of 12 May 1905 and letter of 13 May (Scott 1973, 372, 373).
7 Peirce to Carus, draft letter of 6 January 1909 in PP L 77. See also Peirce's letter of January 8 to William James in the James Papers in The Houghton Library at Harvard University.
8 Actually in a draft of his Carnegie Application of 1902 he had used the phrase 'four stages or degrees of clearness' (PP L 75, on a p 9 about 125 pages from the beginning).
9 For example, in drafts other than those used in CP 5.464-96 of 'Pragmatism' (see note 1 above) – the paper on which Thompson and Fitzgerald chiefly rely – Peirce says 'But pragmatism is plainly, in the main, a part of methodeutic' (PP 320, 23) and 'Pragmatism is, thus ... a mere rule of methodeutic, or the doctrine of logical method' (PP 322, 12). Of course methodeutic depends on speculative grammar and on critic, and the way to pragmatism will have been cleared in these first two parts of semeiotic. That is, they will have made their contributions to the 'proof.' But, as Peirce puts it in a draft of a letter to C.A. Strong in 1904 (PP L 427), it is to 'this third part,' namely methodeutic, that pragmatism 'belongs.' See note 3 above.
10 It should be added that Fitzgerald is preparing a revised edition of his book, which will take account of the unpublished manuscripts, and that Don D. Roberts has in progress a series of articles on the proof, which will also make use of Peirce's unpublished papers. Since writing the present essay, I have tried in another to place

Peirce's pragmatism within his semeiotic (Fisch 1978, 38–40, 49–51). As of February 1980 a new edition of Peirce's writings, in a single chronological order, in twenty or more volumes, with a high proportion of previously unpublished papers, is being prepared; the galley proofs of volume I (1857–66) have been corrected; and typescripts of volumes II, III, and IV are nearly ready for the printer.

HERZBERGER / PEIRCE'S REMARKABLE THEOREM

1 'Triadic Relations' undated PP 543. To promote a better historical sense of the problems treated in this paper, Peirce's works will be cited for the most part from original sources, with dates where known. Numbers, preceded by 'CP,' provide cross-references in the conventional way to paragraphs in Peirce's *Collected Papers*; manuscript numbers assigned in Robin (1967) are preceded by 'PP'.
2 'The Three Categories and the Reduction of Fourthness' undated PP 915.
3 Although Peirce's categories involved much more than formal structures, the context fully justifies treating the reduction thesis as a claim about the formal properties of monadic, dyadic and triadic relations. Peirce tells us that, having spent 'the two most passionately laborious years of my life' searching for a classification of elements 'according to their matter' he abandoned the attempt as unsuited to his genius and turned to formal classifications ('Prolegomena to an Apology for Pragmaticism' draft $\pi\lambda$ c 1908 CP 1.288).
4 A recent study of the problem 'aims to develop Peirce's argument for his Thesis and to show that this argument is invalid' (Skidmore 1971). Skidmore's treatment is valuable on many points, and I think his conclusions are sound relative to the logical framework within which he chose to work. My departures from that treatment all flow from the decision to try out a different logical framework.
5 A complete refresher course in fallibilism can be obtained from Peirce's *Minute Logic* c 1902 (CP 1.277) where Peirce admonishes the reader to abandon all prejudice before proceeding: 'if [my reader] is still in that stage of intellectual development in which he holds that he has already reached infallible conclusions on certain points, e.g. that twice two makes four ... that he exists ... etc. ... he cannot yet gain much from the perusal of this book, and had better lay it aside.' Parity of treatment suggests that no less stringent standards be applied to our opinions concerning the history of logic.
6 Peirce's work on logic has the reputation of being suggestive but difficult, and in its later stages of being eccentric. See Lewis (1918) 106: 'The most valuable of [Peirce's papers] make tremendously tough reading, and they have never received one-tenth the attention which their importance deserves.' See also Roberts (1973 ch 1): 'Authors who do acknowledge the existence of EG [Peirce's systems of existential graphs] consider it too complicated to be of any value.'

7 'Lowell Lectures' 1903 CP 1.525.
8 A partial list of these formulations is given in the Appendix.
9 'The Basis of Pragmatism' c 1905 CP 1.298.
10 'Pragmatism (Editor [3])' c 1906 CP 5.469.
11 Around 1899 Peirce wrote that he had worked out a proof already in 1867 and that this proof was 'duly published' but he did not say where ('That Categorical and Hypothetical Propositions are One in Essence' c 1899 CP 1.565). An editorial footnote to that passage cites CP 3.93ff in which no such proof is to be found. Probably the citation is the wrong passage from the right source, judging from 'The Three Categories and the Reduction of Fourthness' undated PP 915 where Peirce writes: 'Professor Sylvester seems very proud of having first enunciated this axiom; but I gave it in the *Memoirs of the American Academy of Arts and Sciences*, n.s. 10 [sic] p. 374 (p. 38 of my memoir).' This could refer to CP 3.144, where Peirce sketches the reduction method discussed in section III of the present paper.
12 'Prolegomena to an Apology for Pragmaticism' c 1908 CP 1.292.
13 'Pragmatism (Editor [3])' c 1906 CP 5.469.
14 See for example CP 1.347, CP 1.371, CP 3.421, CP 3.483, CP 4.309, CP 4.445.
15 'The List of Categories: A Second Essay' c 1894 CP 1.293.
16 'Description of a Notation for the Logic of Relatives' *Memoirs of the American Academy of Arts and Sciences* n.s. 9 (1870) 317–78 CP 3.63.
17 'Lowell Lectures' 1903 CP 1.347.
18 See CP 4.581 and further references cited therein. One gathers that Peirce's feelings in this connection were not unmixed, nor did they prevent him from making substantial contributions to algebraic logic. Another factor may have been a general disinclination to prove theorems. Peirce was known to give his results 'summarily with little or no explanation and only infrequent demonstration' (Lewis 1918, 106).
19 'The Regenerated Logic' *The Monist* 7 (1896) 19–40 CP 3.451.
20 In a recent paper Quine (1972) surveys some of the main approaches including his own elegant treatments of algebraic logic – in which the influence of Peirce's work can be discerned, and which in turn have strongly influenced the present investigation.
21 'The Basis of Pragmatism' c 1905 CP 1.294.
22 This representation, along with the spirit and many details of formulation, are adapted from Quine (1936a, 1971, 1972). For the extensional viewpoint in Peirce, see his coefficient notation, 'On the Algebra of Logic' *American Journal of Mathematics* 3 (1880) 15–57 CP 3.220); and also 'The Logic of Relatives' *The Monist* 7 (1897) CP 3.329.
23 Adapted from 'The Basis of Pragmatism' c 1905 CP 1.294.
24 Where subscripts do not intervene, the concatenation sign will be suppressed in expressions for n-tuplets.

299 Notes to pages 46–53

25 From *A Guess at the Riddle* c 1890 CP 1.363, with Peirce's upper-case letters *A, B, C, D, E* changed to lower-case to conform to my notation, and with Peirce's '*a* sells *c* to *b*' corrected to read '*a* sells *b* to *c*'. Other versions of Peirce's reduction method can be found in 'Notation for the Logic of Relatives' 1870 CP 3.144, 'The Logic of Relatives' 1897 CP 3.464, 'Detached Ideas' 1898 PP 439 and 'On Topical Geometry' undated CP 7.524ff.
26 Here I extend Peirce's method, which he held to essentially involve 'hypostatic abstraction,' and I depart from Skidmore (1971) who holds that 'The important thing is to see that in both of Peirce's "reductions" abstraction is employed.'
27 In 'The First Part of an Apology for Pragmatism' 1908 PP 296 Peirce enunciates 'the principle of using but a single mode of combination.' This reference is due to Roberts (1973, 117).
28 See 'The Basis of Pragmatism' c 1905 CP 1.294. Peirce's graphical notation displayed correlates in two dimensions; but some conventions of order (e.g. clockwise arrangement) or indexing are indispensable and were implicit in his practice.
29 See 'The Basis of Pragmatism' c 1905 CP 1.294.
30 See 'On the Algebra of Logic' *American Journal of Mathematics* 3 (1880) 15–57 CP 3.223 and also CP 3.317, CP 3.330, CP 3.420, and CP 3.466.
31 Quine (1972 sec 7).
32 'The Critic of Arguments' 1892 CP 3.421; 'On Logical Graphs' 1896 PP 482; 'The Logic of Relatives' 1897 CP 3.469–71 esp. 3.471; also 3.484; 'Pragmatism (Editor [3])' c 1906 CP 5.469; and 'Prolegomena to an Apology for Pragmaticism' c 1908 CP 1.292.
33 The diagram on the left is intended to represent a triadic relation **R**, with its three argument places; and the diagram on the right is intended to represent the result of bonding that triad so as to produce a monad, as in the formation of [-- betrays someone to himself] from [--betrays--to--].
34 'The Logic of Relatives' 1897 CP 3.471.
35 'The Critic of Arguments' 1892 CP 3.421.
36 'The Logic of Relatives' 1897 CP 3.469 and CP 3.471.
37 This theorem, which he called the 'second census proposition', governed configurations of a kind Peirce called 'valental graphs': 'the valency of a graph equals the sum of the valencies of its spots less twice the number of connecting bonds' ('On Logical Graphs' 1896 PP 402). The reference is due to Murphey (1961, 197), whose chapter 9, 'Topology,' discusses Listing's Theorem and its importance for Peirce. It may be mentioned that Robin (1967) dates 'On Logical Graphs' as c 1896–8, leaving it open whether it was written before or after 'The Logic of Relatives' *The Monist* 1897, in which the valency rule was published.
38 'The Logic of Relatives' 1897 CP 3.484.
39 The doctrine of surrogate constructions will be the central theme of a sequel to this

paper which uses Peircean methods to develop the full resources of bonding algebra and to prove even stronger reduction theorems.

40 Peirce's notation, equivalent bonding formulas, and sources in Peirce's writings are:

involution	$R^S = cp(cR, S)$	CP 3.77
backwards involution	$R_S = cp(R, cS)$	CP 3.113
transaddition	$RoS = p(cR, cS)$	CP 3.242
relative sum	$R\dagger S = cp(cR, cS)$	CP 3.332

41 See 'Notation for the Logic of Relatives' 1870 CP 3.68; 'On the Algebra of Logic' 1880 CP 3.198; and 'The Logic of Relatives' Note B to *Studies in Logic, By Members of the Johns Hopkins University* (Boston 1883) 187–203 edited by C. S. Peirce CP 3.331.

42 Nonrelative product can be extended to polyadic relations in more than one way, as discussed in Quine (1972 sec 7); but all of these extended operations agree with the calculations in the text in respect of algebraic type. Quine's own version, which he calls 'intersection,' has the algebraic type [max(r, s)].

43 'Notation for the Logic of Relatives' 1870 CP 3.68–76, described by Quine as follows: 'By an elegant device Peirce contrives furthermore to construe the ordinary logical product of classes ... as a case of the general multiplication described' (Quine 1934, 285–97).

44 Letter to O.H. Mitchell 21 December 1882 PP L 294 quoted from Roberts (1973, 20): 'We thus do away with the distinction of relative & non-relative operations by discarding the latter altogether.' Another relevant source is 'The First Part of an Apology for Pragmatism' 1908 PP 296 where Peirce writes that 'the analysis represented in the algebra of Dyadic Relations by the non-relative product ... must be wrong.'

45 'The Critic of Arguments' 1892 CP 3.423.

46 'The Critic of Arguments' 1892 CP 3.424.

47 Skidmore (1971, 18). A related criticism can be found in Braithwaite (1934).

48 In the work mentioned in note 39, bonding-algebraic surrogates will be provided for all the definitional resources of first-order quantification theory with identity.

49 'On Logical Graphs' c 1896 PP 482.

50 An interesting position of this kind results from one of the simplest possible departures – relaxing the valency rule to allow the limiting case $\lambda = 0$. Compounds of the algebraic type [r +s] thereby admitted, would violate the reduction principle in allowing, for example, some tetrads to be defined from pairs of dyads. Peirce mentions a related case in which a dyad is constructed as the cartesian product of two monads, and he calls the dyad so formed 'a degenerate form of relation' ('On the Algebra of Logic' *American Journal of Mathematics* 7 [1885] 180–202 CP 3.392). From the standpoint of the modified valency rule, some qualification

would then have to attach to the reduction thesis – such as 'no further reduction is possible *for genuine (nondegenerate) triads.*' And the formal demarcation of the genuine-degenerate boundary would be a substantial complication and possibly even an obstacle to the reduction program.

51 The recent publication of Peirce's mathematical papers makes it possible now to identify the manuscript cited in note 1 ('Triadic Relations') as a fragment of a letter written in August 1905 to William James. The letter is reprinted in Eisele (1976 III/2, 809–35). This letter contains extended remarks on the reduction problem and related logical matters.

MADDEN / SCIENTIFIC INFERENCE: PEIRCE AND THE HUMEAN TRADITION

1 See, for example, CP 5.180 ff, 5.265, 5.28, 5.371, 5.372, 5.375, 5.376, 5.376 n, 5.394, 5.397, 5.398, 5.416, 5.417, 5.63, 6.469, 7.194. Cf Broyles (1964, 82 ff).
2 Much of the literature relevant to this exposition is to be found conveniently gathered together in Tomas (1957). Cf also PP 1269, 1273, 1277, 1288, 1289, 1290, 1291, 1582, 1604, 1274–87.
3 Peirce thus was allied with Hertz and Boltzmann against Mach. See the preface to Hertz (1899).
4 Cf. CP 2.703, 2.729, 2.749, 1.485, 1.93, 5.575 ff, 1.608, 1.67, 1.68, 1.71–4, 1.121, 2.781, 2.784, 2.269, 2.769, 6.35–65.
5 Cf 'The Fixation of Belief' and 'The Doctrine of Necessity Examined,' in Wiener (1958).
6 Letter to Josiah Quincy, 26 June 1909.

ROBERTS / THE LABELING PROBLEM

1 I use the OED and the *Century Dictionary* as my authorities on definitions, unless a term is to be used in a new or special or technical sense. For the verb 'label' the appropriate definition is given by the OED as figurative: 'To describe or designate as with a label; to set down in a category (*as* so and so).' For the noun 'label' the *Century Dictionary* gives 'A slip of paper or any other material, bearing a name, title, address, or the like, affixed to something to indicate its nature, contents, ownership, destination, or other particulars.'
2 I have in mind such things as Enrico Fermi's 'neutrino' of 1932 as a label for Wolfgang Pauli's postulated particle of 1931. The association of this label with something other than mathematical formulas was not accomplished until 1955 or 1956.
3 Lewis Carroll's 'Jabberwocky,' for example.
4 Jorge Luis Borges' Aleph, for example. See Borges (1970) 15–30.

5 Contrary to a frequently expressed objection, this would no more reduce philosophical 'creativity' than the common terminologies of other sciences – physics, chemistry, biology, mathematics – have reduced creativity there. But this is matter for another paper.
6 Dr Lorraine Beattie Roberts first pointed this out to me.
7 This is from the first chapter of *Through the Looking-Glass* (Gardner 1960, 197).
8 In a letter to T.W. Ward in 1868 James wrote 'I have been growing lately to feel that a great mistake of my past life – which has been prejudicial to my education, and by telling me which, and by making me understand it some years ago, someone might have conferred a great benefit on me – is an impatience of *results*. Inexperience of life is the cause of it, and I imagine it is generally an American characteristic. I think you suffer from it. Results should not be too voluntarily aimed at or too busily thought of. They are *sure* to float up of their own accord, from a long enough daily work at a given matter; and I think the work as a mere occupation ought to be the primary interest with us' (James 1920 I, 133).
9 This is a Peircean notion of philosophy; see for instance CP 1.11–14. Thomas A. Goudge gives a splendid account of the scientific features of Peirce's philosophy in TP. See especially 3, 25–7, 75, 252ff.
10 Learning seems to me to depend partly on what Peirce called 'one of the most wonderful features of reasoning ... namely, that reasoning tends to correct itself, and the more so, the more wisely its plan is laid. Nay, it not only corrects its conclusions, it even corrects its premisses.' (CP 5.575) A detailed development of my oversimplified model would make that clear, and would make clear the similarity between learning of almost any kind and scientific method in general. I have profited from Goudge's discussion of the self-corrective feature of reasoning in his TP 41, 188–94.
11 This is one of the many points worth noting in Professor Goudge's sensitive and insightful introduction to Bergson's essay (Bergson 1949).
12 Or, one of the instruments of knowledge; the photographer who arranges for the points of view should probably be acknowledged as well.
13 'William James: The thought is already complete at the beginning of the sentence. How can one know that?' (Wittgenstein 1967, 2).
14 Kierkegaard (1941 'Prelude').

SCHOULS / PEIRCE AND DESCARTES: DOUBT AND THE LOGIC OF DISCOVERY

1 Hereafter called *Regulae*.
2 Unless otherwise indicated all quotations from Descartes are from the translation of Haldane and Ross *The Philosophical Works of Descartes* (New York 1955) vols I

and 2 (abbreviated as HR1 and HR2). The second reference, in each case, is to the Adam and Tannery *Oeuvres de Descartes* (abbreviated as AT).
3 This statement is from Descartes' *La Géometrie* (AT6, 376). The translation is from Olscamp (1965, 182).
4 'While possible existence indeed attaches to the ideas of all other natures, in the case of the idea of God that existence is not possible but wholly necessary. For from this alone and without any train of reasoning they will learn that God exists ... ' (HR2, 55; AT7, 163); 'after becoming aware of the existence of God, it is incumbent on us to imagine that he is a deceiver if we wish to cast doubt upon our clear and distinct perceptions; and since we cannot imagine that he is a deceiver, we must admit them all as true and certain' (HR2, 41; AT7, 144).
5 The presence of 'surprise' (in e.g., 'surprising ... phenomenon,' CP 2.776; and in 'surprising fact,' CP 5.189) has been stressed a good deal by commentators. The element of surprise should not be over-stressed. The facts to be explained need not be 'astonishing' or 'jarring.' See, e.g., CP 2.623, 5.443, 5.451, 5.512.
6 All quotations from Descartes' correspondence are from *Descartes Philosophical Letters* (abbreviated as DPL) translated and edited by Anthony Kenny (Oxford 1970).
7 See, e.g., Boethius' comments in the First Book of the second *Commentaries* on the *Isagoge* of Porphyry, in Richard McKeon's *Selections from Medieval Philosophers* vol 1 (Modern Student's Library, Scribner's) 74.

SPRIGGE / JAMES, SANTAYANA, TARSKI, AND PRAGMATISM

1 Even James qualified it somewhat, though not enough, surely, for Santayana. Cf James (1909, 42).
2 Both reprinted in James (1909).
3 It may be noted that if the interpretation of James advocated here is correct, the suggestion of H.S. Thayer that 'cognitive truth' (conceived in terms of the correspondence theory) is, for James, an essential condition of pragmatic truth distinct from its useful workings would lose its force, since correspondence can only be explained by reference to 'working' and 'adjustment.' See Thayer's fine introduction to James (1975b). Chapter 4 seems especially to support my interpretation.

THAYER / PEIRCE ON TRUTH

1 Thus in the articles on 'truth' for Baldwin (1902, 718–20) the definition, which I give as (B) below, is shown to apply to logical, normative, and practical subject-matters. For these passages, see CP 5.565–73.
2 The term 'would' is a very important item in Peirce's discussions of dispositional

traits, metaphysical possibility, his scholastic realism, and his theory of probability. I have commented on this in Thayer 1968, 113-20.
3 But points of doctrine are, in this case, open to question; for a problem ever present to a commentary on any of Peirce's ideas is how to establish connections and priorities between the published and the manuscript material, and to determine which (of often several drafts and revisions) are the authoritative statements, i.e., most representative of Peirce's intention. Perhaps the controlling context for the determination of Peirce's ideas is his general theory (or theories) of signs, with its relation to his tripartite categorial scheme and the existential graphs. Here, to aggravate our problem, there is a vast collection of manuscript material raising difficult questions of chronology, stages of devolopment, and the scope and direction of Peirce's philosophizing. For these reasons one must applaud the efforts now being taken, under Professor Max Fisch's direction, to make more of these materials available in chronologically accurate and complete texts.
4 Quine's objection on this score has been carefully and critically discussed by Robert Almeder (1975).
5 Letter to Lady Welby, 23 December 1908 (Lieb 1953, 26).
6 This has been pointed out by Max Fisch in 'The "Proof" of Pragmatism,' this volume.
7 It should be noticed that proposition signs are not confined to linguistic signs. A portrait of a man with his name under it, or a photograph, can be propositions. (CP 2.320; 5.569)
8 On *representamen* see also CP 1.564.
9 See CP 5.435 where this is developed as the 'essential proposition of pragmaticism.' See also CP 2.316.

THOMPSON / PEIRCE'S CONCEPTION OF AN INDIVIDUAL

1 I also have not had space to discuss the work of others who have written at length on this part of Peirce, though I have learned much from their work.
2 'For since the species is always a concept, containing only what is common to different things, it is not completely determined. It cannot, therefore, be directly related to an individual, and other concepts, that is, subspecies, must always be contained under it. This law of specification can be formulated as being the principle: *entium varietates non temere esse minuendas*' (*Critique of Pure Reason*, CPR A656 = B684). This reference to Kant is given by Peirce at CP 3.612. In his Lowell Lectures of 1903 Peirce mentions the view that 'the analysis of signification must be capable of [being] pushed on further and further, without limit' as 'an opinion which Kant expresses in a well-known passage but which he did not develop' (CP 5.177). The passage is certainly A656 = B684, as the editors of CP say.

In the preceding sentence Peirce says Kant 'consistently neglected the logic of relations.' With this neglect he could not have developed his view of the analysis of signification in the direction Peirce wanted. He could not see that the ultimate reference to an individual is via pronouns (indices or quantified variables) and not via predicates or common nouns. But it took Peirce a long time to reach this position himself, and, as I will try to bring out, he was not very close to it in his 1870 paper on the logic of relatives.

3 '"Singular" and "individual" are equivocal terms. A singular may mean that which can be but in one place at one time ... When an image is said to be singular, it is meant that it is absolutely determinate in all respects' CP 5.299). The word 'individual' does not appear in this paragraph after the first sentence, and what is said about 'singular' is clearly intended to apply to 'individual' as well.

4 The claim that an *infirm species* is determined only by a convention not to seek further differences (*nicht tiefer zu gehen*) occurs in Kant's *Logic* (Jäsche's ed, 1800) §11. However, unlike Peirce in 1870 Kant did not regard the differences neglected as 'differences of time and the differences that accompany them.' Logical division for Kant, no matter how far it proceeds, yields only conceptual differences; any reference to time introduces intuition and results in differences irrelevant to logic. Cf Kant's objection to adding the qualifying phrase 'at one and the same time' to the principle of contradiction (CPR A152–3 = B192–3). I consider below at the end of section 2 Peirce's later view of space and time in relation to individuation and how it contrasts with the Kantian view.

5 Although in CPR Kant classifies intuitions together with concepts as cognitions (A320 = B376–7), the precise cognitive character of a Kantian intuition is a disputed point. In a fragment c 1890 Peirce says Kant's 'sharp discrimination of the intuitive and discursive processes of the mind' was 'what enabled him to see that no general description of existence is possible, which is perhaps the most valuable proposition the *Critic* contains' (CP 1.35). Yet in allowing intuitions a cognitive character, Kant could not hold that positions and dates are distinguished only by a brute sense of reaction. Actually, of course, Kant sets off Space and Time by themselves under the heading of *pure* as distinct from empirical intuition, a distinction Peirce ignores. His futher comments in CP 1.35 concerning Kant's 'too hard a line between the operations of observation and ratiocination' raise important issues, but they are beyond the scope of the present paper.

WILSON / GOUDGE'S CONTRIBUTION TO PHILOSOPHY OF SCIENCE

1 Cf Mill (1872 bk3 ch12). The term 'gappy' is from Mackie (1965); this paper contains a discussion of these matters that is very much in the tradition and style of Mill. The term 'imperfect' is from G. Bergmann (1957 ch2. sec 4); this discussion

306 Notes to pages 153-9

puts such knowledge in the context of a discussion of process knowledge, and is also in the tradition, though not the style, of Mill. Bergmann's discussion is the best available. Strangely, it is ignored by Helmer and Rescher (1959), who deal with some of the simpler aspects of the issues involved; these authors use the term 'quasi-laws.' See also Wilson (1981 sec 2).

2 '$(\exists !x)$' abbreviates 'there is at least one and at most one x such that ... '
3 Scriven (1962) has wrongly concluded that, since the symmetry of explanation and prediction sometimes thus breaks down, since (in other words) we can sometimes explain ex post facto where we cannot predict, therefore explanation is not by deduction from laws. Brodbeck (1962) criticizes Scriven's position. See also Wilson (1981 sec 4).
4 Goudge quotes this passage from Simpson (1950).
5 Goudge (AL 61) points out that such ex post facto explanations are more than common in biology; and that their basis is the existence of mixed quantificational laws of the sort (E). Goudge speaks of 'reading back into the historical record' rather than of 'ex post facto explanations' but the point is the same. And he makes the point that the laws like (E) can be taken to be instantiations of a more general 'uniformitarian principle.'
6 Compare the 'paradoxical' emphasized in the passage quoted just two sentences ago.
7 Ruse (1973c 89-92) suggests, wrongly I think, that Goudge is here arguing that *all* generalities involved in narrative explanations should be construed as 'inference-tickets' and that therefore narrative explanations do not involve deduction from general premises. I do not think Goudge is rejecting law-deduction as a condition of explanation, but rather only the idea that when we explain 'E because s' the relevant law for the deduction is the generality 'Whenever an event exactly like s occurs then an event exactly like E occurs.' For, as Scriven (1959b 471-5) points out, E and s are unique: only they are *exactly* like E and s. In which case the generality *is* tautological, and not a law. But it does not follow from this – nor (I think) does Goudge suggest it follows – that no law and no deduction from a law is involved. Rather, what Goudge suggests is that we need an account of how the law-deduction occurs which is both more subtle and more complicated.
8 (AL 74); his italics. Compare the 'could' with the 'seem' emphasized in the passage quoted at the beginning of this paragraph.
9 Cf the discussion of various senses of 'hypothesis' in Bergmann (1951) and Wilson (1981 sec 4, 7).
10 For this point in connection with integrating explanations, cf AL 68.
11 Two theories can always be put into one axiomatic system simply by conjoining their separate axioms into one by means of 'and.' The requirement that the two theories have a shared form or generic structure eliminates this trivial case. It is also

necessary, however, to exclude genera of the 'gruesome' sort designed by Goodman. Eliminating these 'arbitrary' predicates is not so easy. But I think Goodman himself has given the essential ingredients for a solution, with his notion of a predicate being 'non-arbitrary' just in case it has become *entrenched*.

12 Ruse (1973c 66–7) does not take seriously the idea that the axiomatic model is a model, minimally, for the generic unification of several laws. If Ruse is correct, then simply to make a deductive inference is to be involved in axiomatics. Upon that criterion, *all* science is in fact at present axiomatized. Yet we can still distinguish classical mechanics and evolutionary theory: as Goudge asserts, in the former a real generic unity of laws is achieved, in the latter such a unity is still a goal, and in fact a distant goal.

13 For the importance of clearly distinguishing definite descriptions from definitions, cf Wilson (1969a).

14 For an excellent discussion of historical laws, see Bergmann (1957 ch 2 sec 5).

15 Ruse (1973c 65) fails to see how historicity complicates the task of axiomatization, that is, complicates it *in fact*, though not in principle – but that is all that Goudge wishes to argue.

16 Ruse (1973c 62) misses this de facto complexity also.

17 Ruse (1973c 88–9) argues that truth is necessary to the success of a narrative explanation: 'If we do not know which are the true conditions, then I fail to see how we can claim to have a narrative explanation either.' Goudge's point is that a narrative explanation can do the job expected of it even if the truth of (some part of) the explanation is not known. (See, for example, Goudge's emphasis in the quotation cited by note 8.) Narrative explanations are in this way distinguished from ordinary causal explanations, which are acceptable in an explaining situation only if the causal statement is known to be true, or, at least (given the limits of induction), if there is reason to believe it to be true.

18 For a discussion of how the location of an hypothesis in the context of research affects the logical status of the concepts appearing in it, see Wilson (1967).

19 There is a tendency too quickly to deploy the 'context of discovery/context of justification' dichotomy to dismiss the process and concentrate on the product; see Wilson (1981 sec 3, 7).

20 The relevance of context for providing laws not explicitly stated is indicated in AL 76.

21 Cf Broad (1925, 82–3). Goudge refers to this section of Broad in AL 195.

22 David Hume, in his *Dialogues Concerning Natural Religion*, has Philo ask the atheist (Part XII) ' ... if it be not probable, that the principle which first arranged, and still maintains, order in this universe, bears not some remote inconceivable analogy to the other operations of nature, and among the rest to the oeconomy of human mind and thought' (Hume 1947, 218). This is indeed the best that can be got

from the argument from design. And to this the atheist agrees: 'The atheist allows, that the original principle of order bears some remote analogy to it' (Hume 1947, 218). This is surely correct; this conclusion follows *soundly* from the Argument from Design. Only, we now *know* (since Darwin) what it is that bears this analogy to the human mind: it is no transcendent God, but simple natural selection. Nor is the analogy anything more than the fact that natural selection and reasoning are processes in which problems of adaptation are solved: nothing, in short, sufficiently strong (as Hume knew) to establish anything about morals or politics, about praying or about aborting.

23 On the role of the composition law, see Bergmann (1957 ch 3). Also Wilson (1969b; 1971a; 1981 sec 5).
24 Kuhn (1970b 153 ff). Campbell (1957, 129–37) suggests an account along the same lines. Kuhn's account has been elaborated in Wilson (1981 sec 7, 10, 11).
25 Cf Feyerabend (1962, 1970). Feyerabend's irrationalism follows from his embracing the thesis that 'all concepts are theory-laden.' For criticism, see Kordig (1971), and Wilson (1971b; 1981 sec 8).
26 See the remark made above on the passage quoted on p 159.
27 Cf the discussion of 'principles of actions' in Dray (1957 ch 5).
28 Toulmin (1961 ch 6) has pursued the evolutionary analogy with some insight, but unfortunately couples it with a version of the thesis that 'all concepts are theory-laden' (cf note 25 above) so that he ends up with a radically distorted view of science; cf Wilson (1969b; 1981 sec 8).
29 See the passage quoted above on p 168.
30 On the situation of evaluation and of the production of values within an evolutionary view of mind, cf Goudge (1973, 141). Goudge there refers to Dewey (1939).

HULL / HISTORICAL NARRATIVES AND INTEGRATING EXPLANATIONS

1 Some years earlier, Karl Popper termed these same two classes of statements 'universal laws' and 'initial conditions' (1959; see also 1966, 117).
2 Stephen Toulmin (1976, 36) sketches a similar distinction in quite another context. 'For the purposes of general knowledge,' Toulmin remarks, 'data are significant only when they reveal correspondingly general facts; empirical data about particular cases exemplifying general truths are spoken of, by contrast, as "boundary conditions," "values of the parameters," or the like. For the purposes of particular knowledge, the contrast is reversed. Any relevant general relationships, laws, and/or correlations merge into the continuous background of "standing conditions": the particular values of all the relevant magnitudes and parameters now become the significant focus of attention.'
3 One devastating argument against the role of covering-law explanation in biology

is the denial that any biological laws exist in the first place, an argument which neither Goudge nor Ruse favours. However, J.J.C. Smart (1963 and 1968) has argued just this, though in the meantime he has tempered his views (1976–7).
4 Considerable disagreement exists among taxonomists over the proper way to define 'monophyly' and related terms. Some biologists argue that all taxa must be strictly monophyletic, i.e., that all species in a taxon must be descended from a single immediately ancestral species. Others are willing to countenance taxa which are only minimally monophyletic, i.e., that all species in the taxon must be descended from species included in a single immediately ancestral taxon of its own or lower rank. Thus, a genus could be minimally monophyletic if it arose from two or more species just so long as all of these species are included in a single immediately ancestral genus. A second disagreement exists concerning the proper treatment of the species descended from a particular species. Certain taxonomists argue that all the descendants of any one species must be included in a single higher taxon. Others are willing to exclude some of the descendant species from certain higher taxa. The cladists (or phylogenetic taxonomists) hold the strictest definition of 'monophyly.' According to the cladists, all higher taxa must arise from a single immediately ancestral species (minimal monophyly will not do), and all the species which result from a single species must be included in a single higher taxon; see David L. Hull (1979).
5 David L. Hull (1979).
6 Michael Ghiselin (1969) was the first biologist to argue explicitly that species are individuals rather then classes.
7 Because of the epistemological role of observable, structural similarities in reconstructing phylogenetic descent, a vast literature has developed concerning the relative priority of evolutionary homology and phenetic similarity. The answer is, I think, rather straightforward; epistemologically phenetic similarity is prior; metaphysically evolutionary homology is prior; see Hull (1968).
8 David L. Hull 'The Units of Evolution: A Metaphysical Essay' in *Studies in the Concept of Evolution* ed by V.J. Jensen and R. Harré (forthcoming).
9 For the distinction between the relationship between ecological and phylogenetic considerations in evolutionary explanations, see Martin L. Cody (1974) and S. Simberloff (1974).
10 Maurice Mandelbaum (1977) finds himself in agreement with much of what I have to say about the role of central subjects in general history but doubts my application of this same analysis to biological species.
11 Biologists currently distinguish between phylogenetic trees and scenarios: see Tattersall and Eldredge (1977, 204–11). Trees are descriptions or diagrams purporting to represent which species gave rise to which. Scenarios are descriptions of the relations between the organisms and their environments which produced these

trees. These biologists are also concerned to emphasize how speculative reconstructions of phylogenetic trees and evolutionary scenarios actually are.
12 Under the influence of Ludwing Wittgenstein, the notion of 'cluster concepts' has become quite popular in philosophy. The names of particular organisms (like Moses) and particular species (like tiger) have been presented as paradigm cases of cluster concepts. However, once biological species are conceptualized as individuals, antother interpretation seems more plausible: 'tiger' like 'Moses' is a proper name rigidly designating the individual to which it refers. I think that it is no accident that the metaphors which Wittgenstein uses to elucidate cluster concepts are family resemblances and a thread consisting of overlapping fibres (1953 sec 67). A thread is quite literally an individual, and family resemblances result from the genetic relationships which integrate organisms into lineages. See David L. Hull 'The Principles of Biological classification: The Use and Abuse of Philosophy,' in Peter. D. Asquith and Ian Hacking ed *PSA 1978* Dordrecht (forthcoming).

MCRAE / LIFE, *Vis Inertiae*, AND THE MECHANICAL PHILOSOPHY

1 'The possibility of living matter is quite inconceivable. They very conception of it involves self-contradiction, since lifelessness, *inertia*, constitutes the essential characteristic of matter' (Kant 1928, 46).
2 I have modified the translation to make it more literal. Kepler's *vis inertiae* was the tendency of a body towards rest. This is plainly not what Newton or Kant are referring to.
3 Cambridge University Library, Add. 3970, fol. 620^r, quoted in McGuire (1968, 170–1).
4 Quoted in McGuire (1968, 171).
5 Newton (1952) Query 31; *The Leibniz-Clarke Correspondence*, Clarke's Fifth Reply.
6 After giving reasons for denying thought to animals, Descartes says to More, 'Please note that I am speaking of thought, and not of life or sensation. I do not deny life to animals, since I regard it simply as in the heat of the heart; and I do not deny sensation so far as it depends on bodily organs.' 5 February 1649 (Descartes 1970, 245).
7 To Plempius from Fromondus, 3 October 1637: 'I believe that the souls of animals are nothing but their blood, the blood which is turned into spirits by the warmth of the heart and travels through the arteries to the brain and from it to the nerves and muscles.' Descartes quotes Scripture in his support.
8 I am not asserting that Descartes initiated the revolution. Galileo in his *Letters on Sunspots* (1613) says 'And therefore, all external impediments removed, a heavy body on a spherical surface concentric with the earth will be indifferent to rest and

to movements towards any part of the horizon. And it will maintain itself in that state in which it has once been placed; that is, if placed in a state of rest, it will conserve that; and if placed in movement toward the west (for example) it will maintain itself in that movement.' (Drake 1957, 113) See also the editor's remarks in the accompanying footnote.
9 *Ethics* III 8. The fact that Descartes speaks of persisting *in the same state*, while Spinoza speaks of persevering *in its being* is not significant, for Descartes also gives his first law in the following form: 'Now, as I said, any given thing which like motion, is not complex but simple, persists in its being so long as it is not destroyed by an external cause' (Descartes 1954 II 41). If there is a significant difference it could lie in Descartes' restriction of his principle to things which are 'not complex but simple.'
10 *Cogitata Metaphysica* part II ch 6
11 *Ethics* II 9 and sch.

PUCCETTI / THE ASCENT OF CONSCIOUSNESS

1 H.J. Jerison 'The Evolution of Consciousness' *Proceedings of the Seventh International Conference on the Unity of the Sciences* vol II, 711.
2 There can hardly be any ontological fixity about *psycho*physical laws, since a few billion years ago there were no such laws, just as a few billion years hence there will, alas, be none again.
3 I have not bothered to bring out far more complex sorts of pain behaviour that seems unintelligible without experienced pain, especially those evoked by *internal* tissue damage: try to imagine Jones impelled to the dentist's or surgeon's office if he develops tooth cavities or a swollen appendix, *not knowing why he is doing these things*. Yet on the theory being examined, he would do so even in our actual world if, for some reason, pain experience simply ceased to be identical with functioning of the postulated neurological mechanism. The familiar counter that this could never happen because there is some kind of 'natural identity' between them akin to that of water and H_2O overlooks the fact that while if the latter identity relation ceased to obtain the physical laws of the universe would be different, such is not true of the former because psychophysical laws are necessarily impermanent (note 2, above); and in fact Jones *could* survive in our world. Nor have I bothered to bring out here that if such anti-interactionist theories were true of our world, random mutation should occasionally produce individuals just like Jones, given that he would take appropriate avoidance-of-tissue-damage action without feeling pain. But there are no such individuals.

ROSENBERG / THE INTERACTION OF EVOLUTIONARY
AND GENETIC THEORY

1 Cf the detailed treatment of Goudge's views in the two most widely adopted textbooks in the philosophy of biology: Ruse (1973c) and Hull (1974).
2 Cf AL and the criticism of this claim in Ruse (1973, 62–8). The adequacy of Williams' axiomatization is of course the best argument against Goudge, and Ruse does not use this argument for reasons which will become clear below. For the axiomatization makes evident the independence of evolutionary and genetic theories.
3 For a general argument to this effect, cf Armstrong (1968) and Martin (1972).
4 The same can be said of one of fitness' cousin concepts, 'reinforcement.' Failure to recognize this fact has led some psychologists to search in vain for some neurophysiological or topographic feature common and peculiar to all reinforcers, and to conclude that, like evolutionary theory, the law of effect is a vacuous tautology. On the notion of functional characterization, cf Fodor (1968 ch 3).
5 Cf for example, Hull (1974, 66–9), and Ruse (1973c 38–41). My own students at various universities have found these passages difficult to follow.
6 For more details of this sort of reduction cf Ruse (1974). Ruse's paper is a critique of objections to reduction lodged by Hull, which turn on an over-narrow characterization of the criterion of connectability in theoretical reduction. Cf Hull (1974).

RUSE / PHILOSOPHICAL ASPECTS OF THE DARWINIAN REVOLUTION

1 Particularly in AL and in his contributions to the *Encyclopedia of Philosophy* (New York 1967) and the *Dictionary of the History of Ideas* (New York 1969).
2 See also Michael Ruse 'The Recent Revolution in Geology', in Peter D. Asquith and Ian Hacking, eds *PSA 1978* Dordrecht (forthcoming).
3 In (1831, 146) Herschel actually used an example from the *Principles* to illustrate the *vera causa* notion.
4 Particularly revealing are some notebooks Lyell kept between 1855 and 1861. See L. Wilson (1970). See also Herschel (1861, 12n).
5 Similar criticisms were made by [Anon.] (1846, 184), Huxley (1854), and Gray (1846).
6 See 'Species notebook B' 101–2. This and other species notebooks of Darwin have been published by G. de Beer and others *Bulletin of the British Museum (Natural History) Historical Series* 2 nos 2–6; 3 no 1.
7 Darwin 'Notebook D' 135.
8 For example, in a letter to G. Bentham, 1863. See Darwin (1887, 3, 25).
9 Most fully in AL.
10 See Ruse (1976b). I draw the same conclusions based on ahistorical grounds.

STROUD / EVOLUTION AND THE NECESSITIES OF THOUGHT

1 For discussion of some of the problems involved in explaining one's own beliefs in a way that does not render them true, probable, or reasonable see, e.g., Hampshire (1971) and Williams (1970).
2 C.I. Lewis has stressed perhaps more than anyone else the need to appeal to possible alternatives to our present ways of thinking in any attempt to support conventionalism, or indeed any view according to which the necessary or the a priori are shown to be the 'contribution of the mind.' See especially Lewis (1929 chs 7, 8).
3 This argument was first directed against conventionalism by Quine (1936b). He there expresses his own indebtedness to Lewis Carroll's classic (1895).
4 For the idea that we must understand the *point* of the notion of truth for us, and not just discover some set of conditions that serves to distinguish the true from the false, see Dummett (1958–9; 1973 ch 13).
5 For an outline of an explanation of the origin of the concept of truth, and with it the concept of belief, without attempting to analyse or define either notion, see Davidson (1975).
6 Quine is one who has especially emphasized the different degrees of 'centrality' or 'revisability' among all our beliefs, and has suggested that this difference in epistemic status is perhaps all that necessity really amounts to: 'Our system of statements has such a thick cushion of indeterminacy, in relation to experience, that vast domains of law can easily be held immune to revison on principle. We can always turn to other quarters of the system when revisions are called for by unexpected experiences. Mathematics and logic, central as they are to the conceptual scheme, tend to be accorded such immunity, in view of our conservative preference for revisons which disturb the system least; and herein, perhaps, lies the "necessity" which the laws of mathematics and logic are felt to enjoy' (Quine 1950 xiii). This might well be the right place to look for an explanation of our ascriptions of necessity without yielding a *definition* of necessity in such 'epistemic' terms as 'centrality' or 'degrees of likelihood of revision in the face of recalcitrant experience.'
7 See, e.g., Braithwaite (1953 ch IX), Nagel (1961 ch 4), and many others.
8 I have elaborated this interpretation in Stroud (1977).
9 Quine seems to be alluding to some such distinction when he acknowledges the possibility of a physical theory 'of radically different form from ours,' but 'empirically equivalent' to it: 'Once this is recognized, the scientific achievement of our culture becomes in a way more impressive than ever. For, in the midst of all this formless freedom for variation, our science has developed in such a way as to maintain always a manageably narrow spectrum of visible alternatives among which to choose when need arises to revise a theory. It is this narrowing of sights,

or tunnel vision, that has made for the continuity of science, through the vicissitudes of refutation and correction. And it is this also that has fostered the illusion of there being only one solution to the riddle of the universe' (Quine 1975, 81). For an expression of Wittgenstein's endorsement of a similar distinction see, e.g., Wittgenstein (1953, 230): 'I am not saying: if such-and-such facts of nature were different people would have different concepts (in the sense of a hypothesis). But: if anyone believes that certain concepts are absolutely the correct ones, and that having different ones would mean not realizing something that we realize – then let him imagine certain very general facts of nature to be different from what we are used to, and the formation of concepts different from the usual ones will become intelligible to him.' I tried to elaborate on the implications of this distinction for conventionalism in Stroud (1965).

THAGARD / THE AUTONOMY OF A LOGIC OF DISCOVERY

1 Cf Wilfrid Sellars' claim that acceptance of a theory is a matter of practical reasoning (Sellars 1964, 209).
2 On abduction, see TP 195ff., and Paul R. Thagard 'Peirce on Hypothesis and Abduction,' paper read at the C.S. Peirce Bicentennial International Congress, Amsterdam, 1976; to appear in the *Proceedings* of the Congress.

TULLY / EMERGENCE REVISITED

1 Besides these two, Whitehead must also be considered a prominent emergentist, but since most specific criticisms of emergentism have been directed at either Broad or Alexander, they alone will be dealt with here. Minor emergentists include C.L. Morgan and G. Lewes (who was not a contemporary of Broad).
2 Alexander was a realist about the traditional secondary qualities, while Broad was partial to views which stressed their dependency on observation, such as the 'Theory of Multiple Inherence,' which he attributed to Whitehead.
3 Broad makes this comment while discussing the first of three views on the status of secondary qualities, but similar criticisms of Pure Mechanism occur in his discussion of the two others (48–52).
4 Nagel (1965) also recognized that certain classes of phenomena would be indefinable in 'physico-chemical' terms, but his overall treatment appears to be based on the assumption that Broad and others were arguing for the unexplainability of these phenomena, given that they are emergent.
5 For the record, Pap's whole critique of the unpredictability argument actually involves a non sequitur. Broad was inclined to regard any instance of a secondary quality to be a sensum – and an emergent – whether it was perceived, imagined,

dreamed, or otherwise experienced. So to ask (as Pap did) whether some particular quality could have been imagined before ever being perceived would be the same for him as asking whether it could have been sensed before being sensed. Ostension, the disclosure of emergent qualities, was construed by Broad to have a much wider field of operation.

6 In speaking of 'natural piety' Alexander was alluding to the writings of another emergentist, C.L. Morgan.
7 Alexander here applies his two senses of 'mechanical' to the emergent quality of life; there is no reason why it cannot be applied to other emergents such as the traditional secondary qualities.
8 The issue of reductive explanation is taken up more fully in (Tully 1976).
9 According to Goudge, a similar view of perceptual qualities was taken by the pragmatist, G.H. Mead. See Goudge (1973, 143).
10 If we accept Broad's distinctions, he himself was to be counted as an Emergent Materialist, Alexander as an Emergent Neutralist (cf. Broad 1925, 647–8). Alexander did not stop at the atomic level in his analysis of matter; he considered the basic stuff of the universe to be pure spatiotemporal 'motions.'

WILLIAMS / IS BIOLOGY A DIFFERENT TYPE OF SCIENCE?

1 I am following the custom of using feminine pronouns for things significantly larger than a man (ships, hurricanes, Earth, etc.). This is, of course, a gender neutral usage and should not be taken as indicating that all planet-sized scientists are female.

List of Works Cited

For works cited by special abbreviations, see p xiii.

Abel, T. 1953. 'The Operation Called *Verstehen*' in Herbert Feigl and May Brodbeck, eds *Readings in the Philosophy of Science* New York
Addis, Laird. 1975. *The Logic of Society* Minneapolis
Alexander, Samuel. 1966. *Space, Time, and Deity* London
Almeder, Robert. 1975. 'Fallibilism and the Ultimate Irreversible Opinion' *American Philosophical Quarterly* Monograph 9
Anonymous. 1845. 'Vestiges, etc.' *British Quarterly Review* 1
– 1846. 'Explanations etc. '*British Quarterly Review* 3
– 1850. 'Geology Versus Development' *Fraser's Magazine* 42
Armstrong, D.M. 1968. *A Materialist Theory of the Mind* London
Ayer, A.J. 1968. *The Origins of Pragmatism* San Francisco
Ayim, M. 1974. 'Retroduction: the Rational Instinct' *Transactions of the C.S. Peirce Society* 10
Babbage, Charles. 1838. *Bridgewater Treatise* London
Baldwin, James Mark, ed. 1901. *Dictionary of Philosophy and Psychology* vol 1, New York and London
– 1902. *Dictionary of Philosophy and Psychology* vol 2, New York and London
Bartholomew, M. 1973. 'Lyell and Evolution: an Account of Lyell's Response to the Prospect of an Evolutionary Ancestry for Man' *British Journal of the History of Science* 6
Beckner, Morton. 1959. *The Biological Way of Thought* New York
– 1969. 'Function and Teleology '*Journal of the History of Biology* 2
Bergmann, Gustav. 1951. 'The Logic of Psychological Concepts' *Philosophy of Science* 18
– 1957. *Philosophy of Science* Madison

Bergson, Henri. 1949. *Introduction to Metaphysics* ed with introduction by T.A. Goudge, New York
Black, Max. 1954. *Problems of Analysis* Ithaca
Blake, R.M., C.J. Ducasse, and E.H. Madden, eds. 1960. *Theories of Scientific Method: the Renaissance through the Nineteenth Century* Seattle
Boler, J.F. 1963. *Charles Peirce and Scholastic Realism* Seattle
Borges, Jorge Luis. 1970. *The Aleph and Other Stories* New York
Bowen, F. 1845. 'Vestiges, etc.' *North American Review* 60
Braithwaite, R.B. 1934. 'Critical Notice of Peirce, *Collected Papers*' *Mind* ns 43
– 1953. *Scientific explanation* Cambridge
Brewster, D. 1844. 'Vestiges, etc.' *North British Review* 3. Published anonymously
Broad, Charlie Dunbar. 1925. *The Mind and its Place in Nature* London
Brodbeck, May. 1962. 'Explanation, Prediction, and "Imperfect" Knowledge' in Herbert Feigl and Grover Maxwell, eds *Minnesota Studies in the Philosophy of Science* vol 3, Minneapolis
– 1963. 'Meaning and Action' *Philosophy of Science* 30
Brody, Baruch. 1975. 'The Reduction of Teleological Sciences' *American Philosophical Quarterly* 12
Broyles, James R. 1964. 'Charles S. Peirce and the Concept of Indubitable Belief' *Transactions of the C.S. Peirce Society* 1
Buchdahl, Gerd. 1963. 'Descartes' Anticipation of a "Logic of Scientific Discovery"' in A.C. Crombie, ed *Scientific Change* London
– 1970. 'History of Science and Criteria of Choice' in Roger H. Steuwer, ed *Minnesota Studies in the Philosophy of Science* vol 5, Minneapolis
Buchler, Justus. 1939. *Charles Peirce's Empiricism* London
Bunge, Mario. 1976. 'The Relevance of Philosophy to Social Science' in William R. Shea, ed *Basic Issues in the Philosophy of Science* New York
Burroughs, Edgar Rice. 1972. *Tarzan of the Apes* London
Buytendijk, F.J.J. 1961. *Pain: Its Modes and Functions* London.
Campbell, N.R. 1957. *Foundations of Science* (formerly *Physics: the Elements*) New York
Carnap, Rudolf. 1939. *Foundations of Logic and Mathematics* Chicago
Carroll, Lewis. 1895. 'What the Tortoise Said to Achilles' *Mind* ns 4
Causey, R.L. 1969. 'Polanyi on Structure and Reduction' *Synthese* 20
– 1972. 'Attribute-Identities in Microreductions' *Journal of Philosophy* 69
Chambers, Robert. 1844. *The Vestiges of the Natural History of Creation* London. Published anonymously
– 1845. *Explanations, a Sequel to 'Vestiges'* London. Published anonymously
– 1853. *The Vestiges of the Natural History of Creation* tenth edition, London

List of Works Cited

Cody, Martin L. 1974. 'Optimization in Ecology' *Science* 183
Cohen, Morris R. 1923. *Chance, Love and Logic* New York
Collingwood, R.G. 1946. *The Idea of History* Oxford
Darwin, Charles. 1872. *The Expression of Emotion in Man and Animals* London
– 1887. *The Life and Letters of Charles Darwin* ed Francis Darwin, London
– 1903. *More Letters of Charles Darwin* ed Francis Darwin and A. C. Seward, London
– 1958. *The Autobiography of Charles Darwin and Selected Letters* ed Francis Darwin, New York
– 1962. *The Origin of Species* New York
– 1969. *The Autobiography of Charles Darwin* ed N. Barlow, New York
Davidson, Donald. 1975. 'Thought and Talk' in S. Guttenplan, ed *Mind and Language* Oxford
Descartes, René. 1911. 'Reply to Objections' in E.S. Haldane and G.R.T. Ross, trans *Philosophical Works of Descartes* vol 2, Cambridge
– 1954. *Principles of Philosophy* in G.E.M. Anscombe and Peter Geach, eds and trans *Descartes: Philosophical Writings* Edinburgh
– 1970. *Descartes' Philosophical Letters* trans by Anthony Kenny, Oxford
Dewey, John. 1938. *Logic: The Theory of Inquiry* New York
– 1939. *The Theory of Valuation* Chicago
Drake, Stillman, ed and trans. 1957. *Discoveries and Opinions of Galileo* New York
Dray, William. 1957. *Laws and Explanation in History* London
Dummett, Michael. 1958–9. 'Truth' *Proceedings of the Aristotelian Society* ns 59
– 1973. *Frege: Philosophy of Language* London
Eisele, Carolyn, ed. 1976. *The New Elements of Mathematics of Charles S. Peirce* The Hague
Feibleman, James K. 1960. *An Introduction to Peirce's Philosophy* London
Feigl, Herbert. 1958. 'The "Mental" and the "Physical"' in Herbert Feigl, Michael Scriven, and Grover Maxwell, eds *Minnesota Studies in the Philosophy of Science* vol 2, Minneapolis
Feyerabend, P.K. 1962. 'Explanation, Reduction, and Empiricism' in Herbert Feigl and Grover Maxwell, eds *Minnesota Studies in the Philosophy of Science* vol 3, Minneapolis
– 1970. 'Consolations for the Specialist' in Imre Lakatos and A. Musgrave, eds *Criticism and the Growth of Knowledge* London
Fisch, Max H. 1954. 'Alexander Bain and the Genealogy of Pragmatism' *Journal of the History of Ideas* 15
– 1967. 'Peirce's Progress from Nominalism toward Realism' *The Monist* 51
– 1971. 'Peirce's Arisbe' *Transactions of the C.S. Peirce Society* 7

- 1974. 'Supplements to the Peirce Bibliographies' *Transactions of the C.S. Peirce Society* 10
- 1975. 'Hegel and Peirce' in J.J. O'Malley, F.G. Weiss, and K.W. Algozin, eds *Hegel and the History of Philosophy* The Hague
- 1978. 'Peirce's General Theory of Signs' in Thomas A. Sebeok, ed *Sight, Sound and Sense* Bloomington

Fisch, Max H., and Jackson I. Cope. 1952. 'Peirce at the Johns Hopkins University' in Philip P. Wiener and Frederic H. Young, eds *Studies in the Philosophy of Charles Sanders Peirce* Cambridge, Mass.

Fisher, Ronald A. 1958. *The Genetical Theory of Natural Selection* New York

Fitzgerald, John J. 1966. *Peirce's Theory of Signs as Foundation for Pragmatism* The Hague

Flew, Anthony. 1966. *God and Philosophy* London

Fodor, Jerry A. 1968. *Psychological Explanation* New York

Gardner, Martin, ed. 1960. *The Annotated Alice* New York

Ghiselin, Michael. 1969. *The Triumph of the Darwinian Method* Berkeley
- 1974. 'A Radical Solution to the Species Problem' *Systematic Zoology* 23

Glaser, B.G., and A.L. Strauss. 1968. *Time for Dying* Chicago

Goodman, Nelson. 1966. *The Structure of Appearance* New York

Goudge, Thomas A. 1965. 'Another Look at Emergent Evolutionism' *Dialogue* 4
- 1966. 'Plausibility of New Hypotheses' (abstract) *Journal of Philosophy* 63
- 1967a. 'Emergent Evolutionism' in Paul Edwards, ed *Encyclopedia of Philosophy* New York
- 1967b. 'Uexküll, Jacob Johann, Baron von' in Paul Edwards, ed *Encyclopedia of Philosophy* New York
- 1973. 'Pragmatism's Contribution to an Evolutionary View of Mind' *The Monist* 57
- 1976a. 'Philosophical Literature' in C. Klink, ed *Literary History of Canada* second edition, Toronto
- 1976b. 'Neodarwinism, Mental Evolution and the Mind-Body Problem' in William R. Shea, ed *Basic Issues in the Philosophy of Science* New York
- Unpub. 'What Makes a Hypothesis Plausible?'

Gray, A. 1846. 'Explanations' *North American Review* 62

Gray, J.A. 1971. 'The Mind-Body Identity Theory as a Scientific Hypothesis' *Philosophical Quarterly* 21

Greenlee, Douglas. 1971. 'Unrestricted Fallibilism' *Transactions of the C.S. Peirce Society* 7

Gutting, Gary. 1973. 'A Defense of the Logic of Discovery' *Philosophical Forum* 4

Hampshire, Stuart. 1971. *Freedom of Mind and Other Essays* Princeton

Hanson, Norwood Russell. 1958a. 'The Logic of Discovery' *Journal of Philosophy* 55
- 1958b. *Patterns of Discovery* Cambridge

- 1960. 'More on the Logic of Discovery' *Journal of Philosophy* 57
- 1961. 'Is There a Logic of Discovery?' in Herbert Feigl and Grover Maxwell, eds *Current Issues in the Philosophy of Science* New York
- 1963. 'Retroductive Inference' in B. Baumrin, ed *Philosophy of Science: The Delaware Seminar* 1
- 1965a. 'The Idea of a Logic of Discovery' *Dialogue* 4
- 1965b. 'Notes toward a Logic of Discovery' in Richard J. Bernstein, ed *Perspectives on Peirce* New Haven

Hardy, J.D. 1953. 'Thresholds of Pain and Reflex Contraction as Related to Noxious Stimulation' *Journal of Applied Psychology* 5

Harman, Gilbert. 1965. 'The Inference to the Best Explanation' *Philosophical Review* 74
- 1973. *Thought* Princeton

Helmer, O., and N. Rescher. 1959. 'On the Epistemology of the Inexact Sciences' *Management Science* 6

Hempel, Carl G. 1942. 'The Function of General Laws in History' *Journal of Philosophy* 39
- 1949. 'The Function of General Laws in History' in Herbert Feigl and Wilfrid Sellars, eds *Readings in Philosophical Analysis* New York
- 1965. *Aspects of Scientific Explanation and Other Essays in the Philosophy of Science* New York

Hempel, Carl G., and Paul Oppenheim. 1965. 'Studies in the Logic of Explanation' in Carl G. Hempel, *Aspects of Scientific Explanation and Other Essays in the Philosophy of Science* New York

Henle, Paul. 1942. 'The Status of Emergence' *Journal of Philosophy* 39

Herschel, Sir John F.W. 1831. *A Preliminary Discourse on the Study of Natural Philosophy* London
- 1841. 'Review of W. Whewell's *History of the Inductive Sciences* and *Philosophy of the Inductive Sciences*' *Quarterly Review* 68. Published anonymously
- 1857. *Essays from the Edinburgh & Quarterly Reviews with Address and Other Pieces* London
- 1861. *Physical Geography* Edinburgh
- 1878. *Outlines of Astronomy* new edition, London

Hertz, Heinrich. 1899. *The Principles of Mechanics* London

Hesse, Mary B. 1963. *Models and Analogies in Science* London
- 1973. 'Logic of Discovery in Maxwell's Electromagnetic Theory' in Ronald N. Giere and Richard S. Westfall, eds *Foundations of Scientific Method: The Nineteenth Century* Bloomington

Hopkins, W. 1860. 'Physical Theories of the Phenomena of Life' part 1 *Fraser's Magazine* 61
- 1861. 'Physical Theories of the Phenomena of Life' part 2 *Fraser's Magazine* 62

Hull, David L. 1968. 'The Operational Imperative – Sense and Nonsense in Operationalism' *Systematic Zoology* 17
- 1974. *Philosophy of Biological Science* Englewood Cliffs, N.J.
- 1975. 'Central Subjects and Historical Narratives' *History and Theory* 14
- 1976. 'Are Species Really Individuals?' *Systematic Zoology* 25
- 1978. 'A Matter of Individuality' *Philosophy of Science* 45
- 1979. 'The Limits of Cladism' *Systematic Zoology* 28
Hume, David. 1947. *Dialogues Concerning Natural Religion* ed Norman Kemp Smith, second edition, Edinburgh
Huxley, Julian. 1942. *Evolution, the Modern Synthesis* New York
Huxley, Thomas Henry. 1854. 'Vestiges, etc' *British and Foreign Medico-chirurgical Review* 13
James, Henry, ed. 1920. *Letters of William James* Boston
James, William. 1907. *Pragmatism* New York and London
- 1909. *The Meaning of Truth* New York and London
- 1975a. *Pragmatism* Cambridge, Mass.
- 1975b. *The Meaning of Truth* Cambridge, Mass.
Jerison, H.J. 1973. *Evolution of the Brain and Intelligence* New York and London
- 1979. 'The Evolution of Consciousness' *Proceedings of the Seventh International Conference on the Unity of the Sciences* II
Kant, Immanuel. 1928. *Critique of Teleological Judgment* trans J.C. Meredith, Oxford
- 1933. *Critique of Pure Reason* trans Norman Kemp Smith, London
- 1970. *Metaphysical Foundations of Natural Science* trans James Ellington, Indianapolis and New York
Kempe, A.B. 1886. 'A Memoir on the Theory of Mathematical Form' *Philosophical Transactions of the Royal Society of London* 177
Kent, Beverly. 1975. *Logic in the Context of Peirce's Classification of the Sciences* PH D dissertation, University of Waterloo
Kierkegaard, Søren. 1941. *Fear and Trembling* Princeton
Kochanski, Zdzislaw. 1973. 'Conditions and Limitations of Prediction-Making in Biology' *Philosophy of Science* 40
Koestler, Arthur. 1975. *The Act of Creation* London
Kordig, C. 1971. *The Justification of Scientific Change* Dordrecht
Körner, Stephan. 1960. 'On Philosophical Arguments in Physics' in Edward H. Madden, ed *The Structure of Scientific Thought* Boston
Kuhar, H.J., C.B. Pert, and S.H. Snyder. 1973. 'Regional Distribution of Opiate Receptor Binding in Monkey and Human Brain' *Nature* 245
Kuhn, Thomas. 1970a. 'Reflections on my Critics' in Imre Lakatos and A. Musgrave, eds *Criticism and the Growth of Knowledge* London

- 1970b. *The Structure of Scientific Revolutions* second edition, Chicago
Lakatos, Imre. 1970. 'Falsification and the Methodology of Scientific Research Programmes' in Imre Lakatos and A. Musgrave, eds *Criticism and the Growth of Knowledge* London
Laudan, Laurens. 1973. 'Peirce and the Trivialization of the Self-Correcting Thesis' in Ronald N. Giere and Richard S. Westfall, eds *Foundations of Scientific Method: The Nineteenth Century* Bloomington
Lavoisier, Antoine. 1862. *Oeuvres* Paris
Layzer, D. 1975. 'The Arrow of Time' *Scientific American* 233
Lenz, John W. 1960. 'The Pragmatic Justification of Induction' in Edward H. Madden, ed *The Structure of Scientific Thought* Boston
- 1964. 'Induction as Self-Corrective' in E. Moore and R. Robin, eds *Studies in the Philosophy of Charles Sanders Peirce* second series, Amherst
Leonard, H.S. 1951. 'Review of T.A. Goudge, *The Thought of C.S. Peirce*' *University of Toronto Quarterly* 21
Levi, Isaac. 1965. 'Hacking, Salmon on Induction' *Journal of Philosophy* 62
Lewis, C.I. 1918. *A Survey of Symbolic Logic* Berkeley
- 1929. *Mind and the World-Order* New York
Lieb, Irwin C., ed. 1953. *Charles S. Peirce's Letters to Lady Welby* New Haven.
Llinás, R.R. 1975. 'The Cortex of the Cerebellum' *Scientific American* 232
Lyell, Sir Charles. 1830-3. *Principles of Geology* London
Lyell, K., ed. 1881. *Life, letters and Journals of Sir Charles Lyell* London
Mackie, J.M. 1965. 'Causes and Conditions' *American Philosophical Quarterly* 2
Madden, Edward H. 1960a. 'Charles Sanders Peirce's Search for a Method' in R.M. Blake, C.J. Ducasse, and E.H. Madden, eds *Theories of Scientific Method* Seattle
- 1968. 'Peirce and Contemporary Issues in the Philosophy of Science' in Raymond Klibansky, ed *Contemporary Philosophy* Firenze
Madden, Edward H., ed. 1960b. *The Structure of Scientific Thought* Boston
Mandelbaum, Maurice. 1977. *The Anatomy of Historical Knowledge* Baltimore and London
Margenau, H. 1950. *The Nature of Physical Reality* New York
Martin, Robert M. 1972. 'A Reason to Believe in Mind-Body Identity' *Personalist* 1972
Martin, Robert M., and A. Rosenberg. 1976. 'Materialism and Evolution: A Reconsideration' *Canadian Journal of Philosophy* 6
McGuire, J.E. 1968. 'Force, Active Principles and Newton's Invisible Realm' *Ambix* 15
McMurray, G.A. 1950. 'Experimental Study of a Case of Insensitivity to Pain' *Archives of Neurology and Psychiatry* 64
Mead, George Herbert. 1934. *Mind, Self and Society* Chicago

Melzack, R. 1973. *The Puzzle of Pain* London
Mettler, L.E., and T.G. Gregg. 1969. *Population Genetics and Evolution* Englewood Cliffs, N.J.
Meyers, Robert G. 1971. 'Truth and Ultimate Belief in Peirce' *International Philosophical Quarterly* 11
Michael, C.R. 1969. 'Retinal Processing of Visual Images' *Scientific American* 5
Mill, John Stuart. 1872. *A System of Logic* eighth edition, London
Mink, L.O. 1970. 'History and Fiction as Modes of Comprehension' *New Literary History* 1
Muntz, W.R.A. 1962. 'Effectiveness of Different Colors of Light in Releasing Phototactic Behavior of Frogs and a Possible Function of the Retinal Projection of the Diencephalon' *Journal of Neurophysiology* 25
Murphey, Murray G. 1961. *The Development of Peirce's Philosophy* Cambridge, Mass.
Nagel, Ernest. 1961. *The Structure of Science* New York
– 1965. 'Mechanistic Explanation and Organismic Biology' in Sidney Hook, ed *American Philosophers at Work* New York
Newton, Isaac. 1946. *Mathematical Principles of Natural Philosophy* Berkeley
– 1952. *Opticks* New York
Olscamp, Paul J., ed. 1965. *Descartes: Discourse on Method, Geometry, and Meteorology* New York
Oppenheim, Paul, and Hilary Putnam. 1958. 'Unity of Science as a Working Hypothesis' in Herbert Feigl, Michael Scriven, and Grover Maxwell, eds *Minnesota Studies in the Philosophy of Science* vol 2, Minneapolis
Pap, Arthur. 1962. *An Introduction to the Philosophy of Science* Glencoe, Ill.
Pepper, S.C. 1926. 'Emergence,' *Journal of Philosophy* 23
Perry, Ralph Barton. 1935. *The Thought and Character of William James* Cambridge, Mass.
Plato. 1956. *Euthyphro, Apology, Crito* trans T.J. Church, revised and introduced by Robert D. Cumming, Indianapolis.
– 1961. *Euthyphro* trans Lane Cooper, in Edith Hamilton and Huntington Cairns, eds *The Collected Dialogues of Plato* New York
Popper, Karl. 1959. *The Logic of Scientific Discovery* New York
– 1966. *Unended Quest* La Salle, Ill.
Quine, Williard Van Orman. 1934. 'Collected Papers of Charles Sanders Peirce – Volume III: Exact Logic' *Isis* 22
– 1936a. 'Toward a Calculus of Concepts' *Journal of Symbolic Logic* 1
– 1936b. 'Truth by Convention' in O.H. Lee, ed *Philosophical Essays for Alfred North Whitehead* New York
– 1950. *Methods of Logic* New York
– 1960. *Word and Object* Cambridge, Mass.

- 1971. 'Predicate-Functor Logic' in J.E. Fensted, ed *Proceedings of the Second Scandinavian Logic Symposium* Amsterdam
- 1972. 'Algebraic Logic and Predicate Functors' in R. Rudner and I. Scheffler, eds *Logic and Art: Essays in Honor of Nelson Goodman* New York
- 1975. 'The Nature of Natural Knowledge' in S. Guttenplan, ed *Mind and Language*, Oxford

Reichenbach, Hans. 1939. 'Dewey's Theory of Science' in Paul Arthur Schilpp, ed *The Philosophy of John Dewey* Evanston
- 1949. *The Theory of Probability* Berkeley
- 1952. *Experience and Prediction* Chicago

Rescher, Nicholas. 1973. *The Primacy of Practice* Oxford
- 1976. 'Peirce and the Economy of Research' *Philosophy of Science* 43

Roberts, Don D. 1970. 'On Peirce's Realism' *Transactions of the C.S. Peirce Society* 6
- 1973. *The Existential Graphs of Charles S. Peirce* The Hague

Robin, Richard S. 1967. *Annotated Catalogue of the Papers of Charles S. Peirce* (Amherst) Mass.
- 1971. 'The Peirce Papers: a Supplementary Catalogue' *Transactions of the C.S. Peirce Society* 7

Rosenberg, Alexander. 1976. *Microeconomic Laws* Pittsburgh

Royce, Josiah. 1885. *The Religious Aspect of Philosophy* New York

Rudner, Richard S. 1966. *Philosophy of Social Science* Englewood Cliffs, N.J.

Ruse, Michael. 1971a. 'Narrative Explanations and the Theory of Evolution' *Canadian Journal of Philosophy* 1
- 1971b. 'Natural Selection in *The Origin of Species*' *Studies in History and Philosophy of Science* 1
- 1973a. 'The Nature of Scientific Models: Formal *v* Material Analogy' *Philosophy of Social Science* 3
- 1973b. 'The Value of Analogical Models in Science' *Dialogue* 12
- 1973c. *The Philosophy of Biology* London
- 1974. 'Reduction to Genetics' in R.S. Cohen et al, eds, *PSA 1974* Dordrecht
- 1975a. 'Darwin's Debt to Philosophy: An Examination of the Ideas of F.W. Herschel and William Whewell on Charles Darwin's Development of his Theory of Evolution' *Studies in History and Philosophy of Science* 6
- 1975b. 'The Relationship Between Science and Religion in Britain, 1830–1870' *Church History* 44
- 1975c. 'Charles Darwin and Artificial Selection' *Journal of the History of Ideas* 36
- 1975d. 'Charles Darwin's Theory of Evolution: An Analysis' *Journal of the History of Biology* 8
- 1976a. 'Charles Lyell and the Philosophers of Science' *British Journal for the History of Science* 9

- 1976b. 'Is Biology Different?' in R. Colodny, ed *Laws, Logic, Life* Pittsburgh
- 1979. *The Darwinian Revolution: Science Red in Tooth and Claw*, Chicago

Russell, Bertrand. 1939. 'Dewey's New Logic' in Paul Arthur Schilpp, ed *The Philosophy of John Dewey* Evanston

Salmon, Wesley. 1966. *The Foundations of Scientific Inference* Pittsburgh

Santayana, George. 1920. *Character and Opinion in the United States* London
- 1923. *Scepticism and Animal Faith* New York
- 1938. *The Realm of Truth* New York

Savan, David. 1952. 'On the origins of Peirce's Phenomenology' in Philip Wiener and Frederic H. Young, eds *Studies in the Philosophy of Charles Sanders Peirce* Cambridge, Mass.
- 1977. 'Questions Concerning Certain Classifications Claimed for Signs' *Semiotica* 19

Schaffner, Kenneth. 1974. 'Logic of Discovery and Justification of Regulatory Genetics' *Studies in History and Philosophy of Science* 4

Schouls, Peter A. 1972a. 'Reason, Method, and Science in the Philosophy of Descartes' *Australasian Journal of Philosophy* 50
- 1972b. 'Descartes and the Autonomy of Reason' *Journal of the History of Philosophy* 10
- 1973. 'The Extent of Doubt in Descartes' *Meditations' Canadian Journal of Philosophy* 3

Scott, Frederick J. Down. 1973. 'Peirce and Schiller and their Correspondence' *Journal of the History of Philosophy* 11

Scriven, Michael. 1959a. 'Explanation and Prediction in Evolutionary Theory' *Science* 130
- 1959b. 'Truisms as the Grounds for Historical Explanations' in Patrick Gardiner, ed *Theories of History* Glencoe, Ill.
- 1962. 'Explanations, Predictions, and Laws' in Herbert Feigl and Grover Maxwell, eds *Minnesota Studies in the Philosophy of Science* vol 3, Minneapolis

Sedgwick, Adam. 1833. *A Discourse on the Studies of the University of Cambridge* Cambridge
- 1845. 'Vestiges, etc.' *Edinburgh Review* 165. Published anonymously
- 1850. *A Discourse on the Studies of the University of Cambridge* fifth edition, London and Cambridge
- 1860. 'Objections to Mr. Darwin's Theory of the Origin of Species' *Spectator*

Sellars, Wilfrid. 1964. 'Induction as Vindication' *Philosophy of Science* 31
- 1967. 'Scientific Realism or Irenic Instrumentalism' in his *Philosophical Perspectives* Springfield, Ill.

Sellars, Wilfrid, and P.E. Meehl. 1956. 'The Concept of Emergence' in Herbert Feigl and Michael Scriven, eds *Minnesota Studies in the Philosophy of Science* vol 1, Minneapolis

Simberloff, S. 1974. 'Equilibrium Theory of Island Biogeography and Ecology' *Annual Review of Ecology and Systematics* 5
Simpson, G.G. 1950. 'Evolution' *Chamber's Encyclopaedia* London
Skidmore, Arthur W. 1971. 'Peirce and Triads' *Transactions of the C.S. Peirce Society* 6
Smart, J.J.C. 1963. *Philosophy and Scientific Realism* London
- 1968. *Between Science and Philosophy* New York
- 1976-7. 'Under the Form of Eternity' *The Walter and Eliza Hall Institute of Medical Research Annual Review*
Smith, J.W. 1952. 'Review of T.A. Goudge, *The Thought of C.S. Peirce*' *Philosophical Quarterly* 2
Sperry, R.W. 1945. 'The Problem of Central Nervous Reorganization after Nerve Regeneration and Muscle Transposition' *Quarterly Review of Biology* 20
Sprigge, Timothy L.S. 1974. *Santayana: An Examination of his Philosophy* London
Stroud, Barry. 1965. 'Wittgenstein and Logical Necessity' *Philosophical Review* 74
- 1977. *Hume* London
Suppe, Fred. 1974. 'Some Philosophical Problems in Biological Speciation and Taxonomy' in J.A. Wojciechowski, ed *Conceptual Basis of the Classification of Knowledge* Munich
Tarski, Alfred. 1944. 'The Semantic Conception of Truth' *Philosophy and Phenomenological Research* 4
Tattersall, Ian, and Nils Eldridge. 1977. 'Fact, Theory, and Fantasy in Human Paleontology' *American Scientist* 65
Thagard, Paul R. 1978. 'The Best Explanation: Criteria for Theory Choice' *Journal of Philosophy* 75
Thayer, H.S. 1968. *Meaning and Action: A Critical History of Pragmatism* Indianapolis and New York
Thompson, Manley. 1953. *The Pragmatic Philosophy of C.S. Peirce* Chicago
Thomson, Sir George. 1969. 'Matter and Radiation' in R. Harré, ed *Scientific Thought 1900-1960* Oxford
Todhunter, I. 1876. *William Whewell D.D.* London
Tomas, Vincent, ed 1957. *Charles S. Peirce: Essays in the Philosophy of Science* New York
Toulmin, Stephen. 1961. *Foresight and Understanding* London
- 1976. 'On the Nature of the Physician's Understanding' *Journal of Medicine and Philosophy* 1
Tully, Robert E. 1976. 'Reduction and Secondary Qualities' *Mind* 85
Vendler, Zeno. 1976. 'On the Possibility of Possible Worlds' *Canadian Journal of Philosophy* 5
Volpe, E.P. 1967. *Understanding Evolution* Dubuque, Iowa
Wartenberg, G. 1971. *Logischer Sozialismus* Frankfurt

Westfall, Richard S. 1971. *The Construction of Modern Science: Mechanisms and Mechanics* New York

Whewell, William. 1831. 'Review of J.F.W. Herschel's *Preliminary Discourse on the Study of Natural Philosophy,*' *Quarterly Review* 45. Published anonymously
- 1837. *History of the Inductive Sciences* London
- 1840. *Philosophy of the Inductive Sciences* London
- 1845. *Indications of the Creator* London
- 1967. *Philosophy of the Inductive Sciences* New York (facsimile reprint of Whewell 1840, second edition)

White, Morton. 1972. *Science and Sentiment in America* New York

Wiener, Philip P., ed. 1958. *Values in a Universe of Chance* Garden City, New York, and Stanford

Williams, Bernard. 1970. 'Deciding to Believe' in H.E. Kiefer and M.K. Munitz, eds *Language, Belief and Metaphysics* Albany

Williams, Mary B. 1970. 'Deducing the Consequences of Evolution: A Mathematical Model' *Journal of Theoretical Biology* 29
- 1973. 'Falsifiable Predictions of Evolutionary Theory' *Philosophy of Science* 40
- 1976. 'The Logical Structure of Functional Explanations in Biology' in F. Suppe and P.D. Asquith, eds *PSA 1976* Dordrecht

Wilson, D. 1974. 'Herschel and Whewell's Version of Newtonianism' *Journal of the History of Ideas* 35

Wilson, Fred. 1967. 'Definition and Discovery' *British Journal for the Philosophy of Science* 18 and 19
- 1969a. 'Dispositions: Defined or Reduced?' *Australasian Journal of Philosophy* 47
- 1969b. 'Explanation in Aristotle, Newton, and Toulmin' *Philosophy of Science* 36
- 1971a. 'Discussion of Achinstein's *Concepts of Science*' *Philosophy of Science* 38
- 1971b. 'Review of Lakatos & Musgrave, *Criticism and the Growth of Knowledge*' *Dialogue* 10
- 1974. 'Why I Am Not Aware of Your Pain' in M. Gram and E. Klemke, eds *The Ontological Turn* Des Moines
- 1981. *Reasons and Revolutions: An Empiricist Account of Paradigms and Research Programmes* Ottawa

Wilson, L., ed. 1970. *Sir Charles Lyell's Scientific Journals on the Species Question* New Haven

Wimsatt, William. 1972. 'Teleology and the Logical Structure of Function Statements' *Studies in History and Philosophy of Science* 3

Wittgenstein, Ludwig. 1953. *Philosophical Investigations* Oxford
- 1967. *Zettel* trans G.E.M. Anscombe, ed G.E.M. Anscombe and G.H. von Wright, Oxford

Wright, Larry. 1973. 'Functions' *Philosophical Review* 82

The Published Works of Thomas A. Goudge (1910–   )

1934
'Some Realist Theories of Illusion' *The Monist* 44:1 (January 1934) 108–25

1935
'The Views of Charles Peirce on the Given in Experience' *The Journal of Philosophy* 32:20 (26 September 1935) 533–44

1936
'Further Reflections on Peirce's Doctrine of the Given' *The Journal of Philosophy* 33:11 (21 May 1936) 289–95

1938
Review of Mortimer J. Adler *What Man Has Made of Man: A Study in the Consequences of Platonism and Positivism in Psychology* in *Queen's Qarterly* 45:2 (Summer 1938) 265–6

1940
'Peirce's Treatment of Induction' *Philosophy of Science* 7:1 (January 1940) 56–68
Review of Justus Buchler *Charles Peirce's Empiricism* in *The Journal of Philosophy* 37:10 (9 May 1940) 274–6

1942
'The Spectator Fallacy' *The Journal of Philosophy* 39:1 (1 January 1942) 14–21

1943
'Science and Symbolic Logic' *Scripta Mathematica* 9:2 (June 1943) 69–80
'Charles Peirce: Pioneer in American Thought' *University of Toronto Quarterly* 12:4 (July 1943) 403–14

1945

Review of W.H. Sheldon *Process and Polarity* in *The Philosophical Review* 54:3 (May 1945) 280–2

1946

Obituary Notice of George Sidney Brett *Proceedings and Addresses of the American Philosophical Association, 1945–1946* 19 (July 1946) 449–50

Review of B. Croce *Politics and Morals* in *Science and Society* 10:4 (Winter 1946) 432–4

1947

'Philosophical Trends in Nineteenth-Century America' *University of Toronto Quarterly* 16:2 (January 1947) 133–42

'The Conflict of Naturalism and Transcendentalism in Peirce' *The Journal of Philosophy* 44:14 (3 July 1947) 365–75

1948

'The Concept of Verifiability' [Abstract] *Bulletin of the Eastern Division of the American Philosophical Association: Published for Its 45th Annual Meeting at the University of Virginia, Charlottesville* (December 1948) 7–8

1949

An Introduction to Metaphysics by Henri Bergson. Authorized translation by T.E. Hulme. (New York: The Liberal Arts Press 1949) xx, 41. Edited with an Introduction

'The Function of Reason' *Dalhousie Review* 28:4 (January 1949) 329–38

1950

The Thought of C.S. Peirce (Toronto: University of Toronto Press 1950) xii, 360. Diagrams and bibliographical footnotes

'The Future of Materialism' *The Philosophical Review* 59:1 (January 1950) 107–12

Review of P.P. Wiener *Evolution and the Founders of Pragmatism* in *University of Toronto Quarterly* 19:4 (July 1950) 441–3

Review of J.-P. Vaudaire *Bases et Profils de la Société de Demain: Essai de Logique Pure* in *International Journal* 5:4 (Autumn 1950) 371–2

1951

Review of Lloyd Morris *William James: The Message of a Modern Mind* in *Canadian Forum* 30:360 (January 1951) 239

Review of John E. Smith *Royce's Social Infinite* in *The Philosophical Review* 60:2 (April 1951) 253–5

331 The Published Works of Thomas A. Goudge

Review of William P.D. Wightman *The Growth of Scientific Ideas* in *University of Toronto Quarterly* 21:1 (October 1951) 101–3

1952

'Peirce's Theory of Abstraction' in Philip P. Wiener and Frederic H. Young, eds *Studies in the Philosophy of Charles Sanders Peirce* (Cambridge, Massachusetts: Harvard University Press 1952) 121–32
'Philosophy and the Physician' *The University of Toronto Medical Journal* 29:5 (February 1952) 224–6
'Kantian Rationalism' review of H.J. Paton *In Defence of Reason* in *University of Toronto Quarterly* 21:4 (July 1952) 434–7
Review of J.K. Feibleman *Ontology* in *The Journal of Philosophy* 49:16 (31 July 1952) 537–9
Review of Richard von Mises *Positivism: A Study in Human Understanding* in *University of Toronto Quarterly* 22:1 (October 1952) 105–6

1953

'Creation and the Astronomers' review of George Gamow *The Creation of the Universe* in *University of Toronto Quarterly* 22:2 (January 1953) 197–9
Review of George Sarton *A History of Science: Ancient Science through the Golden Age of Greece* in *University of Toronto Quarterly* 23:1 (October 1953) 100–1
'Organismic Concepts in Biology and Physics' *The Review of Metaphysics* 7:2 (December 1953) 282–9

1954

'The Concept of Evolution' *Mind* ns 63:249 (January 1954) 16–25
'Physical Cosmology and Philosophical Physics' *The Review of Metaphysics* 7:3, Issue No 27 (March 1954) 444–51
Review of W.B. Gallie *Peirce and Pragmatism* in *Mind* ns 63:250 (April 1954) 279–81
'Some Philosophical Aspects of the Theory of Evolution' *University of Toronto Quarterly* 23:4 (July 1954) 386–401

1955

An Introduction to Metaphysics by Henri Bergson. Authorized translation by T.E. Hulme. Second revised edition. (Indianapolis: Bobbs-Merrill; New York: The Liberal Arts Press 1955) 62. Edited with an Introduction
Notice of N.W. DeWitt *Epicurus and His Philosophy* and *St. Paul and Epicurus* in 'Letters in Canada: 1954' *University of Toronto Quarterly* 24:3 (April 1955) 287–90
'Broad-minded Philosophy' review of C.D. Broad *Religion, Philosophy and Psychical Research* in *University of Toronto Quarterly* 24:4 (July 1955) 433–5
'What is a Population?' *Philosophy of Science* 22:4 (October 1955) 272–9

1956

Notice of Frederick Creedy *The Next Step in Civilization* and Rupert C. Lodge *Plato's Theory of Art* in 'Letters in Canada: 1955' *University of Toronto Quarterly* 25:3 (April 1956) 346–7

1957

'Progress and Evolution' in E.G.D. Murray, ed *Studia Varia: Royal Society of Canada Literary and Scientific Papers* (Toronto: University of Toronto Press 1957) 86–94
'Is Evolution Finished?' *University of Toronto Quarterly* 26:4 (July 1957) 430–42
'Causal Explanations in Natural History' [Abstract] *The Journal of Philosophy* 54:24 (21 November 1957) 780–1

1958

Review of Charles S. Peirce *Values in a Universe of Chance* edited by Philip P. Wiener *The Journal of Philosophy* 55:14 (3 July 1958) 609–10
Notice of E.G.D. Murray, ed *Our Debt to the Future (Presence de Demain)* Royal Society of Canada Studia Varia Series, vol 2, in 'Letters in Canada: 1958' *University of Toronto Quarterly*, 28:4 (July 1958) 408–9
Review of Arthur W. Burks, ed *Collected Papers of Charles Sanders Peirce, Volume VII, Science and Philosophy* and *Volume VIII, Reviews, Correspondence and Bibliography* in *Dalhousie Review* 38:3 (Autumn 1958) 407, 409
'Causal Explanations in Natural History' *British Journal for the Philosophy of Science* 9:35 (November 1958) 194–202

1959

'Explorations Across the Great Divide' *University of Toronto Quarterly* 29:1 (October 1959) 85–90

1960

Review of John C. Greene *The Death of Adam: Evolution and Its Impact on Western Thought* in *Queen's Quarterly* 67:2 (Summer 1960) 317–18
Notice of George P. Grant *Philosophy in the Mass Age* and R.C. Chalmers and J.A. Irving *Challenge and Response* in 'Letters in Canada: 1959' *University of Toronto Quarterly* 29:4 (July 1960) 486–8

1961

The Ascent of Life; A Philosophical Study of the Theory of Evolution (Toronto: University of Toronto Press; London: George Allen & Unwin Ltd 1961) 236. Includes bibliography
'The Genetic Fallacy' *Synthèse* 13:1 (March 1961) 41–8
'Darwin's Heirs' *University of Toronto Quarterly* 30:3 (April 1961) 246–50

333 The Published Works of Thomas A. Goudge

Notice of T.M. Cameron, ed *Evolution: Its Science and Doctrine*, H.H.J. Nesbitt, ed *Darwin in Retrospect*, and C. de Koninck *The Hollow Universe* in 'Letters in Canada: 1960' *University of Toronto Quarterly* 30:4 (July 1961) 427–9

1962
Review of Murray G. Murphey *The Development of Peirce's Philosophy* in *Dalhousie Review* 42:1 (Spring 1962) 111
'The Evolutionary Vision of Teilhard de Chardin' *University of Toronto Quarterly* 32:1 (October 1962) 70–80
'Salvaging the "Noosphere"' *Mind* ns 71:284 (October 1962) 543–4

1963
Review of Vincent E. Smith, ed *Philosophy of Biology* in *Philosophy and Phenomenological Research* 23:3 (March 1963) 457–8
Review of William Howells, ed *Ideas on Human Evolution: Selected Essays* in *Queen's Quarterly* 69:4 (Winter 1963) 631

1964
'Peirce's Evolutionism, After Half a Century' In Edward C. Moore and Richard S. Robin, eds *Studies in the Philosophy of Charles Sanders Peirce; Second Series* (Amherst: The University of Massachusetts Press 1964) 323–41
Review of Ludwig von Bertalanffy *Modern Theories of Development* and *Problems of Life* in *Dialogue* 2:4 (March 1964) 474–5
Review of Murray G. Murphey *The Development of Peirce's Philosophy* and Hjalmar Wennerberg *The Pragmatism of C.S Peirce* in *Mind* ns 73:292 (October 1964) 602–3

1965
'Peirce's Index' *Translations of the Charles S. Peirce Society* 1:2 (Fall 1965) 52–70
'Another Look at Emergent Evolutionism' *Dialogue* 4:3 (December 1965) 273–85. Presidential Address to the Canadian Philosophical Association at the annual meeting, University of British Columbia, 16 June 1965

1966
De Wijsgerige Aspecten van de Evolutie [Vertaald door S.Y.B. de Glas] (Utrecht and Antwerpen: Het Spectrum 1966) 285 (Aula-Boeken 277) Bibliography. Translation of *The Ascent of Life*
'Plausibility of New Hypotheses' *The Journal of Philosophy* 63:20 (27 October 1966) 621–4
Review of A.G. van Melsen *Evolution and Philosophy* in *Philosophy and Phenomenological Research* 27:3 (December 1966) 293–4

1967

Encyclopedia of Philosophy, edited by Paul Edwards (New York: Macmillan & The Free Press 1967). Eight volumes.
Articles:
Henri Bergson, vol 1, 287–95
L. von Bertalanffy, vol 1, 306–7
Samuel Butler, vol 1, 434–6
Charles Darwin, vol 2, 294–5
Erasmus Darwin, vol 2, 295–6
Emergent Evolutionism, vol 2, 474–7
Asa Gray, vol 3, 377–8
Thomas Henry Huxley, vol 4, 101–3
Chevalier de Lamarck, vol 4, 376–7
Pierre André Lecomte du Noüy, vol 4, 417–18
Origin of Life, vol 4, 477–9
Jacques Loeb, vol 4, 503–4
C. Lloyd Morgan, vol 5, 392–3
George John Romanes, vol 7, 205–6
Jan Christiaan Smuts, vol 7, 464–5.
Teilhard de Chardin, vol 8, 83–4
Jacob Johann, Baron von Uexküll, vol 8, 173–4
Alfred Russel Wallace, vol 8, 276–7
John Henry Woodger, vol 8, 346–7
'Existentialism and Biology' *Dialogue* 5:4 (March 1967) 603–8

1968

'Fulton Henry Anderson, 1895–1968' *Proceedings of the Royal Society of Canada* 4th series, vol 6 (1968) 77–8
'A Century of Philosophy in English-Speaking Canada' *Dalhousie Review* 47:4 (Winter 1967–8) 537–49
'Memorial for Fulton Henry Anderson, M.A., Ph.D, Ll.D., D.Litt, F.R.S.C.' *Dialogue* 7:1 (June 1968) 91–3

1969

The Thought of C.S. Peirce (New York: Dover Publications 1969) xii, 360. Unaltered reprint of the 1950 edition
Preface to Roy Wood Sellars *Evolutionary Naturalism* (New York: Russell and Russell 1969) 1–3
'Fulton Henry Anderson, 1895–1968' *Proceedings and Addresses of the American Philosophical Association, 1968–1969* 42 (October 1969) 162–3
Review of H.S. Thayer *Meaning and Action: A Crtical History of Pragmatism* in *Dialogue* 8:3 (December 1969) 508–10

The Published Works of Thomas A. Goudge

1970
'Henri Bergson' in C.C. Gillespie, ed *Dictionary of Scientific Biography* vol 2 (New York: Charles Scribner 1970) 8–12

1971
'Does Evolution Manifest Progress?' in *Man and Nature: Philosophical Issues in Biology*, edited and with introductions by Ronald Munson (New York: Dell Publishing Co. 1971) 255–67. An unaltered reprint of chapter v, section 5, of *The Ascent of Life*

Review of Giorgio Tagliacozzo, ed *Giambattista Vico: An International Symposium* in *Philosophy of the Social Sciences* 1:4 (December 1971) 350–2

1972
'Ryle's *Collected Papers*' *Dialogue* 11:4 (December 1972) 596–601

Review of Stephen Toulmin *Human Understanding* volume 1, *The Collective Use and Evolution of Concepts* in *Queen's Quarterly* 79:4 (Winter 1972) 577–8

1973
'Evolutionism' in Philip P. Wiener, editor-in-chief *Dictionary of the History of Ideas: Studies of Selected Pivotal Ideas* vol 2 (New York: Charles Scribner's Sons, 1973) 174–89

'Pragmatism's Contribution to an Evolutionary View of Mind' *The Monist* 57:2 (April 1973) 133–50

Review of Morton White *Science and Sentiment in America: Philosophical Thought from Jonathan Edwards to John Dewey* in *Philosophy of the Social Sciences* 3:3 (September 1973) 270–2

1974
Review of Michael Ruse *The Philosophy of Biology* in *Dialogue* 13:1 (March 1974) 176–9

1976
'Neodarwinism, Mental Evolution, and the Mind-Body Problem' in William R. Shea, ed *Basic Issues in the Philosophy of Science* (New York: Science History Publications 1976) 91–106

'Philosophical Literature: 1910–1960' and 'Philosophical Literature' in Carl Frederick Klinck, ed *Literary History of Canada* second edition (Toronto: University of Toronto Press 1976) vol 2, 95–107 and vol 3, 84–103

Review of Don D. Roberts *The Existential Graphs of Charles S. Peirce* in *Dialogue* 15:1 (March 1976) 150–5

1979

Review of Nicholas Rescher *Peirce's Philosophy of Science: Critical Studies in His Theory of Induction and Scientific Method* in *Transactions of the Charles S. Peirce Society* 15:2 (Spring 1979) 176–9

Review of Nicholas Maxwell *What's Wrong With Science? Towards a People's Rational Science of Delight and Compassion* in *Philosophy of the Social Sciences* 9:2 (June 1979) 241–4

FORTHCOMING

Review of P.H. Barrett, ed *The Collected Papers of Charles Darwin* in *Dialogue*

'Peirce and Rescher on Scientific Progress and the Economy of Research' (critical notice for *Dialogue*, invited by the editor)

Notes on Contributors

JOHN BURBIDGE is Professor of Philosophy and Master of Champlain College at Trent University. In addition to several articles on the logic of Hegel he has written two books of philosophical theology: *One in Hope and Doctrine* (1968) and *Being and Will* (1977). He was a student of Goudge, both as undergraduate and graduate.

MAX H. FISCH is Professor of Philosophy, Emeritus, at the University of Illinois and Adjunct Professor of Philosophy at Indiana University – Purdue University at Indianapolis. His main areas of scholarly interest are the history of philosophy and of science. He is writing a biography of Peirce and is General Editor of the Peirce Edition Project at Indianapolis.

HANS G. HERZBERGER is Professor of Philosophy at the University of Toronto. His scholarly interests include philosophy of language, philosophy and history of logic, and value theory, and he is the author of many papers in these areas.

DAVID L. HULL is Professor of Philosophy at the University of Wisconsin – Milwaukee. His main research interest is in the philosophy of biology and he is the author of *Philosophy of Biological Science* (1974) and *Darwin and His Critics* (1973).

EDWARD H. MADDEN is Professor of Philosophy at the State University of New York at Buffalo. He has written articles and books in metaphysics, philosophy of science, and the history of American philosophy, including *Chauncey Wright and the Foundation of Pragmatism* (1963) and *Causing, Perceiving and Believing* (1975, with P.H. Hare). He is past president of the C.S. Peirce Society and in 1967 taught a seminar in the philosophy of science at the University of Toronto, at Goudge's invitation.

Notes on Contributors

ROBERT MCRAE is Emeritus Professor of Philosophy at the University of Toronto. He has written widely in the history of philosophy in the seventeenth and eighteenth centuries, including *The Problem of the Unity of the Sciences, Bacon to Kant* (1961) and *Leibniz: Perception, Apperception and Thought* (1976). He has been both a colleague and a friend of Goudge throughout his teaching career.

ROLAND PUCCETTI is Professor of Philosophy at Dalhousie University. He works primarily in the philosophy of mind and the philosophy of religion and has published *Persons* (1967) as well as many papers on the implications of neurophysiology for philosophical problems of personal identity and materialism. He was a student of Goudge in 1949–50. His new book, *Brain and Mind*, will be published by Macmillan of London in late 1980.

DON D. ROBERTS is Associate Professor of Philosophy at the University of Waterloo. His major areas of interest are logic and American philosophy. He is an associate editor of the Peirce Edition Project at Indianapolis and is the author of *The Existential Graphs of Charles S. Peirce* (1973). He is currently working on a series of articles on Peirce's proof of pragmaticism.

ALEXANDER ROSENBERG is Professor in the Departments of Philosophy and Social Science, Syracuse University. His main areas of interest are the philosophy of natural and social science and the nature of causation. He is the author of *Microeconomic Laws: A Philosophical Analysis* (1976) and *Sociobiology and the Preemption of Social Science* (1980), and co-author of *Hume and the Problem of Causation* (1980).

MICHAEL RUSE is Professor of History and Philosophy at the University of Guelph. He is the author of many papers on the history and philosophy of biology and of *The Philosophy of Biology* (1973), *Sociobiology: Sense or Nonsense?* (1979), and *The Darwinian Revolution: Science Red in Tooth and Claw* (1979).

DAVID SAVAN is Professor of Philosophy at the University of Toronto. His chief areas of scholarly interest are Peirce, Spinoza, Plato, epistemology, semiotics, and theory of the emotions. His many papers on those topics include discussions of Peirce in both volumes of *Studies in the Philosophy of C.S. Peirce* (1952, 1964) and in the *Transactions of the C.S. Peirce Society*. For many years he has been both a colleague and friend of Goudge.

PETER A. SCHOULS is Professor of Philosophy at the University of Alberta. His research interests have focused mainly on Descartes and Locke and his book *The Imposition of*

Method: A Study of Descartes and Locke appeared in 1980. He was an undergraduate and a graduate student at the University of Toronto where Goudge both taught the first philosophy course he took and supervised his PH D dissertation. He also lectured at that university from 1964 to 1967 when Goudge was chairman.

TIMOTHY SPRIGGE is Professor of Logic and Metaphysics at the University of Edinburgh. His publications include *Facts, Words and Beliefs* (1970) and *Santayana: An Examination of his Philosophy* (1974). He has a special interest in the history of American philosophy.

BARRY STROUD is Professor of Philosophy at the University of California, Berkeley. He is the author of *Hume* (in the Arguments of the Philosophers series) and of philosophical papers on Wittgenstein, Quine, transcendental arguments, scepticism, logical necessity, and related topics. He was a student of Goudge as an undergraduate at the University of Toronto and a visiting professor at that University in 1967–8.

PAUL THAGARD is Assistant Professor of Philosophy at the University of Michigan – Dearborn. He received his PH D in 1977 from the University of Toronto, where he was a student of Goudge. His main areas of interest are philosophy of science and nineteenth-century philosophy. He is the author of 'The Best Explanation: Criteria for Theory Choice' (*Journal of Philosophy* 1978), 'Why Astrology is a Pseudoscience' (*PSA* 1978, vol 1), 'The Unity of Peirce's Theory of Hypothesis' in *Transactions of the C.S. Peirce Society* (1977), and other articles.

H.S. THAYER is Professor of Philosophy at the City College of the City University of New York. His research interests embrace the history of philosophy and the theory of knowledge and his published work includes *The Logic of Pragmatism* (1952) and *Meaning and Action: A Critical History of Pragmatism* (1968, 1980). He is also the author of the entries on 'Pragmatism' in *The Encyclopedia of Philosophy* (1966) and the *Encyclopedia Britannica* (15th edition, 1974).

MANLEY THOMPSON is Professor of Philosophy at the University of Chicago. He has published widely in epistemology, metaphysics, Kant studies, and American pragmatism, including *The Pragmatic Philosophy of C.S. Peirce* (1953, 1963) and 'Peirce's Verificationist Realism,' *The Review of Metaphysics* (1978). From 1946 to 1949 he taught with Goudge at the University of Toronto.

R.E. TULLY is Associate Professor of Philosophy at St. Michael's College, University of Toronto. His special interests include theory of knowledge, philosophy of perception, logic, and twentieth-century British philosophy. His published papers include 'Reduction and Secondary Qualities' (*Mind* 1976) and 'Sense-Data and Common Knowledge' (*Ratio* 1978).

MARY B. WILLIAMS is Associate Professor in the University of Delaware School of Life and Health Sciences. She received her PH D in Mathematical Biology from the University of London for a thesis on the axiomatization of evolutionary theory. She is the author of several papers in philosophy of biology and her main research interest continues to be the structure of evolutionary theory.

FRED WILSON graduated in mathematics and physics from McMaster University in 1960, and completed his PH D in philosophy at the University of Iowa in 1965, writing his dissertation under Gustav Bergmann. He then joined the Philosophy Department at the University of Toronto, when Goudge was chairman, and has been Tom's colleague ever since. He has done work in ontology, philosophy of science, philosophy of logic, and the history of philosophy. Among his publications are (with A. Hausman) *Carnap and Goodman: Two Formalists* (1967) and *Reasons and Revolutions: An Empiricist Account of Paradigms and Research Programmes* (forthcoming).

Index of Names

Abel, T. 166, 317
Achinstein, P. 328
Adams, J.C. 256
Addis, L. 166, 317
Adler, M. 329
Alexander, S. 151, 261, 262, 264, 267, 271, 275, 276, 277, 314, 315, 317
Algozin, K. 320
Almeder, R. 304, 317
Anderson, F.H. viii, 334
Anscombe, G.E.M. 319, 328
Aristotle 34, 36, 135, 161, 328
Armstrong, D.M. 205, 261, 276, 312, 317
Arnauld, A. 94
Asquith, P.D. 310
Ayer, A.J. 124–9, 317
Ayim, M. 7, 317

Babbage, C. 223, 227, 228, 317
Bain, A. 30, 35, 319
Baldwin, J.M. 134, 135, 139, 142, 143, 146, 303, 317
Barlow, N. 319
Barrett, P.H. 336
Bartholomew, M. 278, 280, 286, 317
Baumrin, B. 321

Beckner, M. 278, 280, 286, 317
Bentham, G. 312
Bergmann, G. 152, 167, 305, 306, 307, 308, 317
Bergson, H. 75, 81–7, 163, 207, 302, 318, 330, 331, 334, 335
Berkeley, G. 34
Bernard, C. 69
Bernstein, R.J. 321
Bertalanffy, L. von 333, 334
Berzelius, J.J. 69
Black, M. 59, 60, 318
Blake, R.M. 70, 318, 323
Boethius 303
Boler, J.F. 13, 318
Borges, J.L. 301, 318
Bosley, R.N. 88
Braithwaite, R.B. 300, 313, 318
Brett, G.S. viii, 152, 330
Brewster, D. 318
Broad, C.D. 151, 163, 164, 261–77, 307, 314, 315, 318, 331
Brodbeck, M. 166, 306, 317, 318, 321
Brody, B. 286, 318
Broyles, J.R. 301, 318
Brunning, J. 41
Buchdahl, G. 104, 165, 260, 318

Index of Names

Buchler, J. 3, 90, 318, 329
Bunge, M. 176, 318
Burbidge, J.W. 15–27, 337
Burks, A. xiii, 332
Burroughs, E.R. 77, 318
Butler, S. 334
Buytendijk, F.J.J. 201, 318

Cairns, H. 324
Calderoni, M. 28
Cameron, T.M. 333
Campbell, N.R. 164, 308, 318
Carnap, R. 73, 153, 318
Carroll, L. 80, 301, 313, 318
Carus, P. 296
Causey, R. 187, 318
Chalmers, R.C. 332
Chambers, R. 226–30, 234, 318
Church, F.J. 324
Cody, M.L. 309, 319
Cohen, M.R. 123
Cohen, R.S. 325
Collingwood, R.G. 16, 17, 19–23, 26, 27, 294, 295, 319
Colodny, R. 326
Cooper, L. 324
Cope, J.I. 320
Creedy, F. 332
Croce, B. 330
Crombie, A.C. 318
Cumming, R.D. 324

Darwin, C. 69, 151, 209, 213, 214, 219, 220, 226, 230–5, 249, 250, 251, 252, 255, 256, 258, 259, 279, 293, 319, 325, 332, 333, 334
Darwin, E. 334
Darwin, F. 319
Davidson, D. 313, 319
de Beer, G. 312

De Morgan, A. 22
Descartes, R. 5, 29, 61, 66, 67, 88–104, 191–8, 302, 303, 310, 311, 324
Dewey, J. 32, 123, 124, 130, 308, 319, 325, 326
Dewitt, N.W. 331
Drake, S. 311, 319
Dray, W. 158, 159, 166, 308, 319
Ducasse, C.J. 71, 318, 323
Dummett, M. 313, 319

Edwards, J. 3
Edwards, P. 320, 334
Einstein, A. 221
Eisele, C. 37, 301, 319
Eldredge, N. 309, 327
Ellington, J. 322
Engle, J.S. 296
Epicurus 331

Feibleman, J. 3, 104, 319, 331
Feigl, H. 60, 66, 261, 262, 317, 318, 319, 321, 324, 326
Fensted, J.E. 325
Fermi, E. 301
Feyerabend, P.K. 166, 271, 276, 308, 319
Fichte, J.G. 236
Fisch, M.H. 13, 28–40, 293, 297, 304, 319, 320, 337
Fisher, R.A. 282, 320
Fitzgerald, J.J. 40, 320
Flew, A. 75, 320
Fodor, J. 312, 320
Frege, G. 319

Galileo 279, 310, 319
Galle, J.G. 256
Gallie, W.B. 331

Index of Names

Gamow, G. 331
Gardiner, P. 326
Gardner, M. 320
Geach, P. 319
Ghiselin, M. 278, 280, 309, 320
Giere, R. 321, 323
Gilbert, W. 69
Gillespie, C.C. 335
Glas, S.Y.B. de 333
Glaser, B.G. 176, 320
Goodman, N. 273, 307, 320, 325
Gorgias 7
Gram, M. 328
Grant, G.P. 332
Gray, A. 229, 320, 334
Gray, J.A. 204, 320
Green, T.H. 236
Greene, J.C. 332
Greenlee, D. 67, 320
Gregg, T.G. 177, 324
Guttenplan, S. 319, 325
Gutting, G. 255, 320

Hacking, I. 310, 323
Haldane, E.S. 319
Hamilton, E. 324
Hampshire, S. 313, 320
Hanson, N.R. 63, 248, 250, 251, 253, 254, 257, 258, 320
Hardy, J.D. 202, 321
Harman, G. 240, 259, 321
Harré, R. 327
Hartshorne, C. xiii, 3
Harvey, W. 69
Hegel, G.W.F. 28, 236
Helmer, O. 306, 321
Hempel, C. 16, 17–24, 27, 153, 156, 158, 166, 173, 181, 261, 264, 268, 270, 294, 295, 321
Henle, P. 263, 264, 321

Herschel, J.F.W. 221, 222, 223–5, 229, 231, 232, 234, 251, 312, 321, 325
Hertz, H. 301, 321
Herzberger, H. 41–58, 337
Hesse, M.B. 164, 165, 255, 321
Hook, S. 324
Hopkins, W. 232, 233, 235, 255, 321
Horace 166
Howells, W. 333
Hull, D.L. 172–88, 278, 280, 284, 286, 309, 310, 312, 322, 337
Hulme, T.E. 330, 331
Hume, D. 15, 17, 59, 66, 67, 74, 245–6, 293, 294, 307–8, 322, 327
Huxley, J. 208, 219, 322
Huxley, T.H. 312, 322, 334
Huygens, C. 249

Irving, J.A. 332

James, H. 322
James, W. 31, 32, 35, 57, 80, 105–20, 123, 134, 296, 301, 302, 303, 322, 324, 330
Jerison, H.J. 200, 201, 311, 322
Johnson, A.H. viii
Jourdain, P.E.B. 58

Kant, I. 104, 134, 135, 139, 145, 146, 189–92, 194, 237, 246, 295, 304, 305, 310, 322
Kemeny, J. 269
Kempe, A.B. 53, 55, 322
Kenny, A. 303, 319
Kent, B. 37, 221, 322
Kepler, J. 63, 69, 70, 71, 190, 310
Kiefer, H.E. 328
Kierkegaard, S. 87, 302, 322
King-Farlow, J. 88
Klemke, E. 328

344 Index of Names

Klibansky, R. 323
Klinck, C.F. 320, 335
Kochanski, Z. 278, 322
Koestler, A. 77, 322
Koninck, C. de 333
Kordig, C. 171, 308, 322
Körner, S. 221
Kuhar, M.J. 202, 322
Kuhn, T. 151, 153, 162, 165, 166, 170, 171, 308, 322, 323

Lakatos, I. 151, 162, 164, 165, 171, 260, 319, 322, 323, 328
Lamarck, J.B. 69, 214, 216, 223, 224, 334
Langley, S.P. 15, 293
Laudan, L. 59–61, 323
Lavoisier, A. 256, 323
Layzer, D. 12, 323
Lecomte du Nouy, P.A. 334
Lee, O.H. 324
Leibniz, G. 29, 41, 134
Lenz, J.W. 60, 72, 323
Leonard, H.S. 293, 323
Leverrier, U.J.J. 256
Levi, I. 168, 323
Lewes, G. 314
Lewis, C.I. 297, 298, 313, 323
Lieb, I.C. 57, 304, 323
Liebig, J. von 249
Llinas, R.R. 199, 323
Lodge, R.C. 332
Loeb, J. 334
Lyell, C. 223–5, 231, 233, 312, 317, 323, 325, 328
Lyell, K. 323

Mackie, J. 305, 323
Madden, E.H. 59–74, 167, 168, 170, 318, 322, 323, 337
Malthus, T.R. 210, 231, 252

Mandelbaum, M. 309, 323
Margenau, H. 221, 323
Martin, R.M. 205, 312, 323
Maxwell, G. 318, 319, 321, 324, 326
Maxwell, J.C. 255
Maxwell, N. 336
Mayr, E. 235
McGuire, J.E. 310, 323
McKeon, R. 303
McMurray, G.A. 201, 323
McRae, R. 189–98, 338
Mead, G.H. 166, 315, 323
Meehl, P. 224, 277, 326
Melzack, R. 210, 324
Mendel, G. 208, 214–16, 218, 219
Mendeleev, D. 69, 268
Meredith, J.C. 322
Mersenne, M. 96
Mettler, L.E. 177, 324
Meyers, R.G. 68, 324
Michael, C.R. 199, 324
Mill, J.S. 63, 305, 324
Mink, L.O. 185, 324
Mises, R. von 331
Mitchell, O.H. 53, 300
Montaigne, M. 93
Moore, G.E. 105, 110, 236
Moore, E. 323, 333
More, H. 192, 195, 196, 197, 310
Morgan, C. Lloyd 314, 315, 334
Morris, L. 330
Munitz, M.K. 328
Munson, R. 335
Münsterberg, H. 296
Muntz, W.R.A. 199, 324
Murphey, M. 4, 299, 324, 333
Murray, E.G.D. 332
Musgrave, A. 319, 322, 323, 328

Nagel, E. 160, 164, 261, 262, 313, 314, 324

Index of Names

Nesbitt, H.H.J. 333
Newton, I. 152, 153, 157, 158, 161, 173, 189–92, 194, 197, 221, 222, 229, 230, 287, 310, 324, 328

Olscamp, P.J. 303, 324
O'Malley, J.J. 320
Oppenheim, P. 153, 156, 173, 174, 262, 264, 268, 269, 270, 324

Paley, W. 233
Pap, A. 262, 266, 267, 270, 314, 315, 324
Pasteur, L. 69
Paton, H.J. 331
Pauli, W. 221, 301
Peirce, C.S. vii, ix; part one, passim; part two, 167–71, 207, 237–8, 248, 250, 251, 252, 253, 254, 258, 314
Pepper, S.C. 263, 264, 274, 324
Perry, R.B. 31
Pert, C.B. 202, 322
Planck, M. 250
Plato 36, 75, 77–81, 87, 324, 332
Plotinus 236
Polanyi, M. 318
Popper, K. 308, 324
Puccetti, R. 199–206, 338
Putnam, G.H. 32, 296
Putnam, H. 261, 262, 324
Pyrrho of Elis 93
Pythagoras, 16

Quetelet, J. 226, 228
Quincy, J. 301
Quine, W.V. 10, 49, 124–9, 236, 298, 299, 300, 304, 313, 314, 324, 325

Reichenbach, H. 59–61, 66, 71–4, 325
Rescher, N. 221, 306, 321, 325, 336
Roberts, D. 13, 75–87, 296, 297, 299, 300, 325, 335, 338

Roberts, L.B. 302
Robin, R. xiii, 296, 297, 323, 325, 333
Romanes, G.J. 334
Römer, O.C. 284, 285
Rosenberg, A. 166, 205, 207–19, 323, 325, 338
Ross, G.R.T. 319
Royce, J. 57, 106, 114, 132
Rudner, R.S. 86, 87, 325
Ruse, M. 173, 175, 179, 182, 207, 208, 209, 212, 214, 219, 220–35, 278, 284, 286, 306, 307, 312, 325, 326, 335, 338
Russell, B. viii, 57, 105, 110, 124–9, 151, 160, 326
Ryle, G. 335

St Paul 331
Salmon, W. 255, 257, 258, 323, 326
Sanchez, F. 93
Santayana, G. 105–20, 303, 326, 327
Sarton, G. 331
Savan, D. 3–14, 326, 338
Schaffner, K. 255, 258, 326
Scheffler, I. 325
Schiller, F.C.S. 32, 123, 326
Schilpp, P.A. 325, 326
Schouls, P. 88–104, 326, 338, 339
Schrödinger, E. 221
Scott, F.J.D. 296, 326
Scotus, John Duns 104, 140
Scriven, M. 154, 156, 158, 278, 283, 284, 289, 306, 319, 324, 326
Sebeok, T.A. 320
Sedgwick, A. 223, 224, 228, 232, 233, 326
Sellars, R.W. 334
Sellars, W. 164, 165, 247, 274, 277, 314, 321, 326
Seward, A.C. 319
Sextus Empiricus 93
Shea, W.R. 318, 320, 335

346 Index of Names

Sheldon, W. 330
Simberloff, S. 309, 327
Simpson, G.G. 235, 306, 327
Skidmore, A. 55, 297, 299, 300, 327
Smart, J.J.C. 261, 262, 278, 282, 289, 309, 327
Smith, J.E. 330
Smith, J.W. 293, 327
Smith, N.K. 322
Smith, V.E. 333
Smuts, J.C. 334
Snyder, S.H. 202, 322
Socrates 77–81
Spencer, H. 28
Sperry, R.W. 199, 327
Spinoza, B. 197, 198, 311
Sprigge, T.L.S. 105–20, 327, 339
Stahl, G.E. 256
Strauss, A.L. 176, 320
Strong, C.A. 296
Stroud, B. 236–47, 313, 327, 339
Stuewer, R.H. 318
Suppe, F. 186, 327
Sylvester, J.J. 298

Tagliacozzo, G. 335
Tarski, A. 105, 110, 327
Tarzan 77
Tattersall, I. 309, 327
Teilhard de Chardin, P. 152, 333, 334
Thagard, P.R. 248–60, 314, 327, 339
Thayer, H.S. 121–32, 303, 327, 334, 339
Thompson, M. 10, 40, 133–48, 327, 339
Thomson, G. 254, 327
Todhunter, I. 221, 327
Tomas, V. 69, 327
Toulmin, S. 308, 327, 328, 335
Tully, R.E. 261–77, 315, 327, 339, 340

Uexküll, J.J. Baron von 320, 334

Van der Steen, W.J. 172
van Fraassen, B.C. 248
Van Melsen, A.G. 333
Vaudaire, J.-P. 330
Vendler, A. 177, 327
Vico, G. 335
Volpe, E.P. 215, 327

Wallace, A.R. 334
Ward, T.W. 302
Wartenberg, G. 4, 327
Watson, J.B. 166
Weiss, F.G. 320
Welby, Victoria Lady 11, 57, 323
Wennerberg, H. 333
Westfall, R.S. 192, 193, 321, 323, 328
Whewell, W. 221–5, 229, 230, 232, 259, 321, 325, 327, 328
White, M. 3, 328, 335
Whitehead, A.N. 314
Wiener, P.P. 294, 320, 326, 328, 330, 331, 332, 335
Wightman, W.P.D. 331
Williams, B. 313, 328
Williams, M.B. 207, 209, 213, 278–89, 312, 328, 340
Wilson, D. 221, 328
Wilson, F. 151–71, 204, 248, 306, 307, 308, 328, 340
Wilson, L. 328
Wimsatt, W. 286, 328
Wittgenstein, L. 87, 302, 310, 314, 327, 328
Wojciechowski, J.A. 327
Woodger, J.H. 269, 334
Wright, G.H. von 328
Wright, L. 286, 328

Young, F.H. 320, 326, 331

Index of Subjects

abduction 10, 15, 24–6, 31, 63–4, 68–70, 73–4, 95–6, 168–9, 170, 248–9, 254; and induction 60–1; and pragmatism 40
absolute, the 81–2, 84, 114
absolute product 53
action 36–7; as the end of man 36
actualism, method of 223, 224
adaptation 157, 163–4, 169–70, 199–200, 213, 225, 229, 230, 232, 236, 239–40, 280, 281, 285
adaptive radiation 281
agapism 35
algebraic logic 43
analogy 63, 164–6, 176–7, 223, 232, 255, 257–8, 259–60
analysis 97, 101–2
anti-evolutionists 223–6, 229, 232, 235
appetite 198
a priori 4, 12
axiomatization 152–3, 159–60, 231, 233, 259, 278, 284

behaviourism, Watsonian 272
being 11, 133, 136, 137, 142–4, 146; modes of 140
belief 5–7, 26, 30, 35, 61, 117–18, 125–6, 129, 238–41, 244, 313; and doubt 89–91, 92; fixation of 29
biology 160–1, 166, 167, 278–89
body 192–8
bonding algebra 42, 48–56
bridge principles 209, 212, 219, 279, 283, 284, 287, 289

categories 11, 28, 39, 41, 134, 297
causality 194, 221, 224, 229, 232, 245–6, 275; final 221, 222, 225, 227–8, 229
cistron 218
cladists 309
classification 75; of sciences 35, 36
clearness 29, 37–9
closed systems 161
cognition 88, 141–2
common sense 5, 10
conatus 197
concepts 51–2, 106; biological 280–2; cluster 310; meaning of 38, 112
conceptual valency 43–8
conclusiveness of a proposition 68
consciousness 113, 114–15, 199–206, 267–8, 276; animal 200–10; as complex information processing 201
consilience 259–60

348 Index of Subjects

continuity 23, 27
contradiction, principle of 134, 143, 144
conventionalism 241-4
Copernican hypothesis 71
cosmology, evolutionary 11-12
critic 40, 296

Darwinian revolution 220-35
deduction 96-7
definability 42
definition 38; as permission 36; constructional 273-4
demonstration 29
desire 35-6
development 227
discovery 95-103
distinctness 38
double-aspect theory of mind 204, 276
doubt 33, 61-2, 64, 88-104; Cartesian method of 66-7, 90-104; genuine vs specious 66-7

economy of hypotheses 254-5
economy of research 63, 74; logic of 257, 259
embryology 208, 232
emergence 261-77
enkephalin 202
entelechy 261
entity, hypothetical 277; observable 209; theoretical 209
epiphenomenalism 205, 263, 277
essence 112, 119
esthetics 35-6, 131; and the *summum bonum* 35-6
ethics 35-6; and esthetics 35-7; and logic 32, 35-7
Euthyphro 77-81
events, historical 19
evidence 21-2, 60

evolution 5, 69, 151, 160, 236-47; cosmic 127; Darwinian 69, 220-35; descent in 175-8, 212-13; Lamarckian 69; of human knowledge 69-70, 237-47; social 5; theory of 151, 175-81, 207-19, 220-35, 261, 282, 284, 312
excluded middle, principle of 134, 143, 144
existence 11, 133, 142-3, 267; and conception 136
experience 29, 64-5, 115, 145
explanation 16, 23-6, 151-71, 252-7, 258, 259-60, 261-4, 267-9, 270, 274, 278, 284, 294, 295; causal 169, 267, 274, 288; covering-law 16, 173-6, 179-81, 186-8, 261, 286-7; deductive-nomological 151-4, 156, 158, 162; definition of 16, 23, 24; evolutionary 151, 236-47; functional 288; genetic 246; historical 15-27; hypothetical-deductive 152, 221, 231, 234-5, 248, 259; integrating 151, 152, 157-66, 172-4, 175-81, 186, 284-6; narrative 151, 152, 157-66, 172-88, 284-6; teleological 286-8
explanation sketches 294
extensional isomorphism, criterion of 273

faith 7
fallibilism 10, 42, 67, 71, 129, 297
feasibility 23, 27
feeling 4, 5-7, 11
finalism 163
firstness 3, 10, 31
fitness 209-13, 215, 217, 218
force 193-6, 197-8; vital 249, 261
function 222, 232, 235, 286-7

Index of Subjects

gemmules 214
genes 177, 212, 214, 216, 218
general terms 147–8
generality 22–3
generals 133, 136, 140, 141
genetic fallacy 255
genetics 175, 207–19, 280; Mendelian 208–9, 214–18, 219; molecular 215, 218; population 177, 203, 209, 212, 214–18, 219; transmission 218
geology, historical 156, 157
God 75, 93, 103, 115, 192, 194–8, 230, 233–4; and Abraham 76; as First Mover 196
grammar, speculative 40, 296
graphs, existential 12–13, 31, 33, 36, 38, 39, 49
gravity 193–4, 227, 229, 254

habit, and belief 61, 129
Hardy-Weinberg law 214–15
hecceities 134, 139, 140, 145, 146
hedonism 35–6
heredity 208–9, 212–19
historical entity 176–7, 178, 183–5
history 15–27, 166, 295
history of science 126
homologies 157, 172, 175–8, 309
human nature 246
hylozoism 190–2, 194, 198
hypotheses 16, 248–60; cosmological 12; explanatory 63; plausibility of 248–50, 252–8; probable 73–4; quantification of 18; selection of 163–6, 168–71, 257–8; 'weight' of 73

ideal 137
ideas 35, 103, 106–7, 114–16, 119–20; as instruments 114–15
identity, logical 272; of indiscernibles 139, 144; of individuals 141–2, 144–6; referential 276; synthetic 272
images 116–17
imagination 99–100
inconceivability 168
incorrigibility 67–8
independence of hypotheses 18–19
independent assortment, principle of 214–16
indesignative subject 148
index 147
individual events 17
individuals 133–48, 305
indubitability 29
induction 25–6, 258; as self-correcting 59, 65–6, 70–2, 167; qualitative 60, 62–3; quantitative 60–1, 62, 73–4; vindication of 59–74, 167–8
inertia, law of 153, 189–92, 194, 197, 310
inference 10, 16, 59, 70, 131, 198, 238
infimae species 134, 135, 305
inner meaning of a poem 83–5
inquiry 61, 125
instinct 4, 232; and scientific theories 20–1, 68–9
intellectual sympathy 85
interactionism 206
interbreeding populations 212, 214, 217
interpretant 10, 39, 130–1; emotional 11
intuition 84–7, 135, 145–6, 305
involution 53, 300; backwards 53, 300

judgment 98–9; perceptual 141–2

Kepler's laws 221
knowledge 61, 75, 78–87, 119–20; scientific 64

labeling 75–87
laws 17–18, 152–7, 173–4, 221, 226, 228–9; biological 278–89; bridge 187–8; emergent 269–70; evolutionary 160–1, 185, 227; general 294; statistical 154–7, 226
learning 302
life 191, 197–8; Kant's definition of 189
linguistic analysis, school of 152
linkage 215
logic 42, 237–8, 253; and psychology 35; ethics and 32, 35–7
logic of discovery 88–104, 248–60
logic of justification 248–60
logical analysis, theory of 38
logical atom 135–6, 142
logical composition 53
logical division 135, 142, 144, 146, 147, 305
logical empiricism 151, 187, 219, 259
love, evolutionary 35
lume naturale 6

matter 189–90, 194, 195–6, 197–8, 265, 275, 276
meaning 9, 123, 131
mechanical philosophy 191–2, 195, 198
mechanics, composition law of classical 165; terrestrial 279
mechanism, pure 263, 264, 266, 273, 274
metaphysics 151, 163, 167
method of Descartes 91–2
methodeutic 40, 296
micro-reductions 187
mind 114, 192–3, 198; behaviouristic view of 167, 206; emergent nature of 166–7, 205; logical function of 16
mind-brain relation 203–6
miracles 223–5
morphology 208, 232

motion 142, 194–7, 198
musement 11
mutation 164, 205, 214–16, 280, 282, 285
muton 218

names 75
narratives, evolutionary 166; historical 172–88
natural kinds 187
natural mechanisms 223, 225
natural theology 222
naturalism 15, 121, 131–2, 133, 134, 238–9, 244, 245–6; Renaissance 192
nature 19, 68
nebular hypothesis 226–8
necessary truth 238–9, 240–5, 247
necessity 245–7, 313
neo-Lamarckianism 216
Newtonian astronomy 221–2, 223–4, 226–8, 230, 231, 233, 234
Newtonian mechanics 189, 190, 221, 222
nociception 201–13
nominalism 13, 34–5, 69, 133, 136, 138, 141
nonrelative multiplication 53
nucleic acid 218

observable property 261, 265, 277
observation 70, 113
occasionalism 205–6
open systems 161
opinion 110–11, 122–5
order 98
ostension 315

paleontology 156, 157, 175, 182, 185, 208
pangenesis 214

Index of Subjects

paradigm, Kuhn's notion of 162, 170–1
parallelism 204, 268, 271
phaneroscopy 39
phenetic resemblance 309
phenomenology 39
philosophy 77; and philosophic method 8–13; as positive science 10
phlogiston theory 256
phylogeny 177–9, 181, 185, 309, 310
physics 102, 160–1, 207, 211, 218, 221, 222, 226, 231–2, 278, 284–6
piety 78–81
pineal gland 196
positivism 17, 18, 69, 112, 151–2, 262
possibility 108–9; the doctrine of real 34
pragmaticism 28, 32–3, 123, 147, 295–6, 304
pragmatic maxim 10, 37, 118
pragmatism 28–40, 43, 105–20, 123, 296
prediction 26, 27, 153, 155–6, 263, 264–7, 272, 278, 280, 283–4, 289
probability 18–19; and relative frequency 73–4
problem-solving capacities 164
process 152–4, 159–60, 171
progress 151
psychics 16, 20, 294
psychology 18, 35, 110, 252, 272–3
purpose 151, 163, 166, 170–1, 235, 286–7

quantification theory 50, 54
quantum theory 279

rationality, essence of 127
reaction 141–6, 148
real, the 137, 139, 143
realism 13–14, 34–5, 133, 136, 138, 140, 143, 152; and Platonism 4, 13; scholastic 13, 136; Scotistic 13
reality 11, 122, 143, 144
reason, cosmic 127
reasonableness 63, 95–6, 126–7, 254, 295; defined 22–3
recon 218
reduction thesis 41–58
reductionism 217–18, 234–5, 262, 269, 270, 271–2, 274, 276
reference 115–16, 118
regulative principles 220, 234, 255, 258
Newtonian 221–35; teleological 221–35
reinforcement 312
relations 41–58
relative vs absolute in Bergson 81–2
relative product 43–8
relative sum 53, 300
relatives, logic of 138
religion 120
replacement, pragmatic criterion of successful 269
representamen 130
reproduction rate 210, 211, 213, 215
reproductive fallacy 86–7
res cogitans 192
res extensa 192
retroduction 63, 168
rheme 56
Roentgen's hypothesis 249

scenarios 309–10
science 36–7, 72, 74, 91–2; history as a 20–1; method of 5–8; theoretical 37
scientific community 67, 171; and inquiry 126; and truth 67
secondary qualities 263, 265–6, 270–7
secondness 3, 10, 31
segregation, law of 214–16

Index of Subjects

selectionist theory in biology 158, 159–60, 164, 165–6, 170, 171
self-consciousness 198, 199, 206
self-transcendence of mental occurrences 116, 119
semeiotic 10–13, 40
semiotic 131
sentiment 4, 7
signification 111–12, 304–5
signs 38, 39, 113, 122, 130–1, 304; hierarchy of 11; pragmatic interpretation of 111
simplicity 168–70, 259–60
singulars 135, 138, 305
singular statements 138
singular terms 147–8
social behaviourism 166
soul 196–8
space 145–6
speciation 177, 225
species 177, 179, 186, 201, 223–4, 227, 229, 232, 234, 236, 251–2, 256; considered as individuals 279–89; evolving 184–5; geographical distribution of 232
Stoicism 36, 141
struggle for existence 231, 232, 237
substantial forms 193
surrogate constructions, theory of 53–5
symbols 119–20
symmetry 257–8
synechism 28, 31, 33, 40
synthesis 97
systematics 208

taxa 309; monophyletic 177–8
taxonomy 156, 177, 280, 209

teleology 163, 171, 222, 232
theory, and practice 4; and science 7
theory evaluation 167, 170–1, 174, 255
thirdness 3, 10, 31
thought 7, 19–20, 36–7, 236–47
time 145–6
transaddition 53, 100
transcendentalism 3–14, 121, 131–2
transpositions 49
trees, phylogenetic 309–10
triadic sign relation 10–11
truth 72, 106–12, 116–20, 121–32; cognitive 303
tychism 40

understanding, empathetic 19, 22; operations of 99–100
uniformitarian principle in geology 249, 252
uniformitarian view of world history 223
uniformity 12; of nature 70
universals 13–14, 23–7
universal statements 147–8

valental graphs 299
valency, doctrine of 51–5, 299, 300
vera causa 221, 224, 225–6, 229, 232
verification 22, 106, 118
vertebrates 200, 202, 206
vestigal organs 175
vis inertiae 190–2, 194, 197, 198, 310
vitalism 163, 262
volition 145–6

will 198
world-soul 196